Building Surveyor's Pocket Book

Building Surveyor's Pocket Book is an accessible encyclopaedia of matters vital to building surveyors. Well-illustrated with diagrams, pictures, tables, and graphs, it covers all essential elements of building pathology, building performance, and building construction terminology in a simple, accessible way for the practitioner and student. This Pocket Book provides a practical and portable reference text, working as a first-stop publication for those wishing to refresh their knowledge or in need of guidance on surveying practice. Working through fundamental principles in key practice areas, the book is not overly bound by the regulation and legislation of one region, and the principles can be applied internationally.

This book is ideal reading for individual surveyors, practitioners, and students in building surveying, facilities management, refurbishment, maintenance, renovation, and services management. It is also of use for those interested in building forensics, building performance, pathology, and anyone studying for their RICS APC. Many other professions in architecture, contracting, engineering, and safety will also find the book of use when undertaking similar practice.

Dr Melanie Smith, FRICS, GIFireE, NRAC Consultant, PGCHE, Leeds Sustainability Institute, Leeds Beckett University. Melanie is a Building Surveyor and Fellow of the Royal Institution of Chartered Surveyors with over 40 years' experience in local authority, private practice, and education sectors. She has been a graduate member of the Institution of Fire Engineers and on the National Register of Access Consultants. Melanie has worked on all types and sizes of buildings for professional services, compliance, fire, access, retrofit, and upgrade purposes, with expertise in conflicting requirements. She has been instrumental in governmental and institutional guidance and codes. Melanie is a member of the Leeds Sustainability Institute, a past Chair of the Yorkshire branch of the RICS Building Surveyors, a Civic Trust Award assessor, and has sat on BSi and RICS working groups. Her doctorate, awarded in 2019, is entitled "A Study of Building Surveying Praxes to Inform Design of Thermal Retrofit". This identifies common shortcomings of retrofit solid wall insulation measures and determines an appropriate approach for pre-design retrofit building surveys.

Prof. Dr Christopher Gorse, MCIOB, MAPM, Leeds Sustainability Institute, Leeds Beckett University. Chris is a Professor of Construction and Project Management at Leeds Beckett University. He is a Chartered Builder with over 30 years' industrial and academic experience in buildings, materials, management, and construction law. Chris has written extensively on the construction of domestic and industrial buildings, and the processes required to deliver them successfully and to measure their performance. As Director of the Leeds Sustainability Institute and Head of the Low Carbon Sustainability Research Group, Chris leads one of the most successful building performance research units in the UK. The group has contributed to changes in the Building Regulations, undertaken research and development of new energy-saving products, and successfully secured patents. Chris is the Chair of the international group Sustainable Ecological Engineering Design for Society (SEEDS), leads research for subgroups at the International Energy Agency, and is past Chair of the Association of Researchers in Construction Management and Visiting Professor at the Central University of Technology, Free State South Africa, in the Unit of Lean Construction and Sustainability.

Building Surveyor's Pocket Book

Melanie Smith and Christopher Gorse

Routledge
Taylor & Francis Group
LONDON AND NEW YORK

First published 2021
by Routledge
2 Park Square, Milton Park, Abingdon, Oxon OX14 4RN

and by Routledge
52 Vanderbilt Avenue, New York, NY 10017

Routledge is an imprint of the Taylor & Francis Group, an informa business

British Library Cataloguing-in-Publication Data
A catalogue record for this book is available from the British Library

Library of Congress Cataloging-in-Publication Data
Names: Smith, Melanie (Building surveyor), author. |
Gorse, Christopher A., author. Title: Building surveyor's
pocket book / Melanie Smith and Christopher Gorse.
Description: Abingdon, Oxon; New York, NY: Routledge, 2021. |
Series: Routledge pocket books | Includes bibliographical references
and index.
Identifiers: LCCN 2020053018 (print) |
LCCN 2020053019 (ebook) | ISBN 9781138307902 (hbk) |
ISBN 9781138307919 (pbk) | ISBN 9781315142647 (ebk)
Subjects: LCSH: Building inspection–Handbooks, manuals, etc. |
Buildings–Evaluation–Handbooks, manuals, etc. |
Buildings–Valuation–Handbooks, manuals, etc. |
Building–Superintendence–Handbooks, manuals, etc.
Classification: LCC TH439.S64 2021 (print) |
LCC TH439 (ebook) | DDC 690/.21–dc23
LC record available at https://lccn.loc.gov/2020053018LC
ebook record available at https://lccn.loc.gov/2020053019

ISBN: 978-1-138-30790-2 (hbk)
ISBN: 978-1-138-30791-9 (pbk)
ISBN: 978-1-315-14264-7 (ebk)

Typeset in Bembo
by Deanta Global Publishing Services, Chennai, India

Contents

Acknowledgements and contributors ix

1 Building surveying: an introduction **1**
MELANIE SMITH AND CHRIS GORSE

Surveying for property investment 4
Surveying for building works 5
Surveying for legislation 5
Surveying activities discussion and summary 6
Models of survey approaches 6
Building surveying and sustainability 9

2 Form and construction **11**
CHRIS GORSE AND MELANIE SMITH

Building form 11
Wind and building form 14
Wind in urban areas 14
Protecting the building envelope from rain and solar 16
Ground floors 18
Walls 21
Walls and upgrade risks 22
Functional requirement of walls 23
Strength and stability of existing walls 24
Wall construction, the positioning of insulation and condensation 26
Comparing different walls and their thermal properties 29
Walls and conservation of energy 29
Foundations 30
Functional requirements 30
Differential settlement 31
Seasonal variation in clay soils 32
Foundation types 32
Strip foundation 33
Pad foundations 35
Combined foundations 36
Raft foundations 37
Pile foundations 38
Placing of piles 39

Jacked piles 40
Ground stabilisation 42
Heat transfer through the foundation structure 42
References 44

3 Legal and regulatory frameworks **45**
MELANIE SMITH

Building works and building regulation 45
Planning and historic listing or registration 48
Adjoining property and neighbourly matters 57
Party walls 57
Acoustics and noise control: Minimising noise transition 62
Contracts 65
Landlord and Tenant 68
Decent Homes 68
Dilapidations 70
Types of survey 73
Health and safety 76
Inclusivity 82
References 86
Guidance, Standards and Codes 87

4 Building pathology **89**
MELANIE SMITH AND B. N. WEST

Introduction 89
Traditional and modern construction 89
Defect management 96
The trouble with water 97
Damp 101
Condensation 102
Penetrating damp 104
Rising damp 105
Mould, decay fungi, and bacteria 106
Everything under the sun 108
Timber infestation: fungal attack and insects 111
All together now! Complex interactions 112
Concrete 113
Movement and stability 118

5 Retrofitting and refurbishment **134**
CHRIS GORSE, MELANIE SMITH, B. N. WEST, CORMAC FLOOD AND LLOYD M. SCOTT

Introduction 134
Considerations for refurbishment 136

Upgrading thermal performance 139
Discussion surrounding thermal upgrades 143
Underpinning 151
Upgrading acoustic performance 152
Upgrading fire protection 155
Fashion 164
References 167

6 Thermal performance **169**

CHRIS GORSE, MELANIE SMITH, MATTHEW BROOKE-PEAT, MARTIN FLETCHER, CORMAC FLOOD, LLOYD SCOTT, AND JOHN SPILLANE

Introduction to thermal performance 169
Thermal comfort 174
Refocusing sustainability assessment methods: A comparative introduction to the Building Energy Rating (BER) Certificate 182
Thermal performance 199
Thermal performance of existing buildings and retrofit 206
Emerging themes: Gaps and needs in building performance simulation for building retrofit 216

7 Fire safety **219**

MELANIE SMITH

Nature and development of fire 219
Effects of a fire: Physiological and behavioural 229
Fire precautions from first principles 238
Means of escape 245
External fire spread 245
Introduction to fire engineering concepts 248

8 Disasters and the built environment **252**

JOHN BRUEN AND JOHN P. SPILLANE

Disasters 252
Disaster management cycle 259
Sustainability 263
Housing in LDCs and post-disaster contexts: Approaches, participation, and barriers 267
References 272

9 Environmental considerations **276**

CORMAC FLOOD, CHRIS GORSE, LLOYD M. SCOTT AND MELANIE SMITH

Environmental issues 276
Ground 277
Flooding 296

Flora and fauna 298
Climate and construction 301
References 306

10 Sustainability **310**
JOHN L. STURGES

Sustainability: An introduction 310
Managing sustainability: Introduction 319
Construction materials and sustainability: An introduction 326
References and further reading 339

11 Glossary and reference section **341**
MELANIE SMITH

Glossary 341
Reference Section 354
References 364

Index 367

Acknowledgements and contributors

We would like to gratefully acknowledge and thank the following for their input to this book:

Ian Dickinson, Leeds Sustainability Institute, Leeds Beckett University, for his huge input, including the illustration production of so many of the figures and hard work in proof-reading this book

Dr Matthew Brooke-Peat, MCIAT, MCIOB, CEnv, Leeds Sustainability Institute, Leeds Beckett University, for his contributions regarding thermal modelling in the chapter on thermal performance

John Bruen, BSc Arch, BArch, MSc Proj. Mgt., RIBA, MRIAI, ARB, RSUA, CEPHD, Chartered Architect and Certified Passive House Designer, Bruen Architects, Belfast, for his contributions to the chapter on disasters and the built environment

Dr Martin Fletcher, Leeds Sustainability Institute, Leeds Beckett University, for his contributions regarding thermal comfort in the chapter on thermal performance

Dr Cormac Flood, BSc, MPhil, Coady Architects, Dublin, and Dublin School of Architecture, for his contributions to the chapters on retrofitting and refurbishment, thermal performance, and environmental considerations

Prof. Julienne Hanson, University College London, for her contribution relating to inclusivity in the chapter on legal and regulatory frameworks

Dr Lloyd M. Scott, Professor, Construction Science Division, University of Oklahoma, and School of Surveying and Construction Management, Technological University Dublin, for his contributions to the chapters on retrofitting and refurbishment, thermal performance, and environmental considerations

Dr John P. Spillane, MCIOB, MAPM, School of Engineering, University of Limerick, Limerick, for his contributions to the chapters on thermal performance and disasters and the built environment

Prof. John Sturges, Leeds Sustainability Institute, Leeds Beckett University, for his chapter on sustainability

Felix Thomas, BSc, Leeds Sustainability Institute, Leeds Beckett University for his contributions

Dr B. N. West, MRICS, Chartered Building Surveyor, Leeds Beckett University (retired), for her contributions to the chapter on building pathology

1

Building surveying: an introduction

Melanie Smith and Chris Gorse

Building surveying is a rare blend of art, science, and management. As many people have observed – assessing building performance and failure is a science, and surveying buildings is an art, so building surveying combines art and science, attracting people interested in many different fields. As such, building surveying is arguably the widest area of surveying practice. Building surveyors are found working on and with buildings in fields of residential, commercial, industrial, educational, institutional, retail, leisure, health, local and central authority, armed forces, public, private, and everything in between. Consequently, workloads include everything from the conservation and restoration of historic buildings to designing and managing new developments.

Building surveying is a profession. Not all building surveyors are chartered, however, Chartered Building Surveyors need to comply with the Royal Institution of Chartered Surveyors (RICS) requirements of the highest levels of integrity, in order that clients can have confidence in the impartial advice given. In this book, "Building Surveyors", with initial capitals refers to surveyors who are chartered, i.e. members of RICS; "building surveyors" without initial capitals refers to all building surveyors, chartered or not.

Core traits which characterise formal professions such as membership of the RICS are:

- Practice is based on theoretical and privileged knowledge.
- Training is rigorous and specialised.
- The proficiency of practitioners is institutionally monitored and examined.
- Self-regulation, the possession of codes of conduct, and the enforcement of standards are the norm.
- The central ethical claim is that the profession exists to serve the interests of society in general, and the client in particular – not itself first.

Figure 1.1 shows the three pillars to building surveying: Technology, Law, and Management.

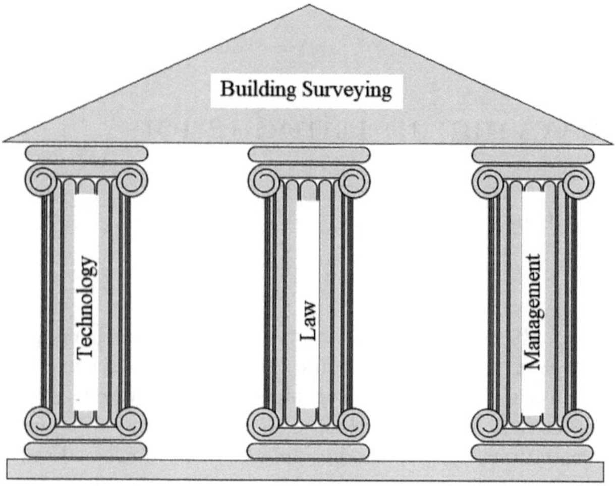

Figure 1.1 Three pillars of building surveying.

Building surveying as an occupation or profession has developed from the earliest times. Figure 1.2 gives a timeline for surveying roles in England. Evidence of surveying as a work process, especially land and property measurement, dates back to the ancients, and it is likely that these skills were brought to Britain by the Romans. More recently, surveying for thermal retrofits only began in earnest in the UK (for more than one-off properties) following the launch of the Green Deal from around 2012, when proposals for retrofitting large numbers of hard-to-treat UK dwellings with external or internal wall insulation began.

Surveys of buildings, or building surveying, predate this by over two thousand years, having been formally recognised as a skill, as a profession for some 400 years, and regulated to some extent (including self-regulation) for approximately 150 years. Because errors can be costly (monetary and/or health-wise) either to clients, occupants, or the country, surveying activities have attracted control. Some surveying aspects became encapsulated into law about AD 1212 when fire safety regulation was introduced for London and compliance control became necessary. Other aspects continued, informally, under the requirements of the clients. In the 19th century, the Industrial Revolution resulted in mass property growth, surveyors grouped themselves into professional alliances, and the RICS originated.

From this time, it has been understood that regulatory control of building construction, and that of the surveyors that oversee or monitor such construction, brings about improved results. The relatively new activity of pre-retrofit surveys can thus be expected to be controlled.

Timeline	Before 1200s	1200s	1500s	1600s	1700s	1800s	1900s	2000s
Commercial and agency (boundaries, landlord & tenant, etc.)	Surveyors (evidence from ancient Babylonia, Greece, Egypt, Rome)				Surveyors Club formed 1792	Institution of Surveyors formed 1868 Chartered Institution of Surveyors (1881 Royal charter given)	Royal Institution of Chartered Surveyors (renamed 1947)	
							Surveyors started to split themselves into types: general, quantity, mining, building, commercial, estate, etc.	Retrofit began earnestly, prompting pre-works surveys
Enforcement inspection		Viewers created to enforce the various 1212 regulations, Assizes: (Nuisance & Building)		Surveyors (London) 1667 London Building Act	Statutory Surveyors (London) 1774 London Building Act	District Surveyors (London) 1844 London Building Act District Surveyors Association, formed 1845	Chief Building Surveyors & Building Control Officers (England & Wales) 1965 national Building Regulations	
							Approved inspectors created 1984.Building Act 1984	
Works' client overseer	Surveyors (evidence from ancient Babylonia, Greece, Egypt, Rome)				Surveyors Club formed 1792	Institution of Surveyors formed 1868 Chartered Institution of Surveyors (1881 Royal charter given)	Royal Institution of Chartered Surveyors (renamed 1947)	
			Surveyor of the King's Works (created 1578)			Office of Works (created 1851)	Ministry of Works (created 1940) Dept of the Environ-ment (created 1970) English Heritage (created 1983)	English Heritage (reorganised 2015) Historic England (created 2015)
				Surveyor to the Fabric (cathedrals) (began around 1600s)				
Performance assessors							Building performance research began around here	
								EPC assessors created 2007
								PAS 2030 (2012) installers

Figure 1.2 A timeline for surveying roles in England.

Mass construction for the Industrial Revolution sometimes resulted in unhealthy conditions due to both construction standards and inadequacy, for example in ventilation and drainage. The consequential Public Health Acts in the 1800s tightened building regulation and compliance control in London and some other cities. However, it was not until 1965 that the first set of national Building Regulations came into force for the whole of England. These regulations required, for the first time, building control surveyors across the country to inspect both building plans and works on-site for compliance with regulations. Statutory site inspection is mandatory for certain stages only, with the emphasis being on observations (1) up to ground level, and (2) at completion.

Running alongside enforcement inspection has been the development of the legal aspects of property, including purchase, occupation, selling, and leases. Landlord and tenant legislation attempts to ensure tenants are able to occupy properties which are not unsafe or unhealthy, and to ensure that the lessor's investment is not diminished by the action or default of the lessee. Surveys for maintenance, fire risk, facilities management, access, party walls, and due diligence are other examples. Surveyors act for, and undertake on-site surveys for, all parties (owners, occupiers, landlords, tenants, and lawyers) in these transactions.

Lately, surveying for the purpose of performance assessment has also been developing. Performance assessment includes testing and measurement, often using new technology and equipment. This type of survey is a relatively new concept as shown in Figure 1.2. Building performance evaluation methods such as air-tightness and co-heating testing tend to be self-regulated by adherence to the methodologies stipulated in industry-accepted standards (such as CEN and ISO). Incoming Smart meter technology also provides metrics on the behaviour and energy efficiency of buildings, and such performance metrics are likely to embed themselves firmly into the practice of assessing and surveying buildings.

Building surveying is therefore a term covering a range of activities that are fundamentally about viewing or observing a building. However, the scope of each survey depends on the purpose of the client and the associated legislation or regulation. Survey purposes can be categorised into three purpose sets, of property investment, building work, and/or legislation, which are further described below.

SURVEYING FOR PROPERTY INVESTMENT

Owners and occupiers usually wish to maintain their financial interest in the property and, thereby, maintain its physical aspects. On-site property surveys are non-mandatory tools used for this purpose, at times of purchase, lease, sale, and periodically in between. They are used to assess condition, find faults (particularly decay or failure), find breaches of contract (dilapidations), assess value, assess and review service charges, budget for future maintenance or building works, and to carry out maintenance. Survey methods for property investment are set out under RICS and RIBA guidance, and although not legally regulated, can attract court proceedings for negligence.

SURVEYING FOR BUILDING WORKS

When building works are to be undertaken, surveys of the site and relevant property are conducted. The surveyor could be the building owner/occupier, architect/building surveyor, or building contractor. Surveys are undertaken to ensure the building work design is appropriate, that the completed building works are as required, and that instalment payments are made appropriately. Such surveys may be casual. Although they are not mandatory, they are practically, contractually, and legally useful. If the surveyor has a profession such as RIBA, RICS, or CIAT, the surveyor is expected to perform their surveying activities in accordance with their professional body's stipulations, and again can attract court proceedings for negligence.

SURVEYING FOR LEGISLATION

When building works are carried out, which fall under the Building Regulations in England and Wales, applicable work must be assessed by building control (local authority or approved inspector) Standards are set by the relevant Approved Document. Those most obviously relevant to retrofit are Approved Documents C (site preparation and resistance to contaminants and moisture), F (ventilation), and L1b (conservation of fuel and power). Building regulation inspections contribute to the health, safety, welfare, and convenience of people in relation to buildings and further the conservation of fuel and power. They are required by the Building Act 1984 and are mandatory.

Plans for the works may be deposited with the local authority or approved inspector, but there is no requirement to visit the site before works start. On-site inspections must be made at certain construction stages which are, as a minimum, the excavations for foundations, the damp-proof course (dpc) and oversite, as the drains are laid but not covered, before plastering, and at completion. At completion, the surveyor is required to confirm that as far as is reasonable to observe, the property is in accordance with the approved deposited plans (if any) or is otherwise in accordance with the regulations. However, much of the detailed construction work is carried out and covered over without inspection being required; for instance, between the dpc and plastering, and between plastering and completion.

When properties are constructed, sold, or leased, they are required to have an Energy Performance Certificate (EPC), which has been produced and registered by an accredited energy performance assessor. The standard assessment procedure (SAP) is the government's tool used to assess the energy and environmental performance of new dwellings. RdSAP is a reduced-data version of SAP used as a lower-cost method of assessing the energy performance of existing dwellings. RdSAP is used to underpin the algorithms in domestic EPCs. EPCs are solely designed to provide an assessment of the energy performance of the property and what can be done to improve it. The assessor must visit the site and conduct a survey using UK government-approved RdSAP software. EPCs and their on-site surveys are mandatory.

SURVEYING ACTIVITIES DISCUSSION AND SUMMARY

The preceding sections have described some of the differing activities and approaches used in building surveying. There are other types of surveying, for example estate/commercial surveying and quantity surveying, which are not covered in this book. Thus, if a survey, assessment, inspection, etc., is required for a building, it is necessary to provide a scope of service or term of reference to define the requirement so that an appropriate survey is conducted.

There is not an examination to certify or guarantee a person's proficiency in undertaking a survey. Traditional types of survey (e.g. dilapidations inspection, condition survey, full building survey, home-buyers reports, due diligence inspections, fire risk assessments) have been developed and refined over many years. Some, for example dilapidations and building surveys, have been honed through the courts by case law. Court action has resulted in surveyors having substantial understanding of what each particular survey entails. Not meeting such accepted understanding can lead to negligence action. Contractual terms of reference do not have to be specific, but can be implied.

Some survey types, such as for thermal retrofit, are a relatively new facet of building surveying (Figure 1.2). They have not, to date, been interrogated in court; however, they could be expected to meet basic, fit-for-purpose methods. There is likely to be some overlap between say retrofit surveys and, for example, building surveys, condition surveys, and EPC inspections. As a new activity, the procurement specification will be important. Using a retrofit survey as an example, because the survey will feed into the success of the retrofit, and the retrofit is required to save energy and money, any failings might attract court proceedings and therefore the scope of a survey will be important, as will limitations of liability and professional indemnity insurance.

MODELS OF SURVEY APPROACHES

The process of survey approach has been modelled. Figure 1.3 shows three comparative approaches of data capture or collection (elemental), diagnostic (problem), and anticipatory (preventative).

Data collection: Observe the element. Identify any manifestations. (Elemental)
If required, can progress to: Identify a cause and determine remedy for the manifestation.

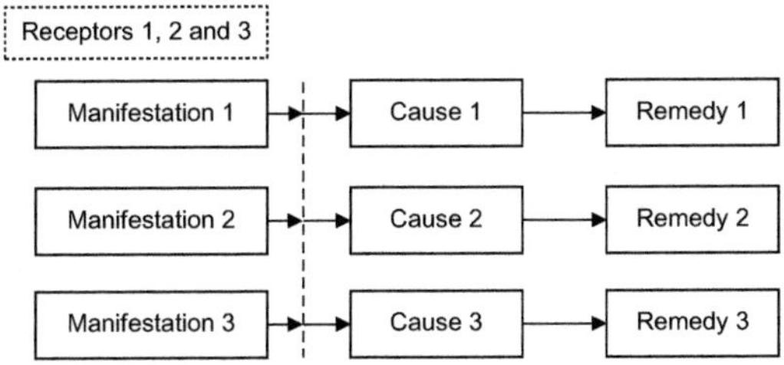

Diagnostic: Decide problem. Look for manifestations (consequences) on individual receptors (elements). Follow pathways from individual receptors back to source(s). Determine cause and thus remedy for problem (if required).

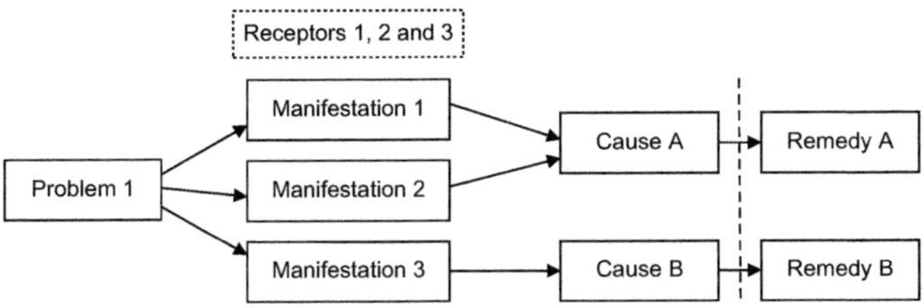

Anticipatory: Observe existing building's aspects. Anticipate retrofit installation(s). Decide problem to interrogate. Anticipate predictable causes of the problem and the possible manifestation(s) on individual receptors (elements). Risk assess problem level. Inform designer of significant risks to enable them to design-out the causes.

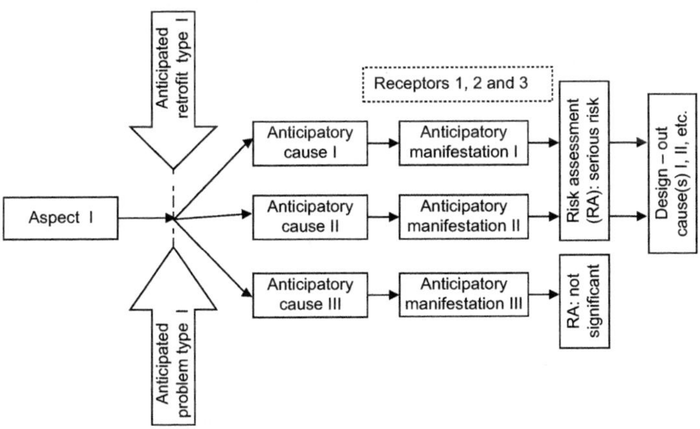

Figure 1.3 Comparative models of Data capture (elemental), Diagnostic (problem), and Anticipatory (preventative) approaches to surveys.

The different approaches can attract differing levels of proficiency and/or speed of undertaking:

The **data collection survey approach** is defined by tick-boxes and a consideration of each element separately and in isolation from other parts of the property. No reflection is required. The choice of observation is chosen from a drop-down menu on the electronic device. Minimal surveying skill is reportedly required. This method can be conducted quickly and easily, with minimal technical equipment. It lends itself to off-site quantitative analysis. The results are of data capture only. This approach identifies existing conditions only, no connections between elements or defects are made, no conclusions are drawn, and only the facts are reported. The time resources associated with this type of survey, coupled with expertise limitations, preclude any technical testing.

The **diagnostic survey approach** incorporates data capture and diagnosis techniques. It is defined by structured observation of an element, simultaneously coupled with corresponding consideration of specific known problems. These problems may involve association with more than the single element observed and may include many elements and other parts of the building. Holistic observation and consideration are therefore intrinsic. Extant problems are searched for, observed, and diagnosed, with possible multi-criteria connections. Reflection is essential. Each element is observed in turn. The order and choice of observations made are surveyor-controlled, usually led by the specific building's characteristics and/or problems found. Extensive surveying skills are required. This method takes longer and costs more than the data capture approach. It results in qualitative assessment. This approach identifies existing conditions and problems only. It ventures into forecasting only when the current circumstances point to an obvious consequence, for example if the roof is not mended, fungal decay of the roof timbers may be expected. The process can provide a holistic view of the property, and adjacent properties where relevant. Expertise limitations can preclude additional testing.

The **anticipatory survey approach** is required for the prevention of future problems. It requires a different approach to either the data collection or the diagnostic approaches. Current guidance for surveys focuses on identifying existing problems, not on anticipating the potential risks associated with undertaking proposed works. The survey is essentially anticipatory in nature, risk-based, and focused on preventing any potential foreseeable and unintended adverse consequences and problems following works. Its main purpose is to identify any potential future problems so that they can be mitigated against as part of the proposed design of the works. The anticipatory survey approach will observe extant elements, aspects, defects, entropy, etc., which have the potential to contribute to failures and shortcomings of the new scheme. It will also anticipate the potential of the prosed works to damage the existing materials and/or construction. Anticipated problems are to be foreseen for the (known or assumed) proposed designs and risk assessed. Reflection will form an essential component of this approach. The choice of observation may be surveyor-controlled, or proforma-controlled, but either way, it will result in an over-riding surveyor intervention. A high level of understanding of retrofit measures and their consequences will be required. It will include data capture, diagnostics, and risk identification. Additional expertise is likely to be required for additional technical testing.

BUILDING SURVEYING AND SUSTAINABILITY

Figure 1.4 shows stages for building surveying involvement in a building's life. Because of their interest in maintenance and existing buildings, building surveyors have historically been involved with sustainable construction. They have chosen materials and practices that will result in longevity and slower entropy. With their understanding of building pathology, they have also looked to construction that fits well in its intended environment. With this understanding comes resilient building. As climate change results in more natural disasters, stronger winds,

Figure 1.4 Stages for building surveying involvement in a building's life.

higher rainfall, increased temperatures, and higher flood risks, then sustainable and resilient buildings are required. Post-disasters, rebuilding is required, which will cope better in the future. Building surveyors' skills and expertise will continue to care for the built environment and sustainability in its many forms is a key issue for the profession.

2

Form and construction

Chris Gorse and Melanie Smith

BUILDING FORM

The shape and type of building, and the materials that make up the building components, have direct impacts on the way the building behaves and performs. The building responds against changes placed on it by the internal and external environment conditions; both environments can change the way the building fabric behaves. Notwithstanding that environmental conditions for each building are unique, there are a few basic assumptions based on a building's form that are useful when characterising the response and behaviour of a building.

Some shapes and assemblies will expose more of the building to the elements, which can result in greater heat loss and gain. An increase in exposure increases the effect that the external environment has on the building. Buildings with greater surface areas are more exposed, tend to get wetter during rain, and transfer more heat energy when it is cold, than those that are less exposed. Equally, more exposed buildings receive more solar energy and have to withstand greater wind loads. Even for very simple shapes, the difference in exposure can easily be seen in the basic assemblies of Figure 2.1.

Assuming the building units in Figure 2.1 are composed of the same materials, then the detached unit has the greatest exposure and will be prone to greater heat transfer to the external environment than the other units. For the single-unit detached building, all five surfaces above ground are exposed. The semi-detached building has one face of the building shared as a party wall with the adjacent attached building and thus has 20% less exposure to the outside elements, when compared to the detached property. The mid-terrace of the group of three units has two walls joined to its neighbouring buildings, and has 40% less exposure. Mid-apartments in a block of flats may just have one exposed elevation and 80% less exposure to the external elements. These are simplified examples, and it is important to note that the interfaces between buildings are not always sealed and may link to the external environment. Where there are cavities within the party walls between buildings, these can act as a heat-loss mechanism in their own right,

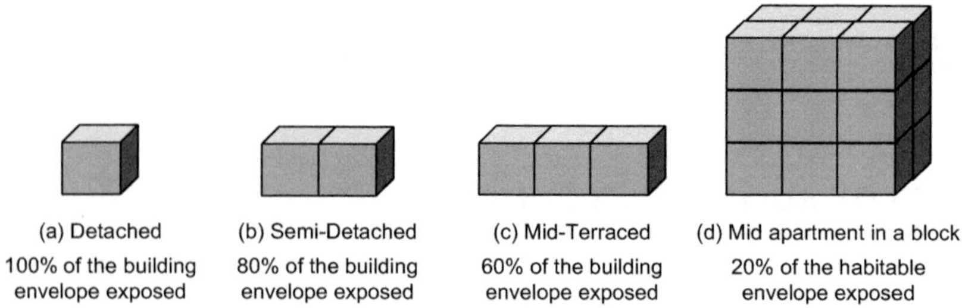

Figure 2.1 Degree of exposure for simplified habitable spaces, relative to the walls and roof of a detached structure. The building units are made up from simple cube shapes, for ease of comparison.

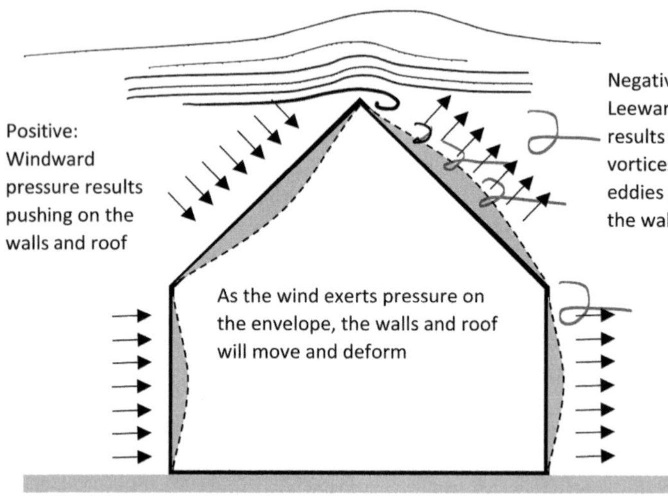

Figure 2.2 Simplified diagram of the effects of wind on the windward and leeward sides.

connecting the internal to the external environment. However, when buildings are connected with a fully sealed wall, the heat loss and gain to neighbours are normally less than those experienced with an external wall, unless adversely affected by a party wall thermal bypass (LABC, 2015).

The solar and wind loads are also heavily influenced by the shape of the building as well as the exposed surface area, as shown in Figures 2.2, 2.3, 2.4, and 2.5.

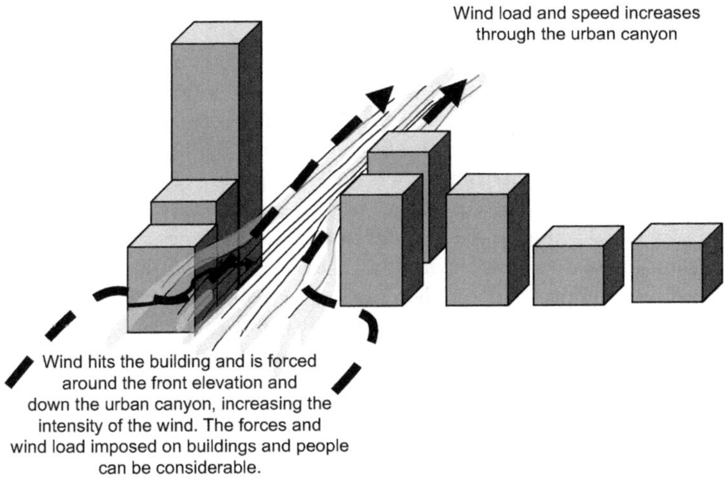

Figure 2.3 Schematic of urban canyon and funnelling of wind.

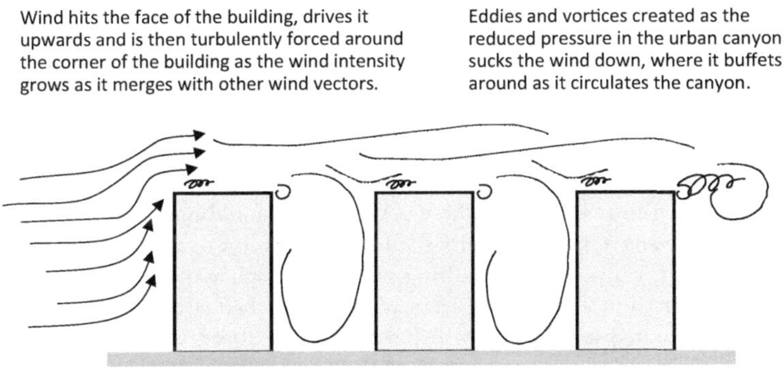

Figure 2.4 Wind driving over buildings, swirling around urban canyons, and creating vortices at the edges and corners.

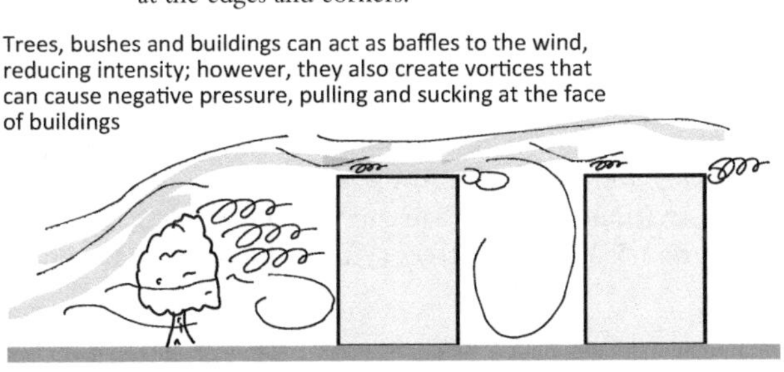

Figure 2.5 Trees and buildings reducing wind intensity at lower levels.

WIND AND BUILDING FORM

Wind has a major impact on the building element. As the wind pushes on the building, it puts pressure on the faces that it hits, and as the wind rolls over and around the building, wind vortices occur. Swirling eddies occur around ridges and corners of the building, pulling and twisting around the building fabric as illustrated in Figure 2.2. The eddies and vortices induce negative pressure and suction, which can lift and displace roof tiles, sheets, trusses, and other loose fabric. This is one reason why many countries require roof timbers to be strapped to the walls.

Where rain, snow, or hail is caught up in the swirling winds hitting the surfaces, the wall surfaces may suffer higher levels of erosion, especially where wind pressures are high. Where vortices result in the lifting of roof tiles or sheets, the negative pressure can also result in water being forced upwards and under eaves, verges, and roof tiles, against the natural pull of gravity. Thus water captured by the wind can penetrate gaps unexpectedly.

WIND IN URBAN AREAS

Trees, buildings, and hedges may offer shelter to buildings, but equally they can create wind tunnels or funnels that concentrate wind pressure and increase its power. Wind is also useful in clearing air; low-lying areas in the cities and built environment "canyons" can become areas of concentrated pollution. Wind passing through canyons can help remove and reduce pollution in these areas. However, if the impact and potential build-up of wind in canyons are not properly considered, this can lead to serious consequences.

In one situation in Leeds, UK, where a 112 m high building was erected, the roads, waterways, and adjacent buildings led to the creation of a significant wind tunnel – an urban canyon. The resulting extremely high winds pushed over a lorry, causing fatal injuries to a pedestrian who was crushed (BBC, 2017a). Large gantries and perforated steel wings, which act as wind baffles, were subsequently placed where the wind tunnel effects were the greatest (BBC, 2017b). The gantry and perforated wings diffuse and disperse the wind, reducing the effect of the wind tunnel at ground level, and are effective at reducing the wind speed at lower levels. Models and wind simulations have further explored the wind forces in the area (Ingrid Cloud, 2018).

Figure 2.3 provides a schematic of how wind can be pushed through urban canyons, increasing the intensity of the wind, causing building damage and presenting risks to those working and living within the area. Figure 2.4 shows how wind drives over buildings, swirling around urban canyons and creating vortices at the edges and corners. Figure 2.5 shows how trees and buildings can reduce wind intensity at lower levels

The shape of buildings can reduce the impact of wind. Tall buildings are often narrower as they increase in height. At lower levels the surrounding buildings and trees create friction and slow down wind speed; however, at higher levels, where buildings and vegetation are negligible, wind speeds increase. Wind speed exposure increases the higher the buildings are; thus the narrowing of buildings at heights reduces the exposure to dynamic wind loads placed on such buildings.

While winds at a higher level are stronger, they tend to be relatively consistent and easier to predict, whereas the winds closer to the ground are changeable and unpredictable. Thus, the turbulent effects of winds lower down the building may result in damage to the building.

Flat or square buildings create the highest impact area and potential pressure from wind, whereas curved surfaces reduce the pressure perpendicular to the surface. Thus, some tall buildings are aerodynamic in shape to reduce wind loads. However, the oscillation and transverse swaying of tall buildings can become a problem with such shapes. Studies have shown that small dimples in the surfaces of buildings can reduce the effect of the vortex, helping the building to slip through the wind, similar to the dimples on a golf ball or on the skin of a shark. See Figures 2.6 and 2.7. While such observation and studies tend to be associated with tall buildings, the principles will also apply to smaller buildings. Rough surfaces can be helpful in diffusing the impact of wind.

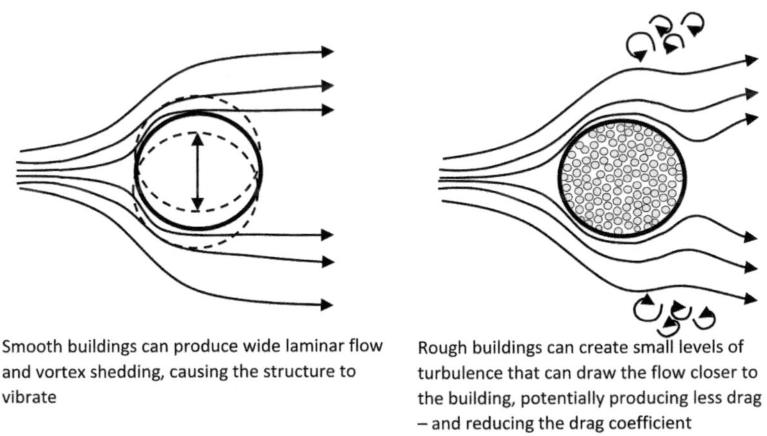

Figure 2.6 Wind flow around buildings shapes: impact of rough or dimpled surface compared with a smooth-surfaced building (adapted from Vongsingha, 2015).

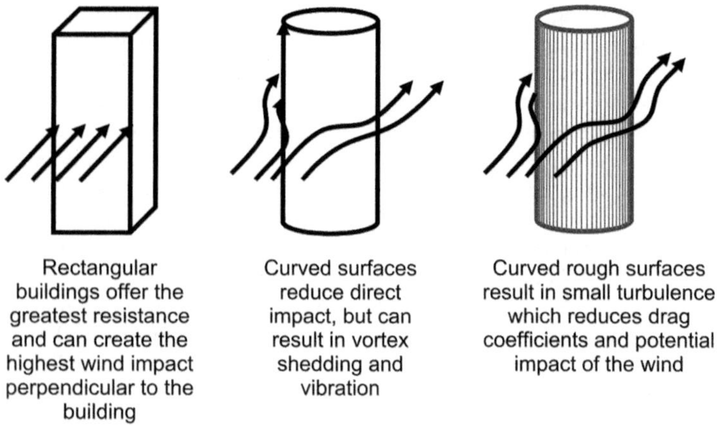

Figure 2.7 Tall building and the impact of shapes on wind resistance.

PROTECTING THE BUILDING ENVELOPE FROM RAIN AND SOLAR

The roof structure is an important feature when considering exposure to the elements. The different shapes and forms of the building can considerably change the building's behaviour and resilience to the elements. The way a roof overhangs the building's envelope walls can change the performance and resilience of those walls. High levels of solar exposure will cause materials to degrade, mainly as a result of the ultraviolet component of the spectrum of light. Plastics, wood, and paints are particularly prone to physical, chemical, and mechanical changes (normally on the surface of the material) when exposed to solar radiation for prolonged periods. While most building materials will have additives and stabilisers to reduce the impact of solar radiation, it is possible to reduce exposure through changes to the shape and form of the building. For example, recessing windows and doors helps to reduce their exposure, thus protecting them from rain ingress.

The exposure of walls can be reduced by increasing the projection of the eaves. Increasing the projection of the eaves not only results in reducing the level of solar radiation impacting on the walls, it also reduces the levels of water and pollutants carried by rain that can hit the face of the building. The pollutants can make the rain more acidic or alkaline, making it chemically aggressive to the masonry. Dust carried by rain can also cause staining of the surface.

Table 2.1 Roof forms and impact on building envelop performance and resilience

Roof form	**Impact**
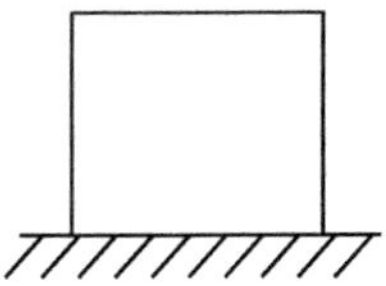	Simple flat roof structure – no overhang: Offers limited protection to the building element, places the walls at highest level of exposure. While gutters prevent immediate run-off it is commonplace to see walls close to the eaves discoloured and affected by high moisture levels.
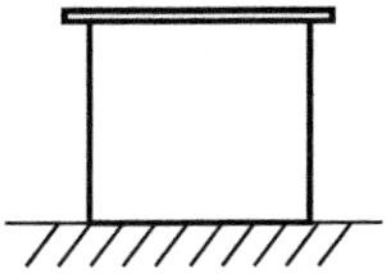	Flat roof structure – with eaves overhang: The projection of the eaves helps to shelter the walls from rain and reduced solar exposure. The protection does help to reduce staining, and reduces the period in which walls are wet (and thus allows greater heat flow).
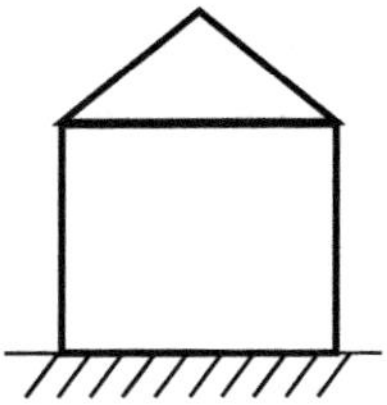	Pitched roof with no overhang: Loft space in the roof, when insulated, creates a buffer to the external environment. With no overhang, the walls are exposed to rain and solar. The finish on the exposed walls tends to discolour and degrade with exposure. Wet walls allow heat flow more readily.
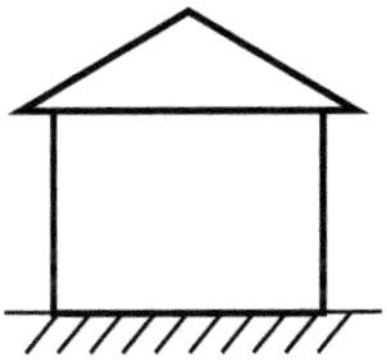	Pitched roof with overhang: Loft space creates a buffer; the eaves offer the walls some protection from the rain and reduce solar exposure, which limits staining, degradation, and reduces wetting.
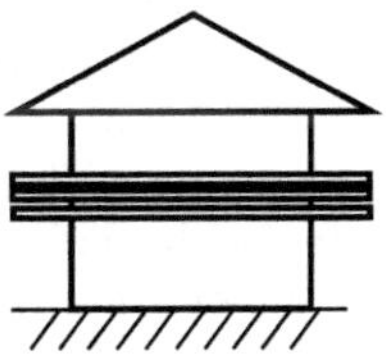	Pitched roof with extended overhang and balconies: Typical of higher-altitude buildings. The extended roof and balconies reduce the wall wetting, limit any freeze-thaw action, and reduce exposure to solar, limiting degradation of the wall. The roof and the balconies are exposed and will require maintenance. These are sacrificial elements helping to protect the main building.
	Room in the roof: Room in the roof maximises the habitable space. These can be both retrofits and incorporated from new. The roof can be complicated in shape; retrofits are often poorly insulated, increasing heat loss and gain in the habitable space.

By reducing the level of water that falls onto the surface of walls, the walls do not become as wet. If walls are wet, the air pockets within the walls fill with water, reducing the wall's ability to act as a thermal insulator. Therefore, saturated walls conduct more heat and reduce the wall's thermal insulation properties. Also, as the water evaporates, further heat energy is lost. Latent heat is required to change liquid to vapour; thus the additional heat energy is transferred from surrounding materials.

When water hits the wall surface, moisture becomes trapped inside the wall's pores. If the temperature drops, this moisture may freeze and, if there is insufficient space for the water to expand as it freezes, the material surrounding it will break and crack. This freeze-thaw cycle of water trapped in the building pores will lead to degradation of the building materials. Where buildings are particularly exposed to the elements and are in elevated or cold climates, it is common to have extended eaves and balconies that protect the main building envelope. Although the exposed balconies and roofs do suffer increased exposure and require more maintenance, they are less costly to maintain and are treated as sacrificial, putting themselves before the main building envelope and offering protection.

GROUND FLOORS

The way a building rests on the ground, transfers its load, and separates the internal and external environment has an impact on the behaviour and performance of the building.

Traditionally ground floors were constructed directly on the ground, with no insulation or damp barrier. If there was a finish (other than compacted earth), tiles, stone slabs, or timbers were laid, either directly onto the ground, on compacted earth, a bed of ashes, or onto pitch. From around 1880, floors began to be sealed with a concrete cover. Ground floors are now insulated and separated from the ground with thermal insulation and a damp-proof membrane. Suspended timber floors are ventilated. The thermal and moisture layers create barriers that prevent the ingress of water and restrict the transfer of heat. In properties with no floor insulation, solid concrete floors are prone to condensation forming on the surface, even when finished with tiles or another surface covering. This can sometimes be hidden, for example below lino or vinyl floor coverings.

Solid floors can and do act as heat reservoirs. The heavy dense structure of a solid floor can have high thermal capacitance and store high levels of heat energy. Thus, floors can take a considerable amount of heat energy to heat. Equally, once the heat energy is contained within the floor, it is steadily released back to the building and surrounding areas, as the supply of heat is reduced. In some buildings, heavy structures such as solid concrete floors are used to slow down and smooth the thermal response of buildings. While solid floors still have high levels of thermal capacitance, most newer builds are now insulated, which will reduce the heat stored. However, the insulation does reduce the heat being lost to the external environment and reduces the potential of condensation in the floor.

Even when solid floors are insulated, there is a tendency for the perimeter of the floor to present a thermal bridge where it interfaces with the wall. This is due to the increased exposure of materials (a geometric thermal bridge – at the corner of the building there are two edges and the underside of the floor all exposed) and due to the interfaces of different structural materials (thermal conduction). Care should be taken to insulate the perimeter of the floor and reduce the risk of condensation.

There are many different types of floor, and Table 2.2 represents just a few of the main configurations. It is useful to consider the way the different floors affect the performance of the building. As discussed above, solid floors provide a direct interface with the ground, which may be prone to heat transfer, moisture ingress, and surface condensation. With insulation and damp-proof barriers many of the limitations of solid floors can be overcome.

Another method of reducing the chance of ground water penetrating the floor is to separate the floor from the ground with a cavity. A suspended floor is traditionally constructed of timber supported by the external walls, with a gap between the ground and the timber. The gap is ventilated to prevent the build-up of damp and stagnant air. With the introduction of air vents in the perimeter wall, the void below the ground is ventilated; however, the movement of air in the floor can lead to draughty floors and high levels of air infiltration into and out of the building. Such draughty floors can be a significant heat-loss mechanism. Suspended floors are now constructed both from timber and concrete beams and blocks; they are insulated and more airtight but are typically prone to air infiltration.

Floors above cellars are separated by a large void; in principle this should mean that they are less prone to damp, however, this is often not the case. Unless tanked, cellars which are below ground and the substructure walls and floors can let small amounts of ground and surface water into them. The rooms below ground tend to be colder than the superstructure, so that cellar walls and floors that are in direct contact with the ground are often cold and prone to condensation forming. Unless cellars are ventilated, damp and stagnant conditions can develop; this can result in moisture forming on the underside of the cellar ceiling and the underside of the floor above the cellar. With tanking and insulated basements, such conditions can be considerably limited, but do still need to be taken into consideration.

The way a floor is constructed affects the building's performance. All floors can be designed to perform and offer habitable conditions; however, equally, all floors present different challenges.

The shape and form of the building, and the construction of the different building components and elements, affect the building's behaviour and interaction with the external environment. At all times, the different elements and their interaction

Table 2.2 Different types of floor and their impact on floor performance

Floor type	Impact
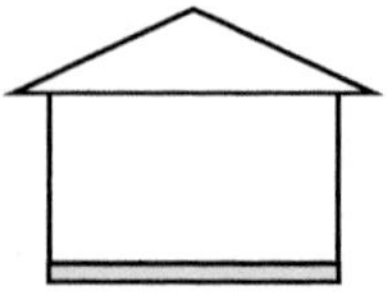	Solid ground floor: the floor rests directly on the ground. There will be heat loss into the ground. The floor is inherently damp commensurate with the type of ground and height of water table. Water vapour evaporates from the surface of the floor, and may condense on any impermeable floor covering, and/or contribute to the internal humidity levels. Any timbers incorporated into, or in close contact with, solid floors may be affected by fungal degradation.
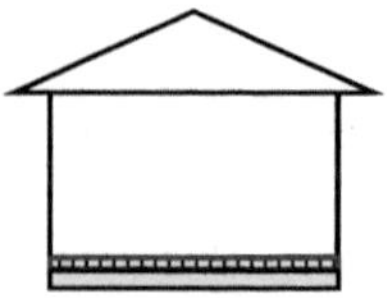	Insulated solid ground floor: the floor is separated from the ground with thermal insulation and a damp-proof membrane. In radon areas, the membrane needs to be gas-tight. If services puncture the floor, there may be a loss of integrity for heat flow and moisture passage.
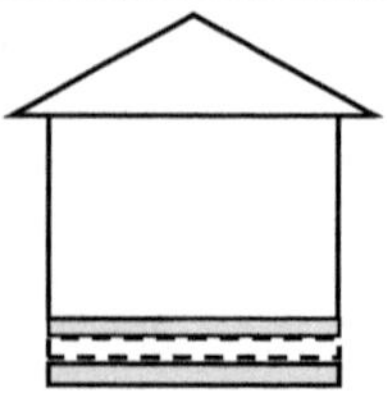	Suspended ground floor: a gap between the floor and ground deters damp penetrating the upper levels of the floor. The cavity below the floor must be ventilated. In renovated homes, there may be insulation inserted between floor joists. There may be a concrete oversite, or not. The joists may be supported by dwarf walls, or be laid directly onto the ground. The dwarf walls may be honeycombed, enabling airflow, or not. There may be DPCs (or slates) effectively positioned to avoid rising damp reaching the timber floor joists both in the external and dwarf walls, or not. Air grates may be sufficient in size, number, and position to provide adequate airflow to avoid build-up of moisture, contaminates (e.g. radon), or not. The ventilated void may be of sufficient height and lacking in obstruction to enable adequate free flow of air, or not.
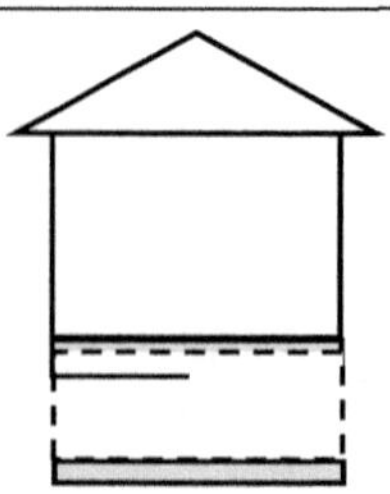	Ground floor suspended over a cellar: floor is separated from the substructure. The cellar will tend to be cooler than the main building. Unless tanked, substructures are likely to experience damp ingress, through floors and walls. The ground floor over the basement may be underdrawn with a ceiling, or otherwise sealed. If not, draughts through the floor, from the basement, are likely. In renovated homes, there may be insulation inserted between ground floor joists. If there is not equal insulation in the stair spandrels and upper flight soffits, there will be large thermal bridges here. Similarly, if there are no or inadequate air-sealants.

with each other need to be properly considered when buildings are designed or changes are made to the existing structure.

WALLS

Building walls perform a number of complex functions including the key requirements of supporting the structure, providing necessary strength, and offering protection from the external environment. To maintain a healthy internal environment, the wall has to respond to external and internal conditions, which fluctuate considerably and lead to complex performance demands being placed on the envelope. Because of environmental changes, differential stresses are imposed. Substantial changes are expected in any season, but during winter conditions, an elevation where surfaces are particularly exposed to the sun can be exposed to freezing temperatures and high solar heat gain at the same time. In such circumstances, if the wall's materials are not durable, cracking due to the differential stresses is likely to occur. When a wall is wet, parts of the surface may quickly dry as the wind and sun induce evaporation, while other shaded areas may remain wet for much of a season, and as a result, deterioration can occur more quickly.

Wind not only exerts positive pressure as it hits a building, but also suction as it flows over and around the building, inducing a negative pressure on the building envelope. The wind can throw rain, snow, and dirt at a wall, causing some areas to suffer from erosion. The external environment is harsh and changeable; thus, walls need to be durable and should be built with materials that give the required performance, given the building's specific environment.

How the building behaves and responds will be a result of the behaviour characteristics of its walls. This is a result of the building form, the protection offered by the shape of the building, and the assembly of components and materials making up the wall. When any new finishes or thermal or acoustic upgrades are applied, the behaviour of the wall will change. Thus, careful consideration is required when upgrading a wall.

The building walls comprise one of the largest building elements and are heavily exposed to the external environment; thus changes to the wall can have a considerable impact on the building's behaviour. In the past, most attention has been given to the wall's resistance to external conditions; however, internal environments also impact on the wall's behaviour and performance. Notwithstanding the structural and environmental loads, the impact of occupants, for example their living habits including washing and ventilation methods, will change the environmental characteristics and can induce stresses on and within the wall.

Different groups tend to use their homes differently. Families with young children and babies often wash and dry clothes more than others and tend to use much of the building space. Affluent professionals traditionally spend minimal time in the home and have relatively little behavioural impact, for example sending laundry out. The occupation patterns of elderly people commonly results in much time spent confined to the home, with the most occupied rooms being heavily heated and other rooms rarely used, heated or ventilation. Where security is a concern, windows may not be opened regularly, often creating stagnant and humid environments. This also happens when occupants are unable to open windows or have ineffective ventilation systems. In humid and poorly ventilated homes, where the internal faces and parts of the building fabric are cold, surface and interstitial condensation are likely, resulting in mould and discolouration of the wall. These traditional behavioural patterns change during COVID and similar lock-downs and the resulting more normalized homeworking may become more prevalent. The way people use their homes affects the type of environmental stresses that walls and buildings are exposed to.

Walls need to be strong and durable, resist the external elements, limit the spread and development of fire, and seal the enclosure whilst enabling ventilation and breathing, in order that the environment enclosed is safe and healthy for occupants to live and work in. Walls are often attached to other structures to create a more stable building. They also interface with other building elements, such as other walls, floors, roofs, stairs, doors, and windows, each transferring loads and creating different environmental conditions that need to be considered. The neighbouring external environment, internal rooms, and vented or unvented voids, with fluctuating temperatures and humidity levels, all create different demands on the wall. In some cases, services are attached and encased within walls. The cables, pipes, and conduits that we place on or within walls need to be properly accounted for so that they can be maintained and do not present hazards. Care must be given to the housing of such services, the surrounding materials, and how the combined risks associated with fire, explosion, or electric shock are reduced and can be managed. Thus, although walls appear to be relatively simple structures, their function and the demands placed upon them are complex.

WALLS AND UPGRADE RISKS

It is vital, for long-term success, to research and identify any unexpected consequences and adverse side effects of any upgrading. Due to their large surface area and the impact that the building element creates, walls are often seen as a building element that can be relatively easily changed. However, upgrading walls so they provide a safe and healthy enclosure can be notoriously complex. Sufficient care and attention should be given when assessing a wall's physical state to ensure it is possible to create the desired characteristics without the manifestation of unintentional consequences such as mould growth or a fire risk. For example implementing external thermal insulation along a row of terraced houses can result in an open route for fire to spread between properties. Thermal upgrades can also result in excessive condensation and mould – see also Chapter 5, "Retrofitting and refurbishment", especially the section "Upgrading thermal performance". Before retrofitting, changing, or upgrading a wall, it is important to consider all of the wall's functions, assessing its condition against the day-to-day requirements and specific performance attributes that are required.

FUNCTIONAL REQUIREMENT OF WALLS

The complexity of walls is a result of the way buildings have developed and the requirements that we have placed upon them over the years. The performance characteristics of walls have evolved; thus in addition to the key requirements, walls have aesthetic functions, need to resist fire, provide compartmentation, and help to provide safe access and egress in the event of a fire. They play a significant part as a passive component for energy efficiency.

Key functions are a good starting point when assessing if a wall is fit for purpose and has the potential to be upgraded. The central functions include providing structural and supporting elements while also enclosing and dividing the space that offers protection from the elements. Beyond the key functions, as the wall encloses and divides much of the building space the performance requirements are more involved.

While relatively inexpensive carbon-rich fuels are being phased out, the harvesting of energy from zero-carbon sources remains expensive, so walls need to help retain and conserve energy used within the home. Outside of conserving heat energy and ensuring thermal comfort, walls assist in providing a secure environment, offering protection against wind, rain, dust, and pollution, and limiting heat lost and/or excess heat gain from solar energy. Importantly, walls must not be made of toxic materials nor become a part of the building where mould and spores grow or create conditions that are hazardous to health. Thus, while walls exist in simple forms, the building requirements and functions of walls are considerable and must be given due attention as buildings are adapted and maintained.

The main functional requirements of walls include:

- Strength (sufficient compressive, tensile strength)
  - Stability (resist distortion and collapse)
  - Sustaining building loads; sustaining dead loads from roofs, floors, walls and live loads of the fixtures, fittings, furniture, and people
  - Resisting eccentric loads, lateral wind forces
  - Sufficiently robust, stiff, and stable (to resist collapse by ensuring slenderness ratio is observed, reinforcement is sufficient, tying into buttressing, ribs, beams, pins, or structural form that offers stability)
- Weather resistance (resisting penetration of rain, snow, and wind)

- Resistance to the passage of heat (resisting thermal transfer from inside and solar heat gain from outside)
- Security (providing a safe and secure enclosure)
- Durability (sufficiently resistant to erosion, allowing for expansion and contraction due to changes in moisture, temperature, and humidity)
- Fire safety (resistance to passage, spread, and propagation of fire, within and over the surface of the wall)
- Resistance to moisture from:
    - Penetration through the structure
    - Penetration from ground moisture
    - Build-up within the structure
    - Surface condensation
- Aesthetics and creating a specific form
- Providing protection from pollution
- Not being made of or releasing toxic substances
- Not creating or harbouring an environment where mould can grow and conditions become hazardous to health
- Safely housing the services placed upon and within them

STRENGTH AND STABILITY OF EXISTING WALLS

When a wall is upgraded, the strength and stability of the existing structure is key and needs to be assessed. It is worth noting that although walls are rigid structures, they are not totally static and will move and adjust. Thus, the stability of a wall is affected as the building moves and adjusts to its surroundings. In most cases, materials and walls will perform for the lifetime of the building; however, surveyors should be aware of factors that can affect buildings and those changes which are more than superficial and require specific attention.

Materials will move as a result of changes in moisture, temperature, and humidity. See also Chapter 4, "Building pathology". Over the course of a year, the wall is exposed to the different seasonal conditions. Patterns of wind, rain, snow, and solar exposure all change with time, although there are seasonal patterns that can be taken into account. The amount of solar heat energy that hits any part of the structure varies as the sun position in the sky changes. Solar exposure is also affected by surrounding vegetation and buildings that may shield the building. The differential temperatures will cause materials to move at different rates and impose stress on the wall. Over the years, movement can increase and cracks develop. Sometimes cracks simply open and close with seasonal change, not particularly affecting the structure. However, it is not uncommon for debris to become lodged within the crack or for the jagged shape of the crack to create a mechanism that locks the crack in an expanded position. Such cracks do not close and can propagate. If during one season a crack becomes locked in an open position, the next year the expansion caused by heat or moisture may continue causing more movement, meaning that year on year the cracks grow and propagate, affecting the structural stability of the building. Such cracks should be monitored with tell-tales and checks, to ensure the cracks do not disconnect structural components and affect the structure's integrity.

The ground under the building also moves with seasonal changes, causing movement of the foundation, which is then transferred to the structure and walls. Eccentric loads, lateral forces (wind forces), and expansion and contraction due to temperature and moisture changes (as a result of solar, rain, external temperature, and humidity) all impose stresses on the wall, which in some cases can render them unstable or lead to collapse.

The strength and stability of the wall will be increased through bonds, ties, straps, buttress walls, and irregular profile (zigzags, offsets, chevron, fins, cross ribs, etc.). Where the wall relies on such mechanisms, it is important to check that they and their connections are still in place and can contribute to the stability of the structure.

The materials selected for walls, when constructed and joined together, must prove to be sufficiently robust and stable, having sufficient compressive and tensile strength, with the required rigidity. When first designed and assembled, test standards and safety factors will ensure that the materials are sufficiently strong and robust to perform their function. Walls should easily be able to carry all loads placed on them. A factor of safety will be included to allow for differences in the designed loads and those encountered during operation. The safety factor will accommodate minor differences in construction, material composition, and workmanship. The addition of a factor of safety means that the actual operation loads are substantially lower than those that would result in material failure under compression and tension. The margin between expected operational stress and failure stress allows for minor changes that occur during the building's lifetime. However, when surveying existing walls, it is essential to determine if the wall is sufficiently stable and no discernible changes have taken place since construction that would compromise safe operation of the wall.

Notable occurrences to consider when assessing the structural components and materials of walls:

- Timber – deterioration as a result of mould, insect infestation, wear, fire, water, movement, or distortion (through wind, impact, or subsidence damage). If any major changes have occurred structural experts should be consulted. Timber will move as it naturally seasons and accommodates changes in temperature and moisture. Checks should be made to ensure that any movement would not impede performance. Obvious distortion and changes, evidenced by cupping, bowing, checking, shakes (splits), and knots, should be considered when assessing if the stability of the structure or wall has been compromised.

- Brick, stone, and masonry – cracking within the brick or mortar, spalling, delamination, and frost attack. Efflorescence and related spalling, wall tie failure, sulphate attack, weathering, erosion, disintegration of joints and masonry. Masonry moves as a result of heat, moisture, and to accommodate any movement in the building. Over the years, it is expected that masonry will experience a degree of notable movement; however if the movement appears excessive or fresh, further investigation may be required.
- Steel – defects such as the protection (paint or other surface treatments) cracking, resulting in corrosion; welding defects, stress cracking, fatigue failure, fretting, erosion, creep, or any changes that suggest the structural component has been compromised should be considered. In most cases, the deterioration of steel and other metals will be obvious, but it is essential that further investigations are conducted, should any changes be noted. Where delamination between steel reinforcement and concrete occurs, expert consultation may be required to undertake structural stability tests on composite structure.
- Concrete – although crazing, blistering, and dusting may be superficial and easily repaired, cracking, delamination, concrete curling, efflorescence, scaling, and spalling may be evidence of more sinister problems. A structural survey may be required when reinforcement has been exposed, or cracks are large or in close proximity to structural members. Where large areas of concrete are missing, the quality of concrete may be suspect and a further investigation is required.

WALL CONSTRUCTION, THE POSITIONING OF INSULATION AND CONDENSATION

The construction of walls has become quite complex, and many different forms of wall can be found. Recently, the most notable change to walls is the level of thermal insulation that is now found within the wall. Thermal insulation can be placed on the external face of the wall, internally, or between the leaves of masonry. Figure 2.8 indicates the risks of interstitial condensation in relation to the position of the insulation.

Interstitial condensation is where the water droplets condensate behind the insulation or within the wall. Where parts of the wall are colder than the rest, there is an increased risk of condensation forming. Internally, air is usually warm and moist, containing vaporised water droplets from cooking, washing, and human activity. This moist air, which can penetrate through the wall, may come into contact with building materials at lower temperatures; as it does the air temperature lowers. Air can hold more water vapour at high temperatures and less at lower temperatures. Moist air at lower temperatures is unable to retain the vaporised water; as the volume of air contracts in size it becomes saturated and reaches what is known as the dew point. Below the dew point, water droplets condense and dew can form

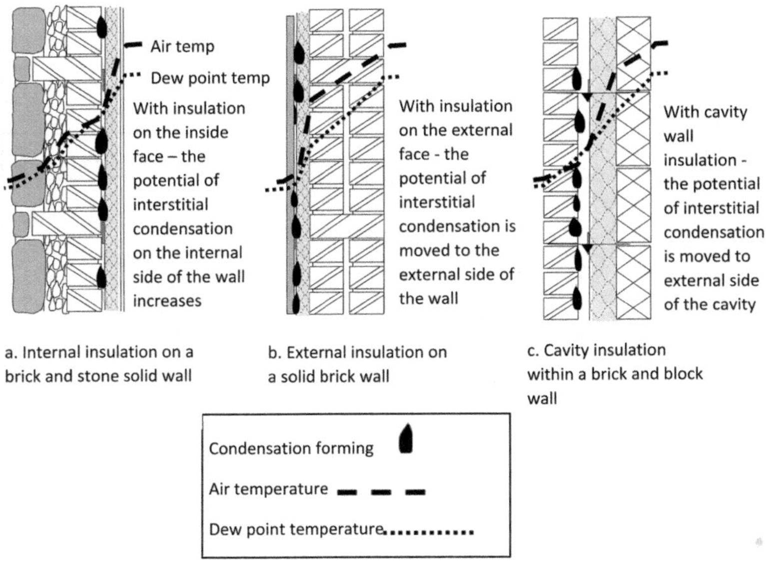

Figure 2.8 Schematic showing the potential of interstitial condensation with thermal insulation in different positions on different walls, without vapour barriers.

on the surface of materials. Thus, in the colder parts of the structures, moist air gives up some of the water that it contains to the wall structure. If the moisture forms within the internal side of a wall's structure and mould grows, spores will be released back into the building, creating unhealthy conditions. For this reason, it is advisable that walls are designed so that the dew point occurs in the external leaf of a cavity wall or more external extremes of solid walls.

Breather papers and vapour barriers are also be used to reduce the risk. Vapour barriers are designed to prevent the passage of moisture; these can be plastic, foil sheets, or foil-backed insulation. Breather papers are different; they are used to allow the structure to breathe but do not allow moisture or moist air from inside the structure to pass. Both breather paper and vapour barriers are used to resist the penetration of moisture. However, services that puncture through the barrier will allow moisture to cross unless they are effectively sealed.

Where insulation is placed on the external face of the wall or within the cavity, the risk of condensation forming within the wall is moved to the external part of the wall. This does not have the same effect on internal environments and presents a lower health risk to the occupants of the building. However, it can still be detrimental to vulnerable timbers present in the wall. Moisture at any part of the wall can cause discoloration, lead to mould growth, and be more prone to material

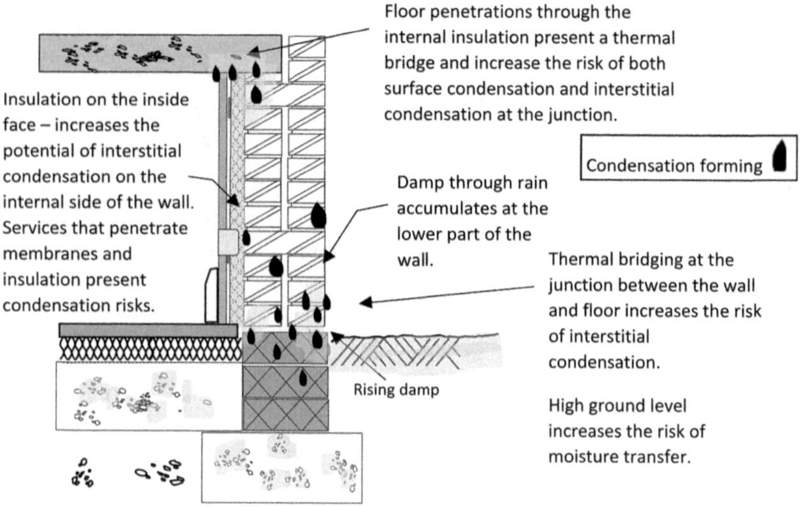

Figure 2.9 Schematic: risks of moisture accumulation and transfer in solid walls due to thermal bridges.

degradation and frost attack. Figure 2.9 shows some of the areas where water and water vapour accumulate.

Where the construction of the building or building features penetrate insulation in solid walls, thermal bridges occur. A thermal bridge allows heat to flow through the fabric more readily, by conduction. Where such penetrations occur, this part of the structure will be colder during the winter and more prone to condensation forming. Typical thermal bridges in solid walls occur at floor junctions, service penetrations, and at the ground floor, wall, and foundation interface. Characteristically, the external part of the wall retains greater levels of moisture at the lower part of the wall. This occurs for a number of reasons. Rainwater hitting the face of the wall naturally percolates down, building up at the base of the wall. Dampness may rise from the ground due to pressure potential differences. Typically, this does not extend beyond 1 m high, although particular conditions may drive the moisture higher. The lower part of the wall can become saturated. The wet part of the structure, due to the presence of moisture, can then become colder, thus causing further condensation and moisture build-up. Reducing moisture build-up in the lower part of the solid walls is difficult. Removing vegetation and materials that shadow the lower parts of the wall can help to improve the wall's ability to shed moisture via evaporation. Where possible, insulation on the external face of the wall below the ground can help reduce the thermal bridge.

COMPARING DIFFERENT WALLS AND THEIR THERMAL PROPERTIES

Figure 2.10 shows a cross-section diagram of a modern masonry wall, with insulation, damp-proof courses, and weep holes, all of which work together to protect the internal environment of the building. The requirements for walls are constantly changing, so it is important that designers and surveyors ensure that they are aware of the latest regulations. When designing or upgrading walls, it is essential that building standards and regulations are observed. The UK Building Regulations can be accessed and downloaded from the Government planning portal (www.planningportal.gov.uk).

WALLS AND CONSERVATION OF ENERGY

The thermal properties of walls have become increasingly important with the need to conserve energy and move to zero-carbon solutions. Depending on their mass and thermal properties, walls help to store and resist the flow of heat energy. The wall's thermal storage property is called the wall's thermal capacitance, and the ability to resist the passage of heat is the wall's thermal insulation. Such properties can help to regulate temperatures, making the internal environment more comfortable, and help to contain heat energy within the building's envelope. Thus, walls are designed to resist the impact of temperature fluctuations resulting from

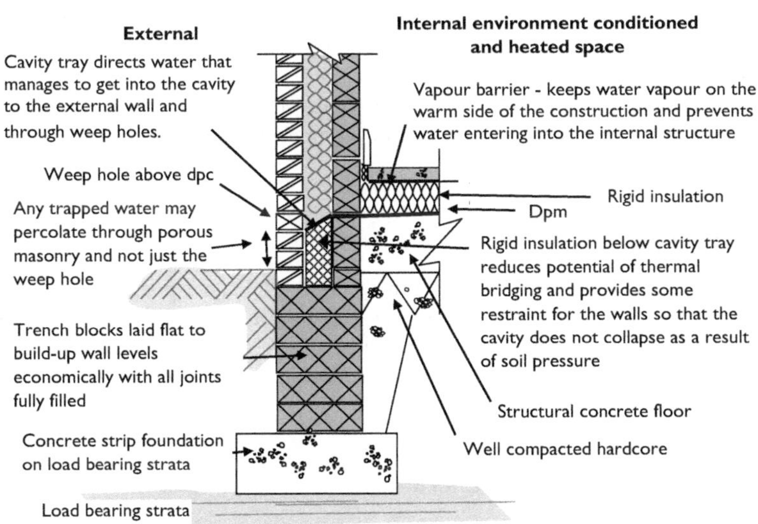

Figure 2.10 Typical detail of a modern masonry brick wall and floor junction.

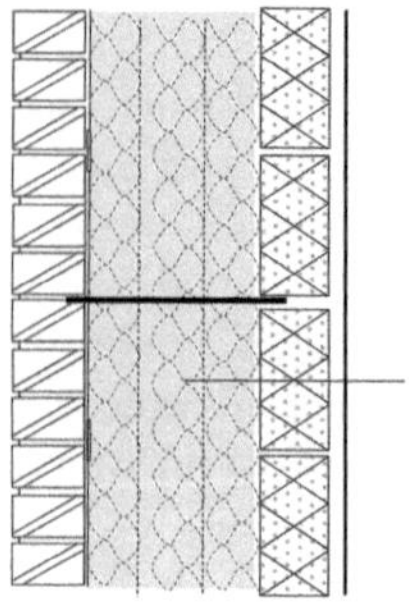

Figure 2.11 Cavity masonry wall built to Passiv standards (adapted from Greenbuildingstore, 2013).

solar heat gain (from the sun's short-wave radiation to the earth) and heat losses (long-wave radiation losses from the earth and buildings to space).

Figure 2.11 shows a highly insulated masonry wall constructed to Passiv standards (see also Chapter 11, Glossary). Together with renewable energy sources and highly insulated floors, roofs, doors, and windows, zero-carbon and energy-plus buildings can be constructed that are highly energy efficient.

While passive construction techniques serve to limit the transfer of heat through thermal bridges, most traditional construction will have a number of heat-transfer mechanisms that need to be considered. Heat energy is transferred between connecting solid materials through conduction, and within cavities, voids, and connecting gaps by convection. When adapting and upgrading buildings, care should be taken so that the flow of heat energy, through connecting elements and ventilation paths, is limited.

FOUNDATIONS

The foundation of a building sits below the ground level and is considered a substructure element. As the interface between the building and ground, the foundation is used to transfer the loads of the building safely to the ground (through the strata) without undue and differential settlement. See Figure 2.12 showing a schematic of foundations and building.

FUNCTIONAL REQUIREMENTS

The primary functional requirement of a foundation is to safely transmit the loads of the building to the ground. The safe transmission of the load to the ground

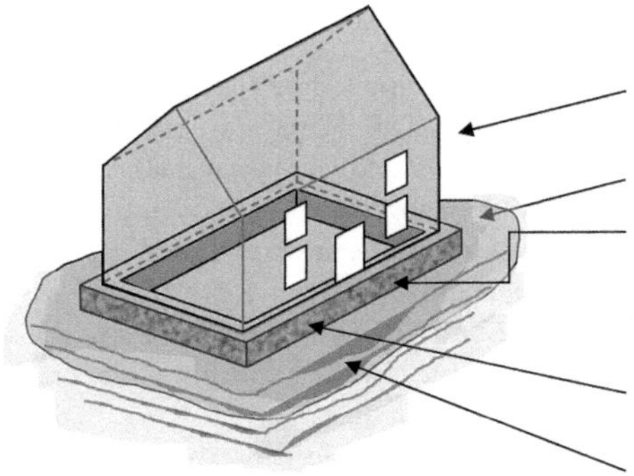

Figure 2.12 Schematic of foundations and building.

should take place without differential settlement or any settlement that could adversely affect the neighbouring or surrounding buildings and structures.

Building foundations are required to:

- Sustain and transfer dead and live loads.
- Resist the movement of soils. They should be designed so that swelling, shrinkage of cohesive soils, or expansion due to the freezing of water in clays does not affect the stability of the building.
- Remain durable, resisting attacks from salts and chemicals in the ground.
- Additionally, foundations can also be designed to resist the passage of heat transfer. As buildings become more energy efficient, the area around the foundations is also designed to resist the flow and transfer of heat energy.

DIFFERENTIAL SETTLEMENT

Differential settlement may sometimes be referred to as relative settlement. Generally, differential settlement is where parts of the foundation of a building settle at different rates due to variations in load on the foundations or variations in the subsoil, movement of the soil, or movement of the underlying strata due to neighbouring loads.

The variations can cause distortion of the building and structure and damage. While different and small magnitudes of settlement are expected, they should not be so great that damage occurs. Where differential settlement has taken place, damage or cracking may be seen in brick or blockwork, rigid infill panels, load-bearing walls, concrete and rigid floors, and finishes such as plaster and render, and where the magnitude is substantial services may be affected.

SEASONAL VARIATION IN CLAY SOILS

During hot, dry periods, the ground dries and vegetation close to the building will attempt to remove any remaining moisture. In clay soils as the moisture is lost from the soil the clay dries and shrinks. In winter, wet seasons, and wet periods, clays become saturated and swell, and can lift buildings. Where movement is not consistent across the whole building, the building will suffer from cracks, as the structure is deformed.

Cracks due to seasonal movement can open and close during wet and dry periods; however, where dirt and debris from the building fill the crack, it is not possible for the crack to close. As the movement of the building continues, the cracks can develop and in extreme cases render the building unstable.

Refer to Figure 2.13 which shows foundations transmitting the loads of the building contending with natural ground conditions and forces caused by other structures.

FOUNDATION TYPES

There are four main classifications of foundation, see Figure 2.14:

- Strip foundations
- Pad foundations
- Raft foundations
- Pile foundations

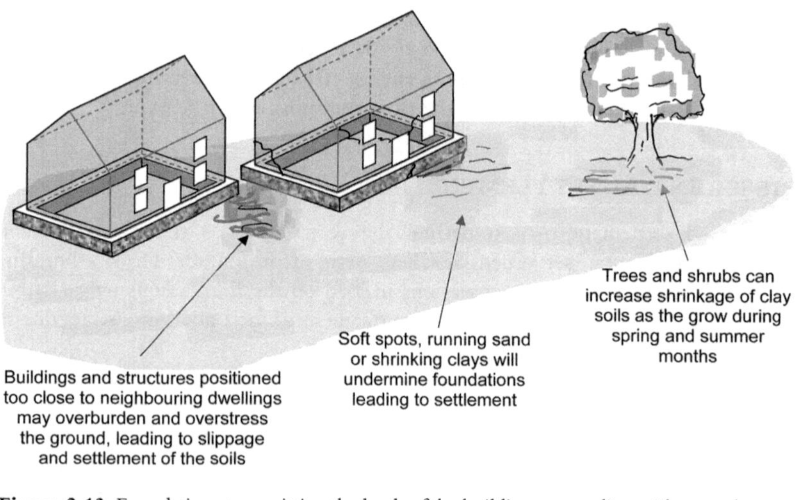

Figure 2.13 Foundations transmitting the loads of the building contending with natural ground conditions and forces caused by other structures.

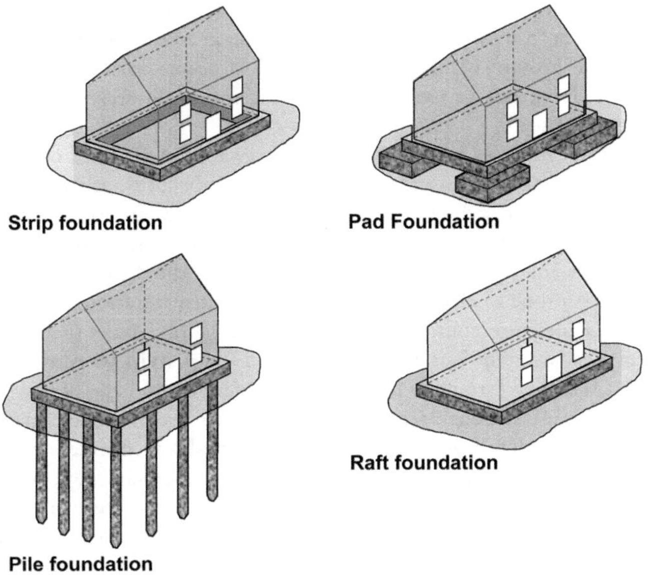

Figure 2.14 Foundation types.

Although the foundation systems can be used in similar circumstances, each foundation system is suited to certain types of buildings and ground conditions.

STRIP FOUNDATION

This type of foundation, in its standard form, is usually considered the most economical foundation to use on good firm ground, under continuous loads, such as masonry walls, where the loads placed on the building are relatively small. Relatively low loads include domestic buildings, such as bungalows and houses and small industrial buildings. As it is a single length of foundation, it is most suited to loads that need continuous support, such as brick and block walls. Masonry walls for houses, garden walls, and outhouses are typically mounted on strip foundations. The loads from the structure typically are transferred down through the inner leaf of the walls to the strip foundation.

Key attributes of strip foundations:

- Economical form of foundation
- Suitable where there are good load-bearing strata near to the surface
- Transfers the loads from the walls, to the strip foundation, to higher level load-bearing strata

Traditionally the strip foundation is formed of a standard concrete mix, with sufficient concrete strength to transfer the loads without the need for any steel reinforcement. Where the ground is sufficiently stiff and of good bearing capacity, offering a reasonable resistance to the forces imposed, an unreinforced concrete strip foundation can be used.

The strip foundation provides a continuous, longitudinal solid platform of concrete; the level base is used to build on and transfer the loads to the load-bearing strata.

Where the conditions under the foundation vary or the loads are increased, it may be necessary to extend the foundation deeper into the ground, to strata that offer a more consistent and higher load-bearing capacity. As the depth increases, the strata are often less prone to changes in moisture content, vegetation, and tree routes that also affect moisture content and freezing. Such foundations that penetrate deeper than standard strip foundations are termed **deep strip or mass fill foundations** and can still be formed of unreinforced concrete.

To avoid changes in moisture content in clay soils (which are prone to expansion and contraction), the depth of the foundation is normally recommended to be at least 750 mm deep. The depth of foundation required does vary depending on local conditions. Where the foundation rests on clay and the moisture content suffers from seasonal variation, the clay will expand when wet and lift or push the foundation, and when dry, the clay will shrink and cause the foundation to settle. Seasonal movement is avoided by constructing the foundation at a suitable depth where seasonal variation is not experienced. See also Figure 2.15 showing schematic sections of deep strip, wide strip, reinforced pad and mass fill foundations.

Alternatively, the width of the strip foundation can be increased to spread the load and reduce the load per unit area placed on the ground. Such foundations are termed **wide strip foundations**.

When the load-bearing ground is relatively consistent and the loads of the building are high, the width of the foundation can be increased in order to reduce

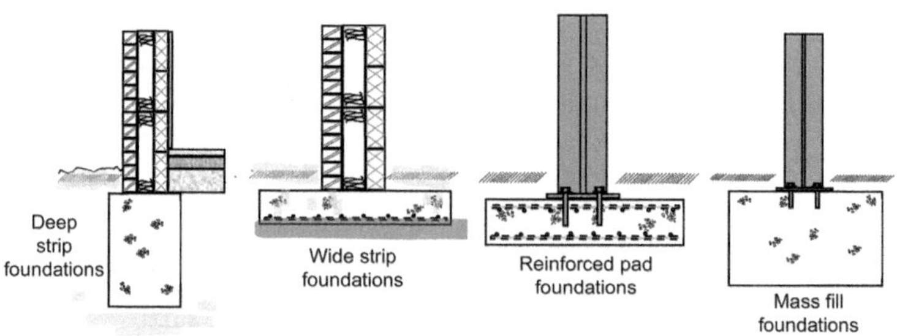

Figure 2.15 Schematic sections of deep strip, wide strip, reinforced pad, and mass fill foundations.

the load-bearing pressure per unit area placed on the ground. The width of the foundation is increased where the bearing pressure of the soil is not quite sufficient to take the loads when distributed across the traditional strip foundation. A wide foundation is not a solution to use when building on non-load-bearing ground.

In order to increase the width of the strip foundation, it is necessary to reinforce the foundation. This distributes the loads across the full width of the foundation. The main reinforcement is positioned towards the base of the foundation and used to withstand the tensile forces across the foundation.

PAD FOUNDATIONS

Pad foundations attempt to distribute the loads placed on them through a block of material. They are often used to take a point load placed on the foundation and disperse the loads over a wider area, reducing the load per unit area placed on the ground. In good load-bearing strata they are economical. While they can be used under small domestic buildings, to carry loads from ground beams, typically they are used to support the columns and ground beams from skeleton-frame commercial buildings, where the intensity of point loads from columns is high. Figures 2.14 and 2.15 show the schematics of a pad foundation. Note that the arrangements are for illustration, and it is not common to find pad foundations used in such proportions under domestic buildings.

Key attributes of pad foundation:

- Economical form of foundation
- Suitable where good load-bearing strata exist near to the surface
- Transfers the loads to the pad, suitable for point loads, pad foundations positioned under columns, or where ground beams transfer loads to the pad and to the load-bearing strata

Pad foundations are used to support and transfer point loads to the ground. Typically, the point loads resting on the foundation are then distributed across the width of the foundation and then to the ground, in a bulb of pressure, through the strata (Figure 2.16). The type of load could be the bearing of a beam, masonry piers (brick, concrete, or stone), or a steel column. Mass fill concrete pad foundations are used where good load-bearing strata are found some distance below the surface of the ground.

The foundations take the point loads placed on them and distribute the load over a large surface area, reducing the load per unit area placed on the ground. The concentrated load placed at the top of the foundation places the base of the foundation in tension; the reinforcement resists the tensile forces and allows the load to be distributed across the base of the foundation. Figure 2.16 shows loads transferred from the wall to the foundation and then to the sub-strata.

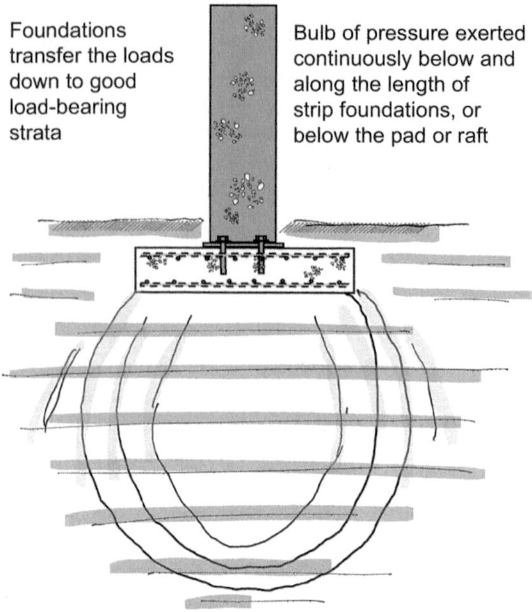

Figure 2.16 Schematic showing loads transferred from the wall to the foundation and then to the sub-strata.

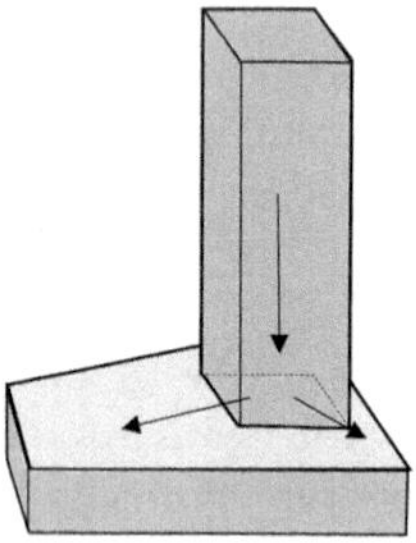

Figure 2.17 Pad foundation with eccentric load.

COMBINED FOUNDATIONS

Pad foundations exist in many forms, shapes, and sizes, including simple, wide, deep, and mass, as well as combinations of irregular shapes. The irregular-shaped foundations are used to withstand and transfer different loads that can be offset from the centre of the foundation. Thus, loads can be placed in eccentric positions on the foundation, Figure 2.17. The plan shape of the foundation can be wider at one end than the other, helping to further reduce stresses and loads under the end of the foundation with the greater width. The loads can be offset or even cantilevered, to avoid causing loads to be spread much beyond the curtilage of the building; this can prevent the loads from one building burdening the ground under neighbouring buildings or structures.

RAFT FOUNDATIONS

Figure 2.14 shows a schematic of a raft foundation.

Key attributes of raft foundations:

- A more costly form of foundation, as the whole of the area under the building is covered with concrete reinforced with steel.
- Raft foundations are suitable for buildings of high loads or where the strata is sufficiently strong to transfer loads directly.
- Rather than directly transferring the loads, the reinforced concrete raft foundations spread and distribute the loads over the whole plan area of the building.
- The composite concrete and steel platform transfers the loads from the walls, columns, and floors to the raft and distributes the loads, reducing the intensity of the loads per unit area.

A raft foundation offers a continuous platform of concrete, normally of the same size or larger than the plan area of the building. The concrete raft effectively floats on the ground by spreading the load over a large area, drastically reducing the loads per unit area. The considerable distribution of load, reducing the load on the bearing strata, means that either greater loads can be placed on the foundation, or a building can be founded on ground with strata of a more limited bearing capacity.

Raft foundations come in many different forms, including:

- **Solid slab raft**: continuous reinforced slab of concrete, normally spanning the full plan area of the building or greater.
- **Beam and slab raft**: reinforced concrete slab with beams or edge thickenings to further strengthen the raft. Up-stand and down-stand beams can be used around the edge of the raft, and where additional rigidity is required the beams can cut across the raft in a grid formation.
- **Cellular raft (buoyant raft)**: where the beams in the raft become so deep that upper and lower reinforced slabs can be used. The structure creates voids between the two slabs, which help reduce the loads on the ground, creating a buoyant structure. The upper and lower slabs, perimeter beams, and ribs running across the foundation make the structure ridged. Cellular rafts are sufficiently suited to buildings with high loads.

PILE FOUNDATIONS

Figures 2.14 and 2.18 show schematics of pile foundations.

Key attributes of pile foundations:

- Pile foundations offer a more costly form of foundation as special plant and equipment are required to bore holes into the ground or drive piles down through the strata.
- Piles are suitable for buildings of high loads or where the strata near to the surface of the ground are not sufficiently strong.
- Rather than directly transferring the loads to the strata near the surface, the loads are transferred down to strata at a lower level.
- The loads are usually transferred from the walls and floors to beams, which carry the loads to the piles and down to the ground at a lower level. Some piles, which rely on friction, transfer the loads along the shaft of the pile as well as at the end of the pile. Other end-bearing piles transfer the loads directly to the end of the pile, when the latter bears on load-bearing strata.

Pile foundations act as columns of material, which are driven or bored into the ground. They work by either transferring the loads to the end (toe) of the pile (end-bearing piles) or through the resistance experienced around the shaft of the pile (friction piles) or can be a combination of both friction and end-bearing. End-bearing piles are used where the strata are generally weak but where good load-bearing strata are found at some distance below the ground. They are often used

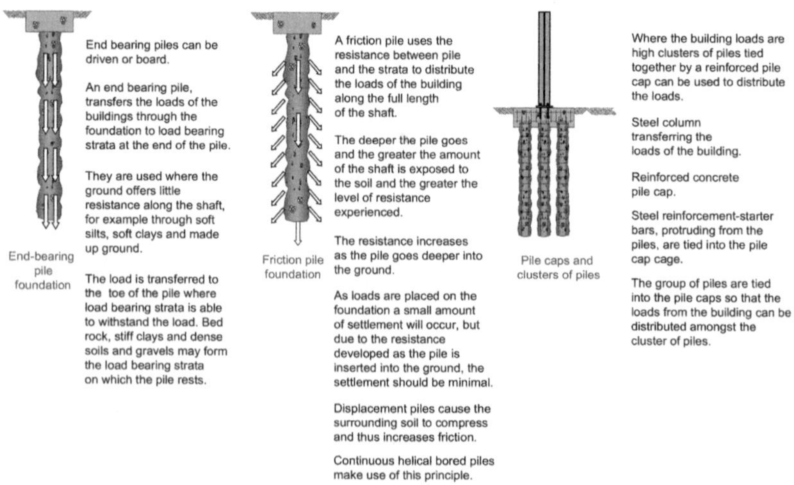

Figure 2.18 End-bearing piles, friction piles, and pile caps and clusters of piles.

where the ground has been built up or filled. Figure 2.18 shows end bearing piles, friction piles, and pile caps and clusters of piles.

Where greater resistance is required, multiple or clusters of piles can be grouped together under a pile cap. Piles may be used in combinations with other foundations, helping to support and transfer loads from pad, strip, and raft foundations.

Where piles are clustered together, they are often formed in close groups that allow pile caps to economically connect them together.

PLACING OF PILES

The way a pile is placed in the ground is often used to classify the pile. Displacement piles are driven into the ground, forced under pressure (percussion driven or hammered into the strata). The columns of steel, timber, or concrete displace the ground as they are forced into the soil. The pile experiences resistance as it penetrates through the strata and further resistance as it comes to rest on load-bearing strata.

In a replacement pile (Figure 2.19), the continuous flight auger continually rotates, drawing the strata to the surface, and replacing the strata with concrete as the auger is withdrawn, leaving a shaft of concrete.

Non-displacement piles excavate the strata by a grab, auger, or bore; as the strata is removed it is replaced with a column of concrete or compressed stone. With non-displacement piles, the toe of the pile can be enlarged by cutting away extra material, providing a larger base to the foundation. Sometimes material at the base of the foundation is driven and rammed into the toe of the excavation, again creating a large and compacted plug that offers additional resistance at the end of the pile.

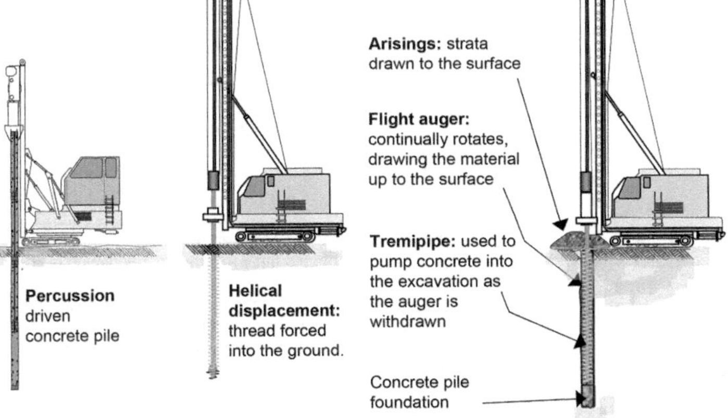

Figure 2.19 Piles: percussion driven and helical displacement piles, replacement piles.

Where non-displacement piles are cut through ground that is loose or unconsolidated, the sides of the pile may be lined with a steel sleeve, tube, or casing to prevent the sides of excavated shaft collapsing. As the concrete or stone is inserted into the shaft, the steel sleeves can be removed, or if the ground is particularly poor, the sleeve may be left in place to contain the pile.

Driven piles can be made of steel, concrete, or timber. Steel piles can be H, I, square, or rectangular in section or have a helical thread. Precast concrete piles come in different sections, often square or rectangle but can also be tubular; they are reinforced and capped with steel heads and toes, allowing the piles to be joined together.

Driven piles are forced into the ground until the "set" resistance of the pile is reached. The set point is the amount of resistance that the pile needs to achieve. It is determined by dropping a pile of a known weight, at a predetermined distance, and measuring the distance the pile drops. Once the set point is reached, the hammering of the pile can stop – as the required resistance is achieved.

Percussion piles, even the more quiet ones, are relatively noisy, cause vibration, and are generally not used when neighbouring buildings have sensitive equipment. Driven piles offer little friction resistance due to the way the soil or clay behaves as the pile drives through the soil.

JACKED PILES

Jacked piles are used where space is restricted and the working area is tight. Where a small pile jack can be inserted, or a small section of pile, then the pile can be

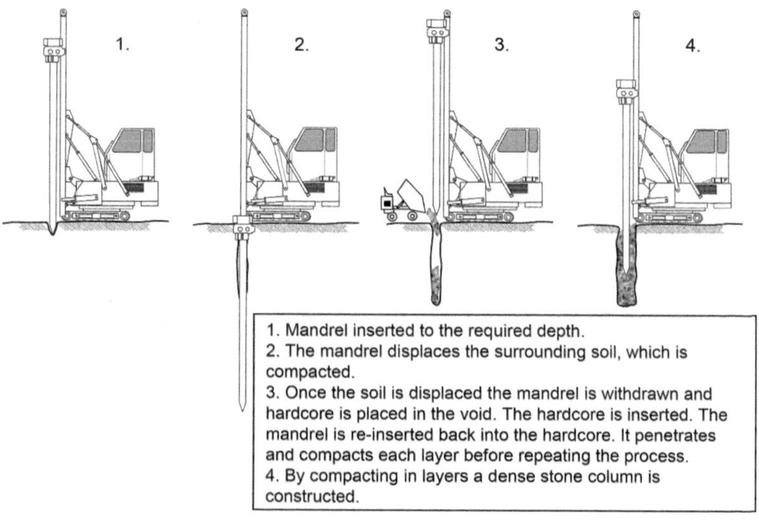

Figure 2.20 Vibro compaction (adapted from www.roger-bullivant.co.uk).

Table 2.3 Methods of ground treatment

Method of ground treatment	**Description**
Dynamic compaction	Ground improved by dropping tampers (weights) that dynamically compact the ground. Used to improve the general ground condition. By improving the strength of the ground, the cost associated with foundations can be reduced. Flat bottom tampers can be used, but it is more common to use a range of shaped tampers that achieve different levels of compaction at different depths. Modern plant use guides and cables to raise and drop tampers. Bearing capacities of 50–150 kN/m^2 can be achieved. Tamper weight are normally between 10 and 12 tonnes. To achieve consolidation multiple passes are used, compacting the ground at different depths.
Vibro compaction (vibro- displacement or vibro- replacement)	Vibrating mandrels (vibrating shafts or rods) are used to penetrate into the strata, displacing the strata and compacting the surrounding soil. When the mandrel is removed, the void is filled with stone, the mandrel is inserted back into the stone and the stone is forced into the surrounding strata – compacting the ground and filling voids. As a guide, bearing capacities of 100–200 kN/m^2 on flat ground and 100 kN/m^2 on sloping sites can be achieved. For non-cohesive soils, the ground becomes compacted on treatment; for cohesive soils the load is transferred through the columns of soil, helping to better distribute the load and making the performance of the ground more consistent.
Vibroflotation	Vibroflotation adopts a similar process to that described by vibro compaction, with the difference being that at the head of the mandrel, and sometimes along the sides, water jets are used to aid penetration. Due to the amount of water involved and the mess created from wet arisings (i.e. waste, detritus) the process is seldom used.
Pressure grouting	Used in soils where there are small cavities, that benefit from being filled with a cement slurry. A hole is bored into the ground, grout is then fed under pressure, created by the head height at which the cement grout is fed into the void. Pressures of up to 70,000 N/mm^2 can be exerted by the grout on the surrounding soil. Slurries are cementitious products with additives, for example, micro-silica, pulverised fuel ash, chemical grouts, cement, or a mixture of cement and the other additives. It is often the cheaper cementitious additives that are used, such as PFA, which, while cost effective, improve the bearing capacity of the ground. The ground becomes stiffer, more consistent and stable, and water resistance is improved.

(continued)

Table 2.3 *(continued)*

Method of ground treatment	Description
Soil modification and recycling	Various systems are available for modifying and recycling the existing strata. Generally, large truck-mounted rotating wheels are used to cut into the soil and break the soil apart. The pulverised soil is then captured in a mixing chamber where cementitious materials are then fed into the soil. When mixed, it can be fed back on to the ground where it is relaid and compacted. The reconstituted material creates a hard stable surface, which seals in contaminants as well as providing a stable platform.

jacked, under hydraulic pressure, into the ground. Once the initial pile is inserted into the ground, the jack is removed and the next section of pile is inserted. This process continues until sufficient resistance is achieved.

GROUND STABILISATION

Different methods are used to improve the ground condition and stability. See Table 2.3. Soils can be compacted from vertical loads and through material forced laterally in the ground, injected with grout and mixed with other materials to create load-bearing strata. Some sites are improved so that plant can simply operate on the ground; others are treated to significantly improve the load-bearing strata, reducing the need for complicated and expensive foundations systems.

HEAT TRANSFER THROUGH THE FOUNDATION STRUCTURE

For zero-energy buildings, there is a need to ensure that the energy required in the heating and cooling of a building is negligible. It is now common to provide a highly insulted enclosure, limiting heat loss and gain. Thermally insulated foundations are becoming more common with insulation products that have structural properties used to transfer building loads.

Not insulating foundations can result in significant thermal bridges when the property is highly insulated. Floors with U-values of less than 0.15 $W/m^2{\cdot}K$ are becoming more commonplace. A major consideration is that the loads transmitted through structural floors and foundations are considerable, whereas standard rigid insulation is not designed, or required, to withstand high loads. Structural

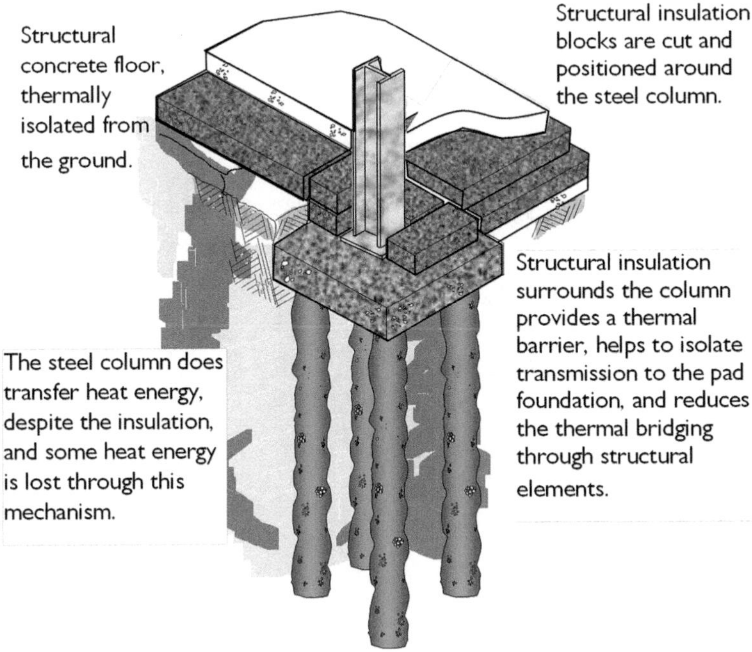

Figure 2.21 Structural insulation under the floor and thermal break around column above pile cap and pile foundation.

cellular glass insulations that do not settle or creep under loads are suitable for highly insulated floors and foundations where a substantial load is transferred through the foundation. Load-bearing capacity of 2400 kN/m^2 has been achieved with some insulation materials (e.g., Foamglas (n.d.)). Such materials are capable of transferring the loads of the floors and buildings directly through the insulation.

For example, Figure 2.21 shows structural insulation surrounding the column bearing on a pile cap, and providing a thermal break between the ground floor and the ground. Figure 2.22 details structural insulation used as part of the basement raft foundation. The insulation can also be used under strip and raft foundations.

As Passivhaus principles work on the basis that fabric U-values are less than 0.15 $W/m^2{\cdot}K$ with the build-up of structural insulation, it is possible in these ways to meet fabric requirements for nearly zero-energy buildings.

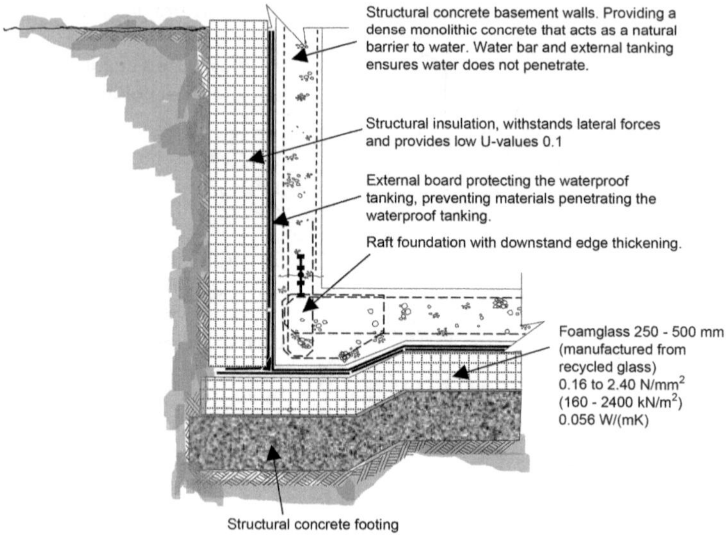

Figure 2.22 Thermal break under raft and around basement – Passivhaus fabric standard.

REFERENCES

BBC (2017a). Bridgewater Place lorry crush death referred to CPS by coroner. https://www.bbc.co.uk/news/uk-england-leeds-16986600

BBC (2017b). Bridgewater Place puts wings in place to block winds. https://www.bbc.co.uk/news/uk-england-leeds-41608461

Foamglas (n.d.). https://www.foamglas.com/en-gb#

Ingrid Could (2018). Wind simulation on bridgewater place by ingrid could. YouTube. https://www.youtube.com/watch?v=4QpUcey0sU8

LABC (2015). Prevent party wall thermal bypass. Accessed from: https://www.labc.co.uk/news/prevent-party-wall-thermal-bypass

3

Legal and regulatory frameworks

Melanie Smith

This chapter is presented in eight sections:

Building works and regulation: meaning of works, and legislative documents of control
Planning and historic buildings: planning permission, listing of historic properties, conservation areas and consents for works
Adjoining property and neighbourly matters: rights over land and property, Party Walls, acoustics and noise control
Contracts
Landlord and Tenant: dilapidations, condition surveys, Decent Homes
Survey types
Health and safety: h&s in construction, CDM regulations, fire safety in occupied buildings, due diligence surveys
Inclusivity: accessibility, Equality Act, Part M building regulations, fire safety and accessibility, typical width requirements
Guidance, standards and codes: British, International and European, HSE guidance, RICS codes and guidance

Although UK legislation is mainly used, if only for illustrative purposes, it has been found that many countries use similar frameworks and have similar aims for legislative devices.

BUILDING WORKS AND BUILDING REGULATION

Building works

Control of building works and refurbishment is held in the country's legal systems. Building surveyors working around the world must obviously be conversant with the legislation in force in the country of work. There is a lot of overlap between the legislation of differing countries – in meaning if not in wording – but compliance with the legislation of the relevant country/dependency is required.

In Great Britain, legislation is different for England and Wales, Scotland, Northern Ireland, the Isle of Man, and each of the Channel Islands. For more information, try:

United Kingdom:	England and Wales: www.legislation.gov.uk/uksi
	Scotland: www.legislation.gov.uk/ssi
	Wales: www.legislation.gov.uk/browse/wales
	Northern Ireland: www.legislation.gov.uk/browse/ni
The Isle of Man:	Isle of Man Government: www.gov.im
Bailiwick of Jersey:	The States of Jersey: www.gov.je
Bailiwick of Guernsey:	The States of Guernsey: www.gov.gg
	The States of Alderney: www.alderney.gov.gg
	Sark Chief Pleas: www.gov.sark.gg

Building acts, regulations, and standards

Most building works are subject to building regulation. In England and Wales, building works are defined in The Building Act, which is the primary, enabling legislation under which secondary legislation (e.g. building regulations) is made. It empowers the Secretary of State (for England and Wales) to make regulations for the purposes of:

- Health, safety, welfare, and convenience of persons in or about buildings and of others who may be affected by buildings or matters connected with buildings
- Conservation of fuel and power
- Preventing waste, undue consumption, misuse, or contamination of water

In Scotland, The Building (Scotland) Act is the equivalent primary legislation. It focuses on:

- Health and safety of buildings
- Conservation of fuel and power within buildings
- Sustainable development

The next legislative level is building regulations or standards, with the third level in England and Wales of Approved Documents, etc., and in Scotland of Building Standards Technical Handbooks, which provide guidance on achieving the standards set out in the building regulations of each country.

Example definitions of building work, for England and Wales and for Scotland, are given in Table 3.1.

Table 3.1 Example definitions of building work

England and Wales (ref: Building Regulations 2010)	**Scotland (ref: Building Standards Technical Handbooks 2017)**
Building work means: • The erection or extension of a building • The provision or extension of a service or fitting (which is controlled by the regulations, e.g. heating system, drains, windows) in or in connection with a building • The material alteration of a building, or a service or fitting (which is controlled by the regulations, e.g. heating system, drains, windows) • Work required by regulations relating to a change of use (as set out in regulations 5 and 6) • The insertion of insulating material into the cavity wall of a building • Work involving the underpinning of a building • Work required by regulations relating to a change of energy status (i.e. now coming under the energy efficiency requirements of the regulations when it did not before, e.g. garage conversion) • Work required by regulations relating to thermal elements (i.e. wall, floor, or roof separating a heated/cooled space from an unheated/uncooled space, e.g. external wall) • Work required by regulations for consequential improvements to energy performance (for an existing building over 1000 m^2) A material alteration is if the work, or any part of it, would at any stage result in the building, service, or fitting: (a) Not complying with a requirement (for structure, fire, or disabled access) where previously it did; or (b) Which did not comply with a requirement before the work commenced (for structure, fire, or disabled access) being more unsatisfactory in relation to that requirement	A building means: Any structure or erection (temporary or permanent), other than: *a structure or erection consisting of, or ancillary to public roads (including any bridge on which the road is carried), private roads, sewers or water mains vested in Scottish Water, aerodrome runways, railway lines, large raised reservoirs, wires and cables, their supports above ground, and other apparatus used for telephonic or telegraphic communication.* Any references to a building include references to: • A prospective building • A part of the building, structure, or erection • So much of the building as is comprised in the extension or the subject of the alteration or conversion (for extensions, alterations or conversions) Work in relation to a building includes work carried out in relation to the enclosure and preparation of the site of the building. Further information including exemptions is set out in The Building (Scotland) Regulations 2004, especially sections 2 and 5, and Schedules 2 and 3.

Approved documents and Technical handbooks

Approved Documents (England and Wales) and Building Standards Technical Handbooks (Scotland) set out to describe how to comply with the relevant building regulation. These set out the lowest acceptable standard.

Contravention of building regulations can lead to enforcement measures, including taking down and replacement of inadequate works, and ultimately fines. It is the regulation that must be complied with, not necessarily the guidance set out in the approved document or technical handbook. If it can be shown that the works meet the regulation in a different way, then this can be deemed acceptable, but proof may be required.

The contents of both Approved Documents and Handbooks are comprehensive and extensive and mostly cover the same aspects under differing headings. See Table 3.2, Sections covered by Approved Documents and Scottish Technical Handbooks. Similar comprehensive advice is provided by other countries.

Enforcement of building control is under two systems in England and Wales: local authority or approved inspectors. There are three types of application:

- Full Plans: the most thorough option. Decisions can be expected within two months of submission. Consents for the works to start are issued when the plans are considered in compliance. Completion certificates are received within eight weeks of completion of the building work if in compliance.
- Building Notice: only for smaller projects. Work can start two days after submission of the notice. The applicant does not receive formal approval.
- Regularisation: retrospective approval for work already carried out without consent – from a local authority only.

In Scotland, the Scottish local authorities act as verifiers. A building warrant is required, giving legal permission to carry out building work, and the planned building has to meet Scottish Building Standards.

PLANNING AND HISTORIC LISTING OR REGISTRATION

Planning

England, Scotland, Northern Ireland, and Wales each have their own primary planning legislation. Each country has a planning system that is "plan-led", meaning that national and local planning policy is set out in formal development plans which describe what developments should and should not get planning permission and how land should be protected and seeks to ensure a balance between development and environmental protection in the public interest. Decisions on individual planning applications are made on the basis of these policies. Each country

Table 3.2 Sections covered by Approved Documents and Scottish Technical Handbooks

Approved Documents for England and Wales	Building Standards Technical Handbooks: • Domestic buildings • Non-domestic buildings
Approved Document	Contents in both handbooks
A Structure B Fire safety: • B1 dwelling houses • B2 buildings other than dwelling houses C Site preparation and resistance to contaminates and moisture: • Approved Document C • Domestic Ventilation Compliance Guide D Toxic substances E Resistance to sound F Ventilation G Sanitation, hot water safety, and water efficiency H Drainage and waste disposal J Combustion appliances and fuel storage systems K Protection from falling, collision, and impact L Conservation of fuel and power in: • L1A … new dwellings • L1B … existing dwellings • L2A … new buildings other than dwellings • L2B … existing buildings other than dwellings M Access to and use of buildings: • M volume 1: Dwellings • M volume 2: Buildings other than dwellings P Electrical safety: Electrical safety, dwellings Q Security in dwellings R High-speed electronic communications networks 7 Material and workmanship	0. General 1. Structural 2. Fire 3. Environment 4. Safety 5. Noise 6. Energy 7. Sustainability App A Defined terms App B List of standards and other publications

defines types of development that are permitted without the need for a planning application and defines "use classes" where change of use within a class is normally permitted. Each country operates an appeal system and enforcement systems. Although the basic structures of the four systems are similar, there are differences in the detail. The system for England will be used as an example.

Planning systems aim to ensure that the right development happens in the right place at the right time, benefitting communities and the economy. These are obviously controversial in practice, as different people will have different opinions about what is "right" and what is meant by "benefitting".

In the UK, the planning system:

- Identifies what development is needed and where
- Identifies what areas need to be protected or enhanced
- Ultimately assesses whether the proposed development is suitable

Key decision-makers in planning:

- Local planning authorities
- Councillors
- Officers
- Secretary of State for Communities and Local Government
- Planning Inspectorate

Main legislation:

- Town and Country Planning Act 1990, for England and Wales
- Town and Country Planning (Scotland) Act 1997
- Planning etc. (Scotland) Act 2006
- Planning (Northern Ireland) Order 1991
- Planning (Wales) Act 2015
- Planning (Listed Building and Conservation Areas) Act 1990
- Planning and Compensation Act 1991
- Planning and Compulsory Purchase Act 2004
- Planning Act 2008
- Localism Act 2011

The National Planning Policy Framework (NPPF) replaced the 44 no. Planning Policy Guidance Notes (PPGs) and their Planning Policy Statements (PPSs). The NPPF sets out the Government's planning policies for England (only). It provides a framework

Table 3.3 Material consideration examples for planning

Overlooking Loss of privacy Loss of light or overshadowing Parking Highway safety Traffic	Noise Effect on listed building and conservation area Layout and density of building Design, appearance, and materials Government policy	Disabled persons' access Proposals in the development plan Previous planning decisions (including appeal decisions) Nature conservation

for locally prepared plans for housing and other development to be produced. It does not contain specific policies for nationally significant infrastructure projects. These are determined in accordance with the decision-making framework in the Planning Act 2008 and relevant national policy statements for major infrastructure, as well as any other matters that are relevant (which may include the NPPF). National policy statements form part of the overall framework of national planning policy and may be a material consideration in preparing plans and making decisions on planning applications. A material consideration is a matter that should be taken into account in deciding a planning application or on an appeal against a planning decision. Examples are listed in Table 3.3 Material consideration examples for planning.

NPPF is a material consideration of significant weight. It should be read alongside the National Planning Practice Guidance (NPPG), a web-based resource bringing together planning guidance on various topics.

NPPF key messages:

- Planning is a plan-led system – local planning authorities must produce plans.
- Retail tests – *Town Centre First*/sequential and impact tests.
- Re-use of brownfield land.
- Local planning authorities must maintain a five-year supply of housing sites.
- Significant weight attached to supporting economic growth.

To gain planning permission for a project, an application is made to the local authority, together with the fee. Different types of planning permission in England:

- **Outline**: Application to establish if the principle of the application is acceptable to the council. Gives permission for the development in principle without detail.
- **Reserved matters**: Following Outline consent, gives permission for the detail required, i.e. for appearance, means of access, landscaping, layout, and

scale. Application for reserved matters must be made within three years of the Outline consent. Following consent for reserved matters, development must occur within two years.

- **Full planning permission**: Application provides all details to the council in order to determine the application. Detailed planning permission. A decision notice can contain conditions which must be discharged prior to a development commencing. Valid for six years.
- **Listed building consent**: Applications to make alterations, extensions, or demolition to a listed building.
- **Advertisement consent**: Permission to put up signs. Some signs do not need permission; others do, e.g. in a conservation area, on a listed building, illuminated signs, certain size.
- **Change of use**: Application to change the use of a building or piece of land to a different use class (e.g. pub Class A4 to shop Class A1). See Table 3.4 for Use classes in England planning.
- **Lawful development certificates**: application to prove to the council a building has been in a certain use and is now lawful (normally happens after enforcement action where the building does not have planning permission for use).

Listed buildings

A "listed building" is a building, object, or structure judged to be of national importance in terms of architectural or historic interest and included on the register named the List of Buildings of Special Architectural or Historic Interest. The list is compiled by the Department for Digital, Culture, Media and Sport (DCMS) under the Planning (Listed Buildings and Conservation Areas) Act 1990. British Listed Buildings is the online database of listed buildings and structures.

Listing identifies a building's special architectural and historic interest and brings it under the protection of the planning system. English Heritage states that the older a building is, and the fewer the surviving examples of its kind, the more likely it is to be listed. The general principles are that all buildings built before 1700 surviving in anything near their original condition are likely to be listed, as are most buildings built between 1700 and 1850. Particularly careful selection is required for buildings from the period after 1945. Buildings less than 30 years old are not normally considered to be of special architectural or historic interest because they have yet to stand the test of time. The descriptions of a building and its special interest in the List are not definitive, however long or short, and should not be treated as an exhaustive survey of features of interest.

Table 3.4 Use classes in English planning

Part	Use class	Comment
Part A	A1 Shops	Shops, retail warehouses, hairdressers, undertakers, travel and ticket agencies, post offices, pet shops, sandwich bars, showrooms, domestic hire shops, dry cleaners, funeral directors, and internet cafes.
	A2 Financial and professional services	Financial services such as banks and building societies, professional services (other than health and medical services), and including estate and employment agencies. It does not include betting offices or pay day loan shops – these are now classed as "sui generis" uses (see below).
	A3 Restaurants and cafes	For the sale of food and drink for consumption on the premises – restaurants, snack bars, and cafes.
	A4 Drinking establishments	Public houses, wine bars, or other drinking establishments (but not nightclubs) including drinking establishments with expanded food provision.
	A5 Hot food takeaways	For the sale of hot food for consumption off the premises.
Part B	B1 Business – offices	Offices (other than those that fall within A2), research and development of products and processes, light industry appropriate in a residential area.
	B2 General industrial	Use for industrial process other than one falling within class B1 (excluding incineration purposes, chemical treatment or landfill or hazardous waste).
	B8 Storage or distribution	This class includes open-air storage.
Part C	C1 Hotels	Hotels, boarding, and guest houses where no significant element of care is provided (excludes hostels).
	C2 Residential institutions	Residential care homes, hospitals, nursing homes, boarding schools, residential colleges, and training centres.
	C2A Secure residential institution	Use for a provision of secure residential accommodation, including use as a prison, young offenders' institution, detention centre, secure training centre, custody centre, short-term holding centre, secure hospital, secure local authority accommodation, or use as a military barracks.
	C3 Dwelling houses – this class is formed of three parts.	

(continued)

Table 3.4 *(continued)*

Part	Use class	Comment
	C3(a)	Use by a single person or a family (a couple whether married or not, a person related to one another with members of the family of one of the couple to be treated as members of the family of the other), an employer and certain domestic employees (such as an au pair, nanny, nurse, governess, servant, chauffeur, gardener, secretary, and personal assistant), a carer and the person receiving the care, and a foster parent and foster child.
	C3(b)	Up to six people living together as a single household and receiving care, e.g. supported housing schemes such as those for people with learning disabilities or mental health problems.
	C3(c)	Allows for groups of people (up to six) living together as a single household. This allows for those groupings that do not fall within the C4 HMO definition but which fell within the previous C3 use class, to be provided for, i.e. a small religious community may fall into this section as could a homeowner who is living with a lodger.
	C4 Houses in multiple occupation	Small shared houses occupied by between three and six unrelated individuals, as their only or main residence, who share basic amenities such as a kitchen or bathroom.
Part D	D1 Non-residential institutions	Clinics, health centres, crèches, day nurseries, day centres, schools, art galleries (other than for sale or hire), museums, libraries, halls, places of worship, church halls, law court. Non-residential education and training centres.
	D2 Assembly and leisure	Cinemas, music and concert halls, bingo and dance halls (but not night clubs), swimming baths, skating rinks, gymnasiums or area for indoor or outdoor sports and recreations (except for motor sports, or where firearms are used).
Sui generis	Certain uses do not fall within any use class and are considered "sui generis"	Including: betting offices/shops, pay day loan shops, theatres, larger houses in multiple occupation, hostels providing no significant element of care, scrap yards. Petrol filling stations and shops selling and/or displaying motor vehicles. Retail warehouse clubs, nightclubs, launderettes, taxi businesses, and casinos.

Different classifications of listed buildings are used in different parts of the United Kingdom:

England and Wales: Grade I, Grade II*, and Grade II:

- Grade I buildings are of exceptional interest; around 2.5% of listed buildings are Grade I.
- Grade II* buildings are particularly important buildings of more than special interest; around 6% of listed buildings are Grade II*.

- Grade II buildings are of special interest; around 91.5% of listed buildings are in this class, and it is the most likely grade of listing for a home owner.

Scotland: Category A, Category B, and Category C:

- Category A are buildings of national or international importance, either architectural or historic, or fine little-altered examples of some particular period, style, or building type.
- Category B are buildings of regional or more than local importance or major examples of some particular period, style, or building type which may have been altered.
- Category C are buildings of local importance, lesser examples of any period, style, or building type, as originally constructed or moderately altered; and simple traditional buildings which group well with others in Categories A and B.

Northern Ireland: Grade A, Grade B+, Grade B1, and Grade B2:

- Grade A: Buildings of greatest importance to Northern Ireland including both outstanding architectural set-pieces and the least altered examples of each representative style, period, and grouping.
- Grade B+: Buildings which might have merited Grade A status but for detracting features such as an incomplete design, lower quality additions, or alterations. Also included are buildings that because of exceptional features, interiors, or environmental qualities are clearly above the general standard set by Grade B buildings. A building may merit listing as Grade B+ where its historic importance is greater than a similar building listed as Grade B.
- Grade B1 and B2: Buildings of local importance and good examples of a particular period or style. A degree of alteration or imperfection of design may be acceptable.

Listed building consent

This form of planning control is intended to prevent the unrestricted demolition, alteration, or extension of a listed building without the express consent of the local planning authority or the Secretary of State. The controls apply to any works to a listed building for its alteration or extension, which are likely to affect its character as a building of special architectural or historical interest, or for its demolition. The control does not depend upon whether the proposed activity constitutes development; nor does it depend on the works normally attracting ordinary planning control.

Works being carried out on a listed building will need consultation with the local authority and possibly an application under the Planning (Listed Building and Conservation Areas) Act. The local planning authority must consult Historic England and the National Amenity Societies on certain listed building consent applications, for example, as shown in Table 3.5 Listed building consent consultation in England and Scotland.

And, additionally, for demolishing a listed building, or alteration of a listed building which comprises or includes demolition of any part of that building, consultation with the following relevant groups may be relevant:

- The Society for the Protection of Ancient Buildings
- The Ancient Monuments Society
- The Council for British Archaeology
- The Georgian Group
- The Victorian Society
- The Twentieth Century Society

Conservation areas

Conservation areas are designated in all the countries of the UK as well as the Isle of Man and the Channel Islands. Legislation for conservation areas exists to manage and protect special architectural and historic interests of a place, that which makes an area unique. Local planning authorities will have parts of their own area that are of special architectural or historic interest, the character and appearance of which it is desirable to preserve or enhance. Local planning authorities are required to designate these areas as conservation areas.

According to Historic England, conservation area designation provides a basis for planning policies whose objective is to conserve all aspects of character or appearance, including landscape and public spaces, which define the area's special interest. It also introduces a general control over the demolition of unlisted buildings.

Table 3.5 Listed building consent consultation in England and Scotland

England	**Scotland**
Consultation to Historic England for works to:	**Consultation to Historic Environment Scotland for works to:**
Any Grade I or II* listed building	A-listed or B-listed buildings and their settings
Any Grade II listed building demolitions	A-listed, B-listed, and conservation area demolitions

Works being carried out in a conservation area will need consultation with the local authority and possibly an application under the Planning (Listed Building and Conservation Areas) Act in England.

ADJOINING PROPERTY AND NEIGHBOURLY MATTERS

This section covers the following:
Rights over land and property
Party walls
Acoustics and noise control: Minimising noise transition

Rights over land and property

See Table 3.6 for various UK rights over land and property. Similar legislation may exist in other countries.

PARTY WALLS

Party wall legislation started in London and then spread to the rest of England and Wales in the 1990s. It aims to provide a framework for preventing and resolving disputes in relation to party walls, boundary walls, and excavations near neighbouring buildings. It involves informal agreements, formal legal notices, all relevant parties (owners, occupiers, etc.). The Party Wall etc. Act gives a procedure when carrying out work to party walls, boundary walls, party structures (such as floors), and excavations within 3 or 6 m of a neighbouring building or structure (depending on the depth of the hole or of the proposed foundations). The Act requires that notices of intentions to do these works need to be sent to adjoining owners and sets out procedures in the event that an adjoining owner does not agree to the proposed works.

What is a party wall?

- Party walls stand on the land of two or more owners and either:
  - Form part of a building.
  - Do not form part of a building, such as a garden wall, termed party fence wall. (Does not include wooden fences).
- Walls on one owner's land which are used by other owners (two or more) to separate their buildings are also party walls.
- Party structures: This could be a floor or other structure that separates buildings or parts of buildings between different owners, e.g. flats.

Table 3.6 Rights over land and property

Occupier's liability	Occupier's Liability Acts 1957 and 1984	Governs relationship between a building occupier and a lawful visitor, imposes private duties, and clarifies common law rules that existed prior to 1957. Occupier's responsibility to ensure that visitors to premises will be reasonably safe in using the premises. In practice this means displaying sufficient warning signs, ensuring that premises are in safe repair, and, where necessary, accompanying the visitor while they remain on the premises. In addition, any proprietor also owes a common duty of care to those using or visiting their premises.
	s2(2) of the 1957 Act defines a duty of care	"... a duty to take such reasonable care as in all the circumstances of the case is reasonable to see that the visitor will be reasonably safe in using the premises for the purposes for which he is invited or permitted by the occupier to be there..."
Defective premises	The Defective Premises Act 1972	Covers leased premises. The landlord has a duty to protect all who might reasonably be expected to be affected by defects in the state of repair of the premises. Not only invited visitors but extends, in a diminished form, to include trespassers. This legislation does not require the owner to take stringent protective measures. In practice – display prominent warnings and take no measures to inflict wilful harm on a trespasser.
Access to neighbouring land	Access to Neighbouring Land Act 1992	Courts have discretionary powers to order access for certain works, e.g. basic preservation works, repairs, decoration, drain clearance, care of trees, etc. The order does not last – each necessary visit will require separate authorisation.
Environmental protection	Environmental Protection Act 1990	Part 3 of the Act has a list of nuisances to which abatement (reduction) procedures apply. These include emission of fumes, smoke, dust, smells, noise, accumulation of noxious waste materials, keeping of animals in unsuitable places or conditions, and any premises in such a state as to be prejudicial to health or a nuisance.

(continued)

Table 3.6 *(continued)*

Rights of light		A "right to light" is an easement that gives a landowner the right to receive light through defined apertures in buildings on his or her land. Generally, people and buildings do not have a right to light. Local planning authorities consider the effect of new buildings upon existing structures, but the planning system can give protection, not rights. When planning permission is applied for, evidence will be required of the effect it will have upon the neighbouring properties, sometimes including the light and other amenities that the properties currently have. For residential property, the BRE publication, "Site layout planning for daylight and sunlight: a guide to good practice" can be consulted. It is often expected that artificial lighting will be used for commercial buildings.

The legislation affects:

- Work on an existing wall or structure shared with another property
- Excavating near a neighbouring building
- Building a free-standing wall, or a wall of a building, up to or astride the boundary with a neighbouring property

Party Wall Surveyors

Party Wall Surveyors are not required to be appointed, if the Building Owner can inform the neighbours what is going in on and gain agreement. If the Building Owner and Adjacent Owners cannot agree, the Building Owner must appoint a surveyor. The Building Owner and Adjacent Owners can appoint a surveyor together or each appoint their own. The Building Owner cannot act as his/her own surveyor. The Building Owner pays for the two or more surveyors. The Building Owner's Party Wall Surveyor draws up the necessary Notices to serve on the Adjoining Owners. The surveyors will then agree on a "Party Wall Award". If they cannot agree, an agreed Third Surveyor is appointed.

What does the Act Cover?

Section 1: New building on line of junction.
Section 2: Repair, etc., of party wall: rights of owner. Work on existing party walls

Section 3: Party structure notices
Section 6: Adjacent excavation and construction. Excavation near neighbouring buildings

Notices:

Line of Junction Notice: Section 1
Party Structure Notice: Section 2 and Section 3 for Notice
Adjacent Excavation Notice: Section 6

- 3m Notice: Section 6(1)
- 6m Notice: Section 6(2)

See Figure 3.1 for Notices in respect of works to a party wall or close to an adjacent structure

The Party Wall Award

This is a legal document which says:

- What work should happen
- How and when it will be carried out
- Who will pay for which part and how much will be paid (including surveyor's fees)

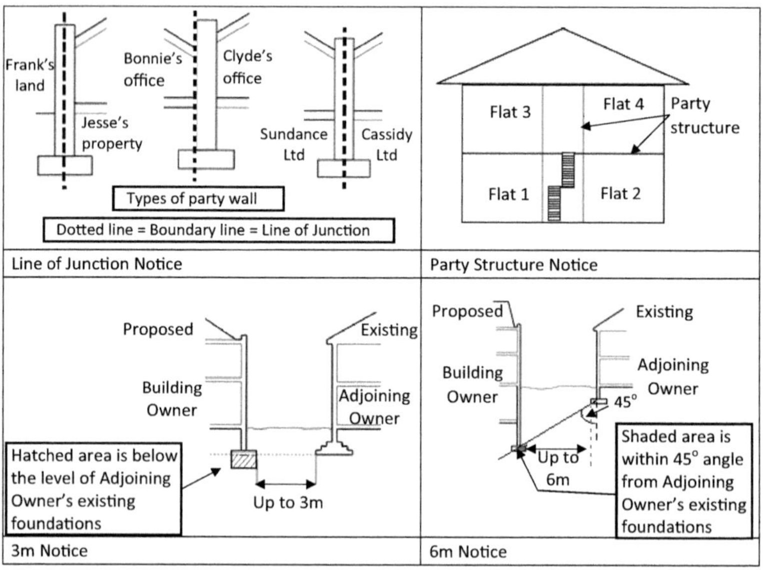

Figure 3.1 Notices in respect of works to a party wall or close to an adjacent structure.

Who should serve the Notices?

The legislation does not prevent a Building Owner from serving a Notice, but any error contained in the Notice can render it invalid. Therefore, to avoid unnecessary delays and costs, Building Owners and Adjoining Owners should seek advice from a Party Wall Surveyor experienced in dealing with party wall matters. See Table 3.7 for periods of time relating to party wall Notices and Awards.

Party walls in Scotland

The proprietor of property must take into account the interest of his neighbours for common structures. For a common dividing wall, either proprietor can object to the other carrying out works which may be injurious to it.

In a tenement building, each flat can be owned separately and the external wall of each flat is also owned by the flat proprietor. Here, each proprietor has a common interest in the whole external wall and should not interfere with their own wall in a manner that might damage the other properties.

Similarly, each proprietor owns their floors and ceilings to the centre point of the joists (unless specified differently). An owner must not cut into the joists or otherwise weaken the structure in a way that damages their neighbour's floor/ceiling.

Table 3.7 Periods of time relating to party wall Notices and Awards

Time factors	
Adjoining Owner receives a notice under the Party Wall Act	14 days to respond
They do not respond or appoint a surveyor within this time period	Further 10 days' notice to be served
Minimum period of notice for construction of a party fence wall or boundary wall where the foundations are to encroach on the Adjoining Owner's land	1 month
Minimum period of notice for construction in relation to Notices of Adjacent Excavation, i.e. 3 m and 6 m Notices	1 month
Minimum periods of notice for repairs, demolition, etc., to a party wall or party structure	2 months
Maximum time to the commencement of the works following the serving of the initial notice	12 months

The roof is owned by the proprietor of the top floor flat(s), but as the lower storey proprietors have an interest in the roof being undamaged and watertight, they may compel the top storey proprietor(s) to keep it in repair.

Feudal titles may contain conditions that override these common law requirements.

ACOUSTICS AND NOISE CONTROL: MINIMISING NOISE TRANSITION

Excess noise can be extremely disruptive and can impact mental health, and result in hearing loss and cardiovascular disease. This is recognised by the creation of statute and guidance to limit and control the passage of sound. Sound is caused by vibrations that transmit through a medium and reach the ear or some other form of detecting device. It is measured in loudness (decibels (dB)) and frequency (Hertz (Hz)). Sound is transmitted in buildings by both airborne sound and structure-borne sound.

Impact sound is a form of structure-borne sound that occurs when an object impacts on or vibrates against another object, resulting in the generation and transmission of sound. Typical examples of an impact sound are footsteps on a floor or banging on a wall, resulting in sound being transmitted respectively through the floor or wall construction and heard in the space below or adjacent. Impact sound can travel through solid structures and through cavities.

Airborne sound (or airborne noise) is sound that is transmitted through the air. Typically, generated by speech, traffic, media devices, animals, e.g. dogs barking.

Air- and structure-borne sound can be considered separate phenomena; however airborne sound can cause an element of the building fabric to vibrate when it comes into contact with a surface, and structural vibrations may radiate from a surface, generating airborne sound.

Noise can be defined as any unwanted sound. Noise can be too loud or repetitive. At certain decibels, it can be hazardous to health. Low-frequency noise can be as damaging as loud noise. Noise accounts for most of the complaints that UK local councils and the Environment Agency receive about environmental pollution and is a major source of stress.

Excessive noise can be a "statutory nuisance" (covered by the Environmental Protection Act) or a lack of compliance with building regulations. For the noise to count as a statutory nuisance it must either:

- Unreasonably and substantially interfere with the use or enjoyment of a home or other premises
- Injure health or be likely to injure health

Exposure to high levels of noise can lead to hearing loss, tinnitus, stress, anxiety, high blood pressure, gastrointestinal problems, and chronic fatigue.

In England and Wales, Approved Document E of the Building Regulations covers:

- Protection against sound from other parts of the building and adjoining buildings
- Protection against sound within a dwelling house, etc.
- Reverberation in common internal parts of buildings containing flats or rooms for residential purposes
- Acoustic conditions in schools

Pre-completion testing, or other approved means, is required to show compliance.

The guidance includes the use of Robust Details. These are published details of high-performance wall and floor constructions with associated construction details, which are expected to be sufficiently reliable not to need pre-completion testing. These can be used for new adjoined dwellings but not for any conversions, home extensions, or building refurbishments. Developers need to register with a scheme for the works. Robust Details Limited is a non-profit distributing company that operates a Robust Details certification scheme, financed by plot registration fees paid by users.

For workplaces, noise exposure is required to be a part of risk assessments. In the UK, guidance is given by the Health and Safety Executive (HSE). The Control of Noise at Work Regulations cover all industry sectors in Great Britain. The aim of these regulations is to ensure that workers' hearing is protected from excessive noise at their place of work, which could cause them to lose their hearing or suffer from tinnitus.

For England, The National Planning Policy Framework, the Noise Policy Statement for England (NPSE), and Planning Practice Guidance (Noise) (PPGN) give guidance for noise limits and control. There may be similar 1st, 2nd or 3rd tier legislation for other countries. The NPPF sets out the Government's planning policies for England. It does not contain any specific noise policy or noise limits but provides a framework for local authorities and people to produce their own local and neighbourhood plans. In respect of health and quality of life, it requires that local planning policies and decisions should aim to:

- Avoid noise, which gives rise to significant adverse impacts as a result of new development
- Mitigate and reduce to a minimum other adverse impacts arising from noise from new development
- Not put unreasonable restrictions on the development of existing businesses because of changes in nearby land uses since they were established
- Identify and protect areas of tranquillity which have remained relatively undisturbed by noise and, thus, prized for their recreational or amenity value

Defra and the World Health Organization use the following terms for noise impact:

NOEL – no observed effect level. Current thinking is that below this level, there is no detectable effect on human health and quality of life due to noise. No specific measures for noise reduction are required at this level.

LOAEL – lowest observed adverse effect level. Adverse effects on health and quality of life can be detected above this level. Above this level noise should be mitigated and reduced to a minimum.

SOAEL – significant observed adverse effect level. Significant adverse effects on health and quality of life occur above this level. Such noise levels should be avoided or prevented.

Choice of design and construction materials used for the building can limit noise emissions and exposure. Poor detailing or poor standards of workmanship can result in airborne sound transmitting directly between spaces, for example through gaps around the edge of doors, and may result in flanking sound, where sound travels around a separating element, even though the element itself might provide very good acoustic insulation. Even very small gaps can cause a significant increase in the transmission of airborne sound. Problems can also occur where openings face onto "noisy" spaces, e.g. a circulation space, road, or outdoor event space. If this deters occupants from opening windows or using trickle vents, natural ventilation can be affected.

There are four basic approaches to reducing sound in construction:

- Distance: Increasing the distance between source and receiver; the greater the distance, the quieter the noise sounds
- Damping: Using damping structures such as sound baffles; dissipates vibrational energy before it can build up and radiate as sound
- Absorption: Using noise barriers to absorb the energy of the sound waves; trapping the sound waves
- Diffusion: Using noise barriers to reflect the energy of the sound waves; scatters sound in different directions

Simple noise control techniques include the use of damping material or sheets incorporated into floors. The efficiency falls off for thicker sheets. Above about 3 mm sheet thickness, it becomes increasingly difficult to achieve a substantial noise reduction. Material for damping includes acoustic absorbent foam, mineral wool, fibreglass, and rubberised materials. The appropriate use of absorption materials within the building can reduce or limit the effects of reflected sound. Specialist help is often advised.

CONTRACTS

Contracts made between parties do not have to be in writing. They are created by an offer and an acceptance, and an intention to enter into a legally binding relationship, and some form of "payment" (e.g. money, action, inaction, or promise).

There are four key elements to a contract procedure:
Offer
Acceptance
Intent to create legal relations
Consideration

The payment does not necessarily have to be directly between the parties of a contract – see Donoghue v. Stevenson [1932] "**the snail in the bottle**" – Duty of care.

A contract may be founded on a conversation or discussions over a period of time. It can be implied from conduct. Surveyors who intend to enter into a contractual relationship need to clarify their terms of engagement and scope of service before entering into an oral contract. The RICS offers a Standard Form of Consultant's Appointment.

It is recommended that care is taken to avoid implying anything that is not intended to be carried out (see Misrepresentation Act and Unfair Contract Terms Act). In the absence of any contractual clause to the contrary, there is the common law implication that the surveyor's work will be carried out with the skill and care reasonably expected of a competent surveyor professing the particular skill enacted. Negligence or breach of duty may otherwise be risked.

Instructions to conduct a building surveying service for the client should be confirmed back in writing, immediately on receipt of the instruction. This written confirmation should include:

- The means by which the instruction was received (e.g. in person, by telephone, by email)
- The name of the person who received the instruction
- The nature of the instruction received
- The name of the person who will carry out the service
- The date, time, and address at and/or about which the service will be conducted
- The date by which the report will probably be issued
- The cost of the service (plus or including VAT)
- The scope of the service to be conducted
- Specific exclusions from that scope
- Any agreement required for additional services, e.g. requirement for a contractor to be present on site, for an articulated lift/aerial work platform, an m+e contractor, etc.
- The provision of the surveyor's complaints procedure

- Requirements for surveyor expenses and/or disbursements
- Dependent on the work and surveyor's insurance company, required professional indemnity insurance (PII) clauses

Main types of construction contract:

Lump sum contracts: The contract sum is agreed before work starts on site. The contractor agrees to carry out specified works for an agreed cost. Can be employed with or without a detailed bill of quantities and drawings. Without such detail, the risk to the contract is increased. Variations to the contract must be separately priced to reduce risk.

Measurement contracts: The contract is assessed and re-measured as the works progress on a formerly agreed basis. Usually involves approximate bills of quantities and drawings. A number of permutations are possible with JCT or ICE contract forms with or without approximate quantities. Such contracts are not normally recommended for work which lasts in excess of three years.

Cost reimbursement contracts: The contractor is reimbursed on the basis of prime cost of labour, materials, and plant, with an agreed percentage added to cover overheads and profit. One risk of this contract is that the project can degenerate into slowness and inefficiency, and measures can be introduced into the contract to deter this.

Design and build contracts: The contractor both designs and builds the project. Can be awarded on a fixed price or lump sum basis, but cost certainty is then dependent on not making any subsequent changes as these have the potential to be expensive (because prices charged by the contractor for those changes will not be subject to competition). Very important for the client to give intense and detailed consideration to the employer's requirements. Independent client advisers may be recommended.

Management contract: Generally involves a management contractor managing the works without undertaking any building works.

Letter of intent: Pre-construction services agreement. Work starts with a letter of intent before a full contract has been agreed. Legal performance varies and may not create a binding contract between parties, with unexpected results. Must be well-drafted and precise – specialist advice recommended.

JCT contracts: The Joint Contracts Tribunal (JCT) produces standard forms of contract for construction, guidance notes, and other standard documentation for the UK construction industry. The main contract forms include:

Building Contract for Home Owner/Occupier (where client deals directly with the builder) (HOB)

Building Contract for Home Owner/Occupier (who has appointed a consultant) (HOC)
Contract for Home Repairs and Maintenance (HO/RMI)
Design and Build Contract (DB)
Intermediate Form of Building Contract (IC or IFC)
Major Project Form (MP)
Management Contract (MC)
Minor Works Agreement (MW)
Standard Form of Building Contract (SBC)

NEC contracts: The New Engineering Contract (NEC) or NEC Engineering and Construction Contract forms of contract for civil engineering and construction projects. Forms include:

Alliance Contract (ALC)
Design Build and Operate (DBO)
Dispute Resolution Services Contract (DRSC) – previously Adjudicator's Contract
Engineering and Construction Contract (ECC)
Engineering and Construction Short Contract (ECSC)
Engineering and Construction Short Subcontract (ECSS)
Engineering and Construction Subcontract Contract (ECS)
Framework Contract (FC)
Professional Services Contract (PSC)
Professional Services Short Contract (PSSC)
Supply Contract/Short Supply Contract (SC/SSC)
Term Service Contract (TSC)

ICC contracts: The Infrastructure Conditions of Contract (ICC) are a suite of related standard forms of contract for infrastructure works that cater for a wide range of contracting strategies. The ICC are published by the Association of Consultancy and Engineering (ACE) and the Civil Engineering Contractors Association (CECA). The ICC replaced the ICE Conditions of Contract. The ICC include:

Archaeological Investigation
Design and Construct
Ground Investigation
Measurement
Minor Works
Network Rail Measurement

Partnering Addendum
Target Cost
Term
With Quantities

LANDLORD AND TENANT

This section covers:

Decent Homes
Dilapidations

DECENT HOMES

Dependent on the relevant country's legislation, landlords and property owners may have a duty to ensure that a residential property for lease/rent provides an adequately safe and healthy environment. The Decent Homes Standard is a UK technical standard for social (i.e. not private owner-occupied) housing. This sets out minimum standards for council and housing association homes: To be free from any hazard that poses a serious threat to occupants' health or safety.

The standards are not prescriptive but tested by a risk assessment procedure – the Housing Health and Safety Rating System (HHSRS). To be "decent", a dwelling should have no Category 1 hazards. Timely remedial action is required for any Category 1 hazards (with exceptions for disproportional cost or disruption), i.e. "trigger action". Category 1 hazard: Serious and immediate risk to a person's health and safety. Category 2 hazard: Less serious or less urgent.

A Decent Home:

- Meets the current statutory minimum standard for housing (i.e. no Category 1 hazards)
- Is in a reasonable state of repair
- Has reasonably modern facilities and services
- Provides a reasonable degree of thermal comfort

There is a suite of associated documents available on the Gov.uk website which expand on these, including, post-Grenfell Tower fire, guidance for the assessment of high-rise residential buildings with cladding systems. The "Housing Health and Safety Rating System Operating Guidance" gives guidance on the risk assessment process. Twenty-nine hazard topics have been identified as shown in Table 3.8.

Table 3.8 Residential 29 hazards identified in the UK "Housing Health and Safety Rating System" for social housing

	Hazards in housing
1	**Damp and mould growth** Includes dust mites, mould, fungal growth, damp, high humidity
2	**Excess cold** A healthy indoor temperature is 18°C to 21°C
3	**Excess heat** (high indoor temperatures) Dehydration, trauma, stroke, cardiovascular and respiratory issues
4	**Asbestos and MMF**
5	**Biocides** Chemicals used to treat timber and mould growth
6	**Carbon monoxide and fuel combustion products** Excess levels of carbon monoxide, nitrogen dioxide, sulphur dioxide, smoke
7	**Lead** Risk of lead ingestion and inhalation
8	**Radiation** Radon gas and its daughters, primarily airborne but also radon dissolved in water
9	**Uncombusted fuel gas** Threat from fuel gas escaping into the atmosphere within a property
10	**Volatile organic compounds** Organic chemicals, including formaldehyde, gaseous at room temperature
11	**Crowding and space**
12	**Entry by intruders** Security
13	**Lighting** (inadequate natural or artificial light) Eye strain, psychological effects associated with the view from the property through glazing
14	**Noise**
15	**Domestic hygiene, pests, and refuse** Poor design, layout, and construction, issues for cleaning and maintaining hygiene, pests attraction, inadequate and unhygienic provision for storing household waste
16	**Food safety** Poor provision and facilities to store, prepare, and cook food
17	**Personal hygiene, sanitation, and drainage facilities**
18	**Water supply** Quality of water supply for drinking household use such as cooking, washing, and sanitation

(continued)

Table 3.8 *(continued)*

	Hazards in housing
19	**Falls associated with baths**, shower, or similar facility
20	**Falls on the level surfaces** Floors, yards, paths, trip steps, thresholds, ramps where the change in level is less than 300 mm
21	**Falls associated with stairs, steps, and ramps, internal and external, and falls over guarding, balustrading, etc.** Change in levels greater than 300 mm
22	**Falls between levels** Falls from one level to another, inside or outside a dwelling, where the difference is more than 300 mm. Including falls from balconies, landings, or out of windows
23	**Electrical hazards** Electric shock and electricity burns
24	**Fire**
25	**Flames, hot surfaces, materials, liquids, and vapours**
26	**Collision and entrapment** In architectural features (e.g. in doors and windows) and colliding with objects such as windows, doors, and low ceilings
27	**Explosions**
28	**Ergonomics** Physical strain associated with functional space and other features at the dwelling
29	**Structural collapse and falling elements** e.g. due to inadequate design, fixing, or disrepair or as a result of adverse weather conditions

DILAPIDATIONS

"Dilapidations" is a part of the process of lease occupation of property. Dilapidations is a specific area of law relating to breaches of a tenant's lease obligations or covenants. It is used for commercial leases, not residential. This section will discuss the process pertinent to England and Wales. Dilapidations in Scotland, Republic of Ireland, Northern Ireland and the Channel Islands are similar to those in England and Wales with the same basic procedures, but some with interesting pitfalls. Other countries, such as the USA, also use dilapidations.

Dilapidations are integral to any commercial lease, as they define tenant obligations for the maintenance and repair of property. A dilapidations claim can be made by the landlord against the tenant during or towards the end of a lease, or after the lease has ended. The landlord's claim document is usually called a

Schedule of Dilapidations. It contains references to breaches of the tenant's lease obligations, mainly relating to redecoration, repair and the reinstatement of any physical alterations.

Dilapidations are breaches of lease covenants that relate to the condition of a property during the term of the tenancy or when the lease ends. They are repairs required during or at the end of a tenancy or lease – but only as stated in the lease.

Covenants are the promises/agreements made between landlord (lessor) and tenant (lessee) when signing the lease. They comprise the clauses of the lease.

The purpose of the lease is to ensure that the lessor's investment is not diminished by the action or default of the lessee.

Interest is how much the premises are worth to the person who has a legal right to the place:

- Landlord's interest – the **reversion**
- Tenant's interest – the **term**

Both these interests have a value and can be sold on, creating other landlord and tenant relationships.
Diminution in the value of the reversion means how much the property's value for the landlord has reduced (because of whatever) at the end of the lease.

A schedule of dilapidations is prepared by the landlord's surveyor, scheduling breaches of reinstatement, repair, decorations, and other legal compliance promises, outstanding from the lease covenants. It identifies the clauses of the lease that have been breached, suggests appropriate remedial works, and, usually, the estimated cost of those works.

There are three types of schedule of dilapidations:

- **Interim**, issued during the course of the lease term
- **Terminal**, issued during the last 3 years to 18 months of the lease
- **Final**, issued after the expiry of the lease, but within a reasonable period

The RICS produces a Guidance Note for Dilapidations in England and Wales. This advises Chartered Surveyors on producing Schedules of Dilapidations, Quantified Demands, Responses, Scott Schedules, and Diminution Valuations. Alongside this, they produce a template for the schedule of dilapidations.

There is a distinction between residential property and commercial in the UK. There is also a difference in the law of property. There are many statutes covering residential renting of buildings – what follows is just a few statutes concerning residential. Common law generally favours the landlords so statute generally redresses the balance. Case law is the law made by cases brought to court where the judge interprets the facts within the law and makes a judgement that may clarify the law. See Figure 3.2 for some of the main UK legislation relating to dilapidations. Further information follows.

Law of Property Act 1925 allows landlords to give tenants a notice demanding that they repair the building in accordance with the terms of the lease they signed.

Landlord and Tenant Act 1927 gives some protection to the tenant by limiting the landlord's claim for breaches at the end of the lease. Section 18 (s18) has two "limbs". It puts a cap on the landlord's claim for damages for disrepair. Damages = a loss that the landlord has suffered. If there is no loss, there cannot be a claim for damages.

First limb: the damages "shall in no case exceed the amount … by which the value of the reversion … is diminished owing to breach".

Second limb: "no damages shall be recovered … if … the premises, in whatever state of repair … would at … the termination of the lease … be pulled down, or such structural alterations made therein as would render valueless the repairs covered by the covenant".

The landlord would not have suffered a loss, therefore no claim. This includes the concept of supersession; i.e. a new tenant's needs would supersede what is there.

Leasehold Property (Repairs) Act 1938 gives some protection to tenants of residential properties with long leases in small houses against the enforcement of repairing covenants in their lease.

This Act identifies the **5-stage test** on whether the disrepair is "damaging" to the landlord's reversion. It builds on the diminution argument. It limits minor repairs and little over-demanding claims unless these are of the "stitch in time" type.

The Landlord and Tenant Act 1954 extends the Leasehold Property (Repairs) Act 1938 to every tenancy (residential or commercial) where the lease is for 7+ years, 3+ years of which remain.

In residential accommodation, the distinction between leases and licences is important because the Rent Acts do not generally apply to licences.

The Defective Premises Act 1972 requires the landlord to keep the property up to a certain standard.

Landlord and Tenant Act 1985 imposes a duty on the landlord to carry out certain repairs of residential property which have been let for a term of less than seven years. It is mainly relevant for the landlord to maintain the premises (generally structure and services) in repair and proper working order. This statute overrides any covenant in a lease.

The Housing Act 1988 offers protection for assured tenancies. These tenants have security of tenure. A licence cannot be an assured tenancy. The rules are complex, and several types of tenancy (as well as licences) are excluded from this protection.

Landlord and Tenant (Covenants) Act 1995 relates to the assignment of the lease to another tenant and the reversion requirements of the various upper tenants.

The Housing Act 2004 relates to private-sector housing with vulnerable tenants being better protected.

Dilapidations Protocol: Pre-Action Protocol for Claims for Damages in Relation to the Physical State of Commercial Property at Termination of a Tenancy. This applies to commercial property in England and Wales and relates to claims for damages for dilapidations against tenants at the termination of a tenancy, i.e. terminal dilapidations claims. It sets out conduct that the court normally expects the parties to follow prior to commencement of legal proceedings. It establishes a reasonable process and timetable for the exchange of information relevant to a dispute, sets standards for the content and quality of Dilapidations schedules and Quantified Demands, and the conduct of pre-action negotiations.

TYPES OF SURVEY

Condition survey: A survey of the condition of a particular property at a particular point in time. A schedule of condition gives a list of each element etc in the property or demise, their material and condition. Can be attached to and referred to in other documents – for example, a lease.

Stock condition survey: The holder of a portfolio of property may want to know the condition of that stock of property, e.g. for maintenance management or estate management purposes. A schedule of condition can be drawn up to prepare a stock condition survey. Uses include for social housing estates.

Residential condition survey: Used for a conventional house, flat, or bungalow, built from common building materials and in reasonable condition. Focuses purely on the condition of the property by setting out the following:

- Clear "traffic light" ratings of the condition of different parts of the building, services, garage, and outbuildings, showing problems that may require varying degrees of attention
- A summary of the risks to the condition of the building
- Advice on replacement parts guarantees, planning, and control matters for the client's legal advisers.

Residential Occupation	Business or Commercial Tenancies
Law of Property Act 1925	Landlord & Tenant Act 1927
Leasehold Property (Repairs) Act 1938	Landlord & Tenant Act 1954
Landlord & Tenant Act 1954	Landlord & Tenant (Covenants) Act 1995
Housing Act 1988	- Occupancy and business purpose
Landlord & Tenant Act 1985	- Security of tenure
Defective Premises Act 1972	- Interim rent
Housing Act 2004	
Licences	

Figure 3.2 Some of the main UK legislation relating to dilapidations.

See Figure 3.3 Traffic light system for condition surveys.

Commercial condition survey: A factual log of the building's condition, not a narrative description of defects, causes, and required action. Normally prepared for legal or contractual reasons. Usually supported by sketches and photographs. For a lease agreement, a photographic schedule of condition examines and details any defects to the property or demise before the lease is signed. Can also be used with a party wall agreement.

Other survey types:

Planned maintenance surveys and **Quinquennial inspections:** Quinquennial inspections are undertaken every five years of, for example, ecclesiastical buildings, in order to assess their condition and plan future maintenance. Planned preventative maintenance (PPM) surveys are an asset management tool, used to proactively maintain, manage, and improve properties over a period of years. The traffic light system can be used with the addition of noting whether the repairs are required in zero, one, two, three, four, five, etc., years.

Compliance audits: These are to check that properties comply with specific legislation, e.g. for fire, or health and safety. Note that it is extremely difficult to prove compliance.

Measured surveys: Often done as a precursor to repairs or redevelopment or may be required for record purposes. All areas are accurately measured and reproduced onto scaled plans.

Residential surveys: The scope and depth of a residential survey depend on the brief and the requirements of the client. The purpose of the survey must be clearly defined at the briefing stage, through questioning of the client. Surveys are about knowledge management, and it is essential to determine the knowledge the client needs and wants to acquire. The survey can then be designed to ensure relevant data is collected, processed, and presented.

Pre-acquisition survey: Reports on the current condition of the property, what defects/issues are present, and remedial repair recommendations, which will assist the prospective purchaser in respect to budgetary planning, etc.

Survey Schedules:

These form part of the reportage of the survey, and can form part of legal documents (e.g. see "Dilapidations"). It is important that where guidance exists this is followed.

Schedule of repair: Includes approximate costings. Includes

- All works considered necessary to bring the property to a reasonable standard
- Any works that may be required to bring the property up to the current standards
- Any enabling works (including, as relevant, scaffolding, waste disposal, storage, highways works, and permissions) that are necessary to allow the feasibility proposals to be undertaken
- The pricing of each element using standard pricing documents

Schedule of condition: For the internal and external elements of the property. Should record the condition of a property accurately and factually. Normally prepared for legal or contractual purposes. Photographs are used to give a record to ensure any property issues are clearly understood. Traffic light systems, as shown in Figure 3.3, are frequently used.

Stock condition surveys: Surveys of large numbers of properties in a portfolio/ estate. Also known as asset condition surveys. Carried out on all building types – institutional, commercial, residential. Commissioned for a range of purposes or levels, e.g. strategic level (informing high-level asset policy and planning), operational level (detailed information on component, location, condition, and costs provides a basis for maintenance plans and implementation). Multifaceted surveys focusing on condition type, legal compliance, facilities, energy, and/or space utilisation. Methods of data collection usually utilise internet (or similar) based technologies. Traffic light systems, as shown in Figure 3.3, are frequently used.

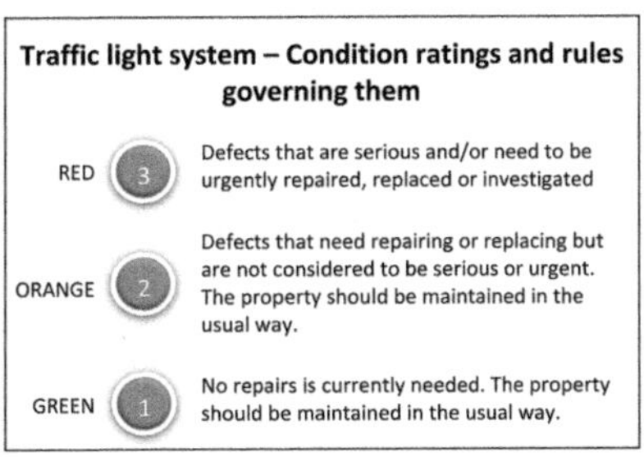

Figure 3.3 Traffic light system for condition surveys.

HEALTH AND SAFETY

This section contains:

1. Health and safety
2. Construction (Design and Management) Regulations
3. Fire safety in occupied buildings
4. Due diligence surveys

1. HEALTH AND SAFETY

All activities in construction and the built environment involve health and safety concerns to a greater or lesser extent. There are legislative issues and guidance concerns. The RICS produces a guidance note on surveying safely. It states that the appropriate management of health and safety is a requirement for all RICS-regulated firms and RICS members, including property-related businesses, practising throughout the world. The principles and practices in the publication are useful for all surveyors and include corporate and personal responsibilities.

Globalisation brings the expectation of common standards, and health and safety legislation is mirrored world-wide. It was recently calculated that every 15 seconds a worker dies from a work-related accident or disease, and every 15 seconds, 160 workers have work-related accidents. This is better than it used to be, because health and safety legislation is becoming more mainstream.

Accidents can result from:

- Tiredness, stress, or inattention
- Faulty construction equipment
- Defective products
- Defective machines
- Inadequate safety or equipment training
- Negligence or recklessness

Accidents are very costly, and not all costs can be reimbursed through insurance. See Figure 3.4 for hidden costs of accidents.

Wherever you work in the world, there is likely to be a similar legislative framework:

- A duty to plan for safety and health
- Duties of employers and responsible people
- Need for risk assessments and safe methods of work
- Duties of us all to ensure safety for ourselves and others

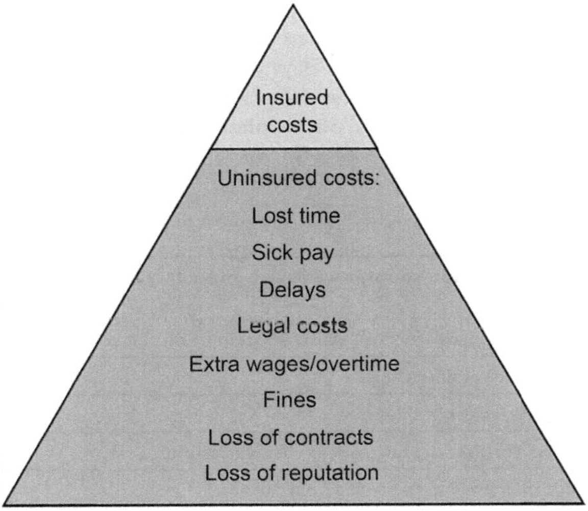

Figure 3.4 Hidden costs of accidents.

2. CDM - CONSTRUCTION (DESIGN AND MANAGEMENT) REGULATIONS

This is the UK version of legislation designed to protect employees who are carrying out construction, maintenance, repairs, and demolition.

This is mirrored world-wide, for example:

In France, health and safety is governed by the French Labour Code and the French Construction Code, which contain similar provisions as the CDM regs. In Russia, health and safety on construction sites is governed by technical regulations, federal laws, and health and safety codes, which set down requirements for the organisation of the construction site, protective clothing and equipment, methods of construction, etc. In Australia, the Commonwealth and each state and territory have legislation to regulate occupational health and safety in each jurisdiction. The occupational health and safety legislation in each jurisdiction imposes duties on employees, those working at the workplace, and those who have sufficient control of the workplace to ensure the health and safety of any person who might be affected by their actions. In Germany, the Ordinance on Health and Safety on Construction Sites sets out the main requirements with regard to health and safety on construction sites. In Italy, health and safety in relation to construction activities is regulated by Legislative Decree 81/2008 (Safety Law). In Mauritius, the Occupational Safety and Health Act 2005 consolidates the legislation on the safety, health, and welfare of employees at work.

The UK CDM regulations are regularly updated, with the current versions and associated guidance available on the HSE website. See Table 3.9 for HSE statements for the requirements of CDM.

At the time of writing, duty holders for CDM are as set out in Table 3.10. Summaries of the duties can be found on the HSE.gov.uk website.

If the work will last longer than 500 person days or 30 working days (with more than 20 people working at the same time), it will need to be notified to HSE. Under the CDM regs 2015, a **construction phase plan** is required for every construction project.

Table 3.9 HSE statements for the requirements of CDM

For all roles in construction, CDM aims to improve health and safety in the industry by requirements to:
Sensibly plan the work so the risks involved are managed from start to finish
Have the right people for the right job at the right time
Cooperate and coordinate individual work with others
Have the right information about the risks and how they are being managed
Communicate this information effectively to those who need to know
Consult and engage with workers about the risks and how they are being managed

Table 3.10 HSE guidance on duty holders and duties

Duty holders	**Comments**	**Main duties**
Client: Domestic	Anyone who has construction work carried out for them but not in connection with any business. Usually work done on their own home or the home of a family member.	CDM 2015 does not usually require domestic clients to carry out client duties as these normally pass to other duty holders.
Client: Commercial	Anyone who has construction work carried out for them as part of their business. This could be an individual, partnership, or company and includes property developers and companies managing domestic properties.	To ensure their project is suitably managed, ensuring the health and safety of all who might be affected by the work, including members of the public.
Designer	Organisation or individual. Preparing or modifying designs, drawings, specifications, bills of quantity, or design calculations. Anyone who specifies and alters designs as part of their work: architects, surveyors, consulting engineers, quantity surveyors, tradespeople, etc.	To eliminate, reduce, or control foreseeable risks that may arise during construction work or in the use and maintenance of the building once built.

(continued)

Table 3.10 *(continued)*

Duty holders	Comments	Main duties
Principal designer	A designer appointed by the client to control the pre-construction phase on projects with more than one contractor.	To plan, manage, monitor, and coordinate health and safety during the pre-construction phase, when most design work is carried out. Identify and eliminate, or control, so far as is reasonably practicable, foreseeable risks to the health or safety of any relevant person during build, maintenance, cleaning, or occupying the building. Assist the client in their duties. Liaise with the principal contractor, sharing relevant information. Prepare and update the **health and safety file** for projects involving more than one contractor. If engagement finishes before end of project, hand this over to principal contractor.
Principal contractor	A contractor appointed by the client to manage the construction phase on projects with more than one contractor.	To plan, manage, monitor, and coordinate health and safety during this phase, when all construction work takes place. Draw up **construction phase plan** during the pre-construction phase, and before setting up a construction site. Update and complete the **health and safety file**, if the principal designer's engagement finishes before end of project.
Contractor	An individual or business carrying out construction work (e.g. building, altering, maintaining, or demolishing), manages such work, or directly employs or engages construction workers. Contractors work under the control of the principal contractor on projects with more than one contractor.	To plan, manage, and monitor the work under their control in a way that ensures the health and safety of anyone it might affect (including members of the public).
Worker	An individual who actually carries out the work (building, altering, maintaining, or demolishing). Plumbers, painters, electricians, decorators, steel-erectors, scaffolders, labourers, foremen, charge-hands, etc. Workers must be consulted on matters affecting their health, safety, and welfare.	Cooperating with their employer and other duty holders, reporting anything they see that might endanger the health and safety of themselves or others.

3. FIRE SAFETY IN OCCUPIED BUILDINGS

The current legislative framework for fire safety considers that most fires are preventable. It requires those responsible for workplaces and other buildings to avoid fires by taking responsibility for and adopting the right behaviours and procedures.

Employers, building owners, and/or occupiers (the responsible person(s)) must carry out a regular fire safety risk assessment, keeping it relevant and current. Based on the findings of this, the responsible person(s) must ensure that adequate and appropriate fire safety measures are in place to minimise the risk of injury or loss of life in the event of a fire.

The risk assessment should identify what could cause a fire to start, i.e. sources of ignition, substances that burn, and the people who may be at risk.

Once the risks have been identified, appropriate action must be taken to control them by elimination, avoidance, risk reduction, risk management, and people protection. Table 3.11 gives legislative guidance for fire safety in occupied buildings.

Table 3.11 Legislative guidance for fire safety in occupied buildings

England and Wales: The Regulatory Reform (Fire Safety) Order 2005	The Department for Communities and Local Government (DCLG) website	Premises-specific guidance documents to help meet responsibilities under the Regulatory Reform (Fire Safety) Order 2005
UK Government	Fire safety law and guidance documents for business	www.gov.uk/government/collections/fire-safety-law-and-guidance-documents-for-business
Scotland: Part 3 of the Fire (Scotland) Act 2005, supported by the Fire Safety (Scotland) Regulations 2006	The Scottish Government website	Information to help meet responsibilities under the Fire (Scotland) Act 2005. https://www2.gov.scot/Topics/Justice/policies/police-fire-rescue/fire/FireLaw/SectorSpecificGuidance https://www2.gov.scot/Topics/Justice/policies/police-fire-rescue/fire/FireLaw/GeneralGuidance
Great Britain	HSE website	Guidance on fire safety in the construction industry http://www.hse.gov.uk/toolbox/fire.htm http://www.hse.gov.uk/construction/safetytopics/generalfire.htm

4. DUE DILIGENCE SURVEYS

Technical Due Diligence refers to the process by which building surveyors investigate, analyse, and report on the physical characteristics, condition, and legal compliance of a building and its essential services, usually prior to acquisition or lease. Conventional condition surveys are often simple visual inspections, but TDD involves more detailed investigations including functional assessment and performance testing of the buildings and plant.

The survey appraises a commercial building for condition, risk management, investment limitations and opportunities, sustainability, legislative compliance, and building service installation systems. TDD surveys for larger or more complex buildings can include a team of building services engineers, structural engineers, flooding consultants, and environmental consultants, alongside the building surveyors.

They usually include the following, dependent on the brief:

- Building services (including heating, cooling, ventilation, building management systems, automation, drainage)
- External areas/buildings/boundaries
- Health and safety matters
- Fire precautions
- Accessibility
- Environmental matters
- Deleterious and hazardous materials
- Sustainability issues
- Cultural heritage

The surveyor needs to have in place:

- Adequate professional indemnity insurance
- Third-party liability insurance

And be aware of:

- Subrogation
- Duty of care
- Negligence
- Contract and limitation
- Civil procedure rules

Guidance is available in the RICS Guidance Note "Building surveys and technical due diligence of commercial property".

INCLUSIVITY

The Equality Act

Many countries have anti-discriminatory laws. In the UK, anti-discriminatory law has developed from domestic law, since 1965, coupled with European directives, since 2000. The current law, the Equality Act, legally protects people from discrimination in the workplace and in wider society. It replaced previous anti-discrimination laws such as the Disability Discrimination Act and Acts relating to race relation and sex discrimination. The Equality Act forms part of the law of England and Wales, and the law of Scotland (with the exception of Section 190 and Part 15), with a few provisions forming part of the law of Northern Ireland (although equal opportunities and discrimination are "transferred matters" under the Northern Ireland Act 1998). Express provision is made for particular provisions (in a limited number of specific cases) to apply outside the United Kingdom.

It sets out the different ways in which it is unlawful to treat someone. The characteristics that are protected are age, disability, gender reassignment, marriage and civil partnership, pregnancy and maternity, race, religion or belief, sex, and sexual orientation. With the main provisions relating to building surveying given in bold, it is unlawful:

- To discriminate against, harass, or victimise a person when providing a service (**this includes the provision of goods or facilities**) or when exercising a public function
- To discriminate against, harass, or victimise a person **when disposing of (e.g. selling or letting) or managing premises**
- To discriminate against, harass, or victimise a person at work or in employment services (includes provisions regarding pay and any employers' requirements for divulging disabilities or health issues)
- For education bodies to discriminate against, harass, or victimise a school pupil, student, or applicant
- For associations (e.g. private clubs, political organisations) to discriminate against, harass, or victimise members, associates, or guests

The Act establishes a general duty on public authorities, when carrying out their functions, to eliminate unlawful discrimination, harassment, or victimisation; to advance equality of opportunity; and to foster good relations. It also contains provisions enabling an employer, service provider, or other organisation to take positive action to overcome or minimise a disadvantage arising from people possessing the protected characteristics. See Table 3.12 for human conditions (short-term, long-term, or permanent) at risk of architectural disability.

Building surveyors provide a service to their clients (and are therefore required to not discriminate, etc., in providing this service), and they also advise clients on

Table 3.12 Human conditions (short-term, long-term, or permanent) at risk of architectural disability

Inclusivity – disabilities or people?		
Traditional disabilities	Hearing	Physical coordination
	Mental ability	Sight
	Mobility	Manual dexterity
	Speech	Continence
Lifestyle disabilities	Adult with a pram, pushchair, and/or toddlers	Children
	People smaller, larger, or taller than average	Adult with large bags or cases
	Knee or leg injuries	Visual/balance disorders, e.g. due to migraine, ear infection, alcohol, drugs, etc.
	Finger, hand, or arm injuries	Back injuries
	Pregnancy	Post-operation
	Dyslexia/reading problems	Digital inabilities
Later-life disabilities (and can occur at any age)	Arthritis	Mobility problems
	Rheumatism	Stiff joints
	Poor hearing	Manual dexterity problems
	Sight problems	Memory loss
	Knees/ankles that won't bend	Excessive simultaneous information

their clients' actions, activities, service provision, etc. A working knowledge of inclusivity is therefore required.

Architecture shapes the choices about where and how to live, work, and play throughout life. However, the built environment can disable anyone by poor design. "**Architectural disability**" occurs when the physical design, layout, and construction of buildings and places confront people with hazards and barriers that make the built environment inconvenient, uncomfortable, or unsafe for everyone to use and even prevent some people from using it at all. On the other hand, "**universal design**" relates to the design of products and environments to be usable by all people, to the greatest extent possible, without the need for adaptation or specialised design. See Figure 3.5 for Benktson's Design Pyramid relating to inclusive design. The intent of universal design is to simplify life for everyone by making products, communications, and the built environment more usable by as many people as possible at little or no extra cost. Universal design should benefit people of all ages and abilities.

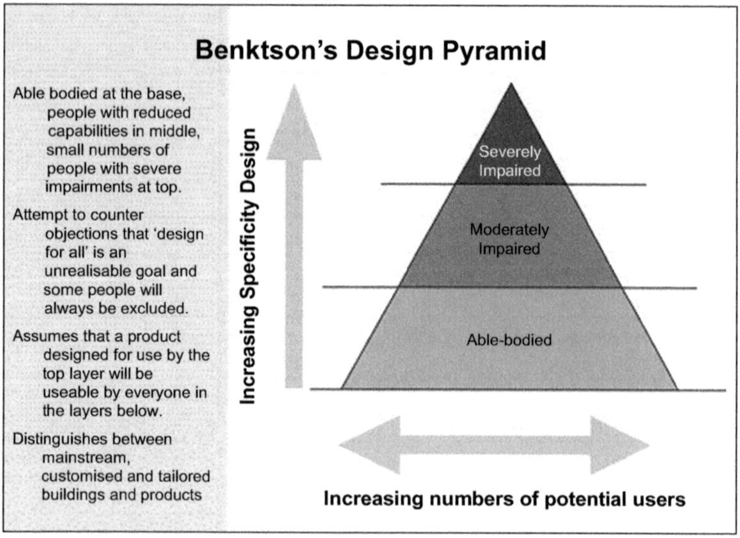

Figure 3.5 Benktzon's Design Pyramid relating to inclusive design (source: Professor Julienne Hanson of University College London (Hanson, 1994), adapted from Benktzon, 1993).

UK Building Regulation Part M requirements

Currently split into two parts: Volume 1 Dwellings, and Volume 2 Buildings other than dwellings.

New dwellings are further split into three types: Visitable, Accessible and adaptable, and Wheelchair user.

"**Visitable dwelling**": Ordinary dwellings where anyone who uses a wheelchair would be able to visit (gain access to premises and a habitable room, use toilet, etc.); people should be able to reach and access a building (implying regardless of ability, etc.), and be able to access facilities in a building (switches, controls, toilets, etc.). New houses should have an entrance-floor toilet that is accessible to some extent.

"**Accessible and adaptable dwelling**": Includes the requirements of a visitable dwelling, and is additionally capable of being used by, or easily adapted in the future to be suitable for, a wide range of occupants including older people, those with reduced mobility, and some wheelchair users. It is broadly similar to "Lifetime Homes", developed by a group of housing experts, including Habinteg and the Joseph Rowntree Foundation.

"Wheelchair user dwelling": Specifically designed for a wheelchair occupant. Access to internal and external spaces, use of all the facilities and rooms.

Volume 2 Buildings other than dwellings: Regardless of ability, people should be able to gain access to, and access within, buildings and use their facilities, both as visitors and as people who work in them. This applies to new buildings and to new extensions of existing buildings. Some parts also apply when a building is altered and if there is a change of use to a hotel, institution, public house, or shop.

If a building is to be extended and sanitary accommodation is provided in the existing building, then the extended building must have accessible provision.

Fire safety and accessibility

Access must also include egress – especially in an emergency. Normal lifts are not to be used as means of escape in a fire situation. Evacuation lifts are required, but Part M and the Approved Documents do not insist on evacuation lifts. Refuges, often used instead, are of no use for safety if good facilities for communication, from and to the refuge, are not also provided. Surveyors will note that these facilities are commonly omitted.

A small rectangle is often left at the top of the stairs (labelled "Refuge" on plans) with no means for the wheelchair user to call for help or tell any managers of the building or fire rescue team that they are there and which floor they are on. Other means of escape designs provide such things as an "evac chair" or sledge – of little use if there is no one with the wheelchair user who is capable of transporting the wheelchair user in this down the stairs.

Current guidance for lifts includes DD CEN/TS 81-76:2011 "Safety rules for the construction and installation of lifts. Particular applications for passengers and goods passenger lifts. Evacuation of disabled persons using lifts". A related publication is BS EN 81-76, "Safety rules for the construction and installation of lifts. Particular applications for passengers and goods passenger lifts. Part 76. Evacuation of persons with disabilities using lifts", published 2019.

Some dimensions

Anthropometric measurements have been provided in various publications to show the general space requirements that people have. Table 3.13 gives typical measurements required width of space, for example of a route, are taken from empirical research.

All of these figures are dependent on the size of the person, the size or shape of any aid they may use, and whether or not they require the presence of another person beside them, adult or child, or a dog.

Table 3.13 Typical width measurements

Typical minimum widths (and other measurements)	
One person walking	0.55 m
A person using a stick	0.85 m
A child	0.5 m
One person pushing a pushchair	0.7 m
One person pushing a double buggy	1.0 m
One person pushing a buggy with a small child alongside	1.2 m
One person pushing a double buggy with a child alongside	1.5 m
One person using a wheelchair, or one person pushing another in a wheelchair	0.85 m
One person in a wheelchair with a small child alongside	1.5 m
One person with crutches	1.2 m
One blind person with a cane (making sweeping movements)	1.2 m
One person in an electric wheelchair	0.8 m
Typical wheelchair size	0.7 m wide × 1.05 m long
Typical larger wheelchair size	1.15 m wide × 1.9 m long
Typical baby buggy	0.75 m wide × 0.81 m long
Typical mobility scooters: Small Medium Large	 0.5 m wide × 1.08 m long 0.59 m wide × 1.22 m long 0.65 m wide × 1.45 m long

When people pass each other, they require the above dimensions plus approximately an additional 100 mm. For example, a person with a stick passing a person using a wheelchair would need 0.85 m plus 0.85 m plus 0.1 m, or 1.8 m.

More space will be required for manoeuvring, turning, using the facilities, etc.

REFERENCES

Benktzon, M. (1993). Designing for our future selves: the Swedish experience, *Applied Ergonomics*, Vol. 24, No. 1, pp. 19–27.

Hanson, J. (2004). The inclusive city: delivering a more accessible urban environment through inclusive design. In: *Proceedings RICS Cobra 2004 International Construction Conference: Responding to Change.* York, UK.

GUIDANCE, STANDARDS AND CODES

Most guidance is not prescriptive and not mandatory. For example, Approved Documents are only guidance for complying with the much more limited wording of the individual building regulations. However, the status, authors, and level of agreement for some publishers mean that their standards are regarded as base lines for acceptable practice. Therefore, non-compliance with standards can lead to accusations of negligence, and robust arguments will be needed as to why the course of action taken was as good as the published standard in the circumstances.

British, International, and European standards

Standards exist at many levels including:

- International standards, e.g. prefixed by "ISO" for international and "EN" for European
- National standards, e.g. prefixed by "BS" for British standards
- Industrial/sector, e.g. RICS guidance notes and Codes of Practice

Standards are defined by the British Standards Institution (BSI) as "an agreed, repeatable way of doing something". Normally the use of standards is voluntary, and they do not impose legal responsibilities. However, in some cases legislation, or for example Approved Documents, may use a specific standard, effectively giving it quasi-legal force as an accepted benchmark.

BSI is the UK's national standards body, publishing all national, European, and many international standards. Standards can be double prefixed, for example "BS EN" where a standard is harmonised for both the EU and the UK.

Standards are also available from databases such as NBS and Construction Information Service.

HSE guidance

The HSE is the UK government agency "responsible for the encouragement, regulation and enforcement of workplace health, safety and welfare, and for research into occupational risks in Great Britain". It is an independent regulator and believes "everyone has the right to come home safe and well from their job" and aims to prevent work-related death, injury, and ill health.

The website http://www.hse.gov.uk is an excellent source of material and guidance.

RICS codes, guidance etc.

For Chartered Surveyors, the RICS sets standards for professional members acting world-wide, and for RICS-regulated firms to follow when conducting their work. These are upheld by the profession through a system of independently led regulation.

Standards of conduct applying to all professional members and RICS-regulated firms include competence, ethical conduct, service standards, and consumer protection.

Sector standards relate primarily to technical competence and ethical conduct. The RICS Building Surveying sector publishes standards for residential and commercial surveys, dilapidations, party walls, and safety. Table 3.14 gives RICS publication status, relative to RICS member Surveyors.

Table 3.14 RICS publication status relative to their member Chartered Surveyors

Status of RICS publications for RICS members	
Rules of Conduct	Mandatory
Professional Statement	Mandatory
Practice Statement	Mandatory
Guidance Note	Areas of good practice and recommendations for professional advice. Not mandatory but could be used by a court as accepted practice
Code of Practice	Mandatory or good practice – will be defined in the specific publication
Information paper	Information and/or explanation

4

Building pathology

Melanie Smith and B. N. West

INTRODUCTION

Pathology: a systematic study of diseases to understand causes, manifestations, and treatments. Applied to buildings – a study of defects, why they arise, what the symptoms are, and what can be done to rectify/ameliorate/reduce, etc.

Three main aspects to building pathology are shown in Figure 4.1.

There is a holistic approach to this. Building pathology requires you to look at the whole building – to study it as a system, and understand how it operates in the wider world. Coupled with this is the requirement for essential underpinning knowledge – building surveyors must know how buildings are designed, constructed, used, and adapted; what they are made from, and how these materials are affected by other factors.

TRADITIONAL AND MODERN CONSTRUCTION

The surveyor needs to consider the whole building, because it is not just about individual components but also about how they interact and how the building interacts with its occupiers and the environment. See Table 4.1, Architectural periods (England), typical features and problems.

The Victorians built in durability through craftsmanship, and an extravagant use of high-quality materials. The technology was simple: a pitched roof to throw off the water, durable materials such as slate and stone, compatible materials that allowed the building to breathe and flex. Durability is compromised through the deleterious effect of industry and pollution. Acids and pollutants in the air cause a degree of deterioration of materials.

Buildings begin to deteriorate from the moment they are constructed. The continual exposure to damaging factors such as weathering and pollutants leads to the breakdown of systems and materials. It can be argued that ultimately all building defects stem from materials' entropy. An understanding of building materials is thus fundamental to an understanding of building defects and consequently building pathology.

Three Main Aspects of Building Pathology

a. Identification, investigation and diagnosis of defects in existing buildings
b. Prognosis and recommendations
c. Design, specification, and supervision of remedies

PLUS monitoring and evaluation of the results in terms of functional, technical and economic performance.

Figure 4.1 Main aspects of building pathology.

Traditional and modern construction are designed and built to meet different standards of performance and behave differently, particularly in relation to moisture and thermal movements. See Table 4.2 Victorian, Post-war, Modern construction.

Traditional – uses natural materials and on-site manufacture. Buildings move and flex. Vapour permeable.

Modern – use of impermeable materials, often prefabricated, using structural frames, modular components, and precision. Movement is designed out. Vapour closed.

Precision and refined tolerances are the norm in modern construction where components are often prefabricated, using very fine tolerances. Compare that to old buildings where timbers and masonry were trimmed and shaped on site – e.g. rubbed brick arches, etc.

Table 4.1 Architectural periods (England), typical features, and problems

Named period	Approximate dates AD/CE	Typical characteristics/features of extant buildings/parts of buildings
Roman	Pre-410	Stone, brickwork. Likely to be Listed.
Anglo-Saxon	410–1066	Ashlar masonry. Reuse of Roman bricks. Likely to be Listed.
Romanesque	500–1200	Heavy, stocky, squat, round arches. Likely to be Listed.
Norman	1066–1150	Stonework. Likely to be Listed.
Medieval	1066–1485	Timber framed. Wattle and daub. Jettied construction. Likely to be Listed. In urban locations, early timber-framed buildings lie behind later brick facades to ape Georgian or Victorian buildings.

(continued)

Table 4.1 *(continued)*

Named period	**Approximate dates AD/CE**	**Typical characteristics/features of extant buildings/parts of buildings**
Gothic	1100–1450	Pointed arches. Mainly ecclesiastical buildings.
Tudor Elizabethan	1480–1600 1550–1600	Timber, stone, or brick. Brickmaking introduced – handmade, narrow, non-standard bricks. Timber frame with separate infill. Brickwork including brick panels. Elaborate chimneys. Likely to be Listed. Walls not plumb and vary in construction. Porous brickwork, eroding stone, disintegrating cob. Woodworm in timbers, including deathwatch and weevil, and fungal attack. Damp floors – necessary to protect the walls.
Jacobean Stuart	1600–1625 1625–1720	Internal timber panelling and plasterwork. More extensive use of brickwork due to birth of the Building Acts. Standardised thicknesses of brickwork (stability) and heights of windows (ventilation).
		Rural properties of indigenous material and underlying geology: cob in the south-west; timber-framing in Hereford, Worcester, Shropshire, Cheshire; pargetting covering timber frames and infills in south-east; tile-hanging and weatherboarding in Kent, Essex, Sussex; ashlar in Bath; stone or slate in West Yorkshire, Cumbria, Wales, Somerset, Lancashire, Scotland, etc.; flint in Norfolk, Suffolk, Hertfordshire; terracotta in Essex.
Baroque	1600–1830	Extravagant decoration.
Rococo	1650–1790	White, graceful, classical, sweeping curves.
William and Mary Queen Anne	1690–1702 1702–1714	Dutch gables. Double-hipped roofs with gabled dormers. London party walls to be 18 inch thick at basements and ground levels, 13½ inch thick upper storeys, and extend 18 inches above the roof. Dwelling houses plain, functional, well-proportioned.
Palladian	1715–1770	Classical forms. Cheap brick and stucco. London party wall thicknesses increased.

(continued)

Table 4.1 *(continued)*

Named period	Approximate dates AD/CE	Typical characteristics/features of extant buildings/parts of buildings
Georgian Regency George IV	1714–1837 1800–1830 1830–1837	Symmetry, elegant, classic features. Lack of detailed ornamentation. Repetitively designed, terraced properties, and streets increasing. "Solid" brick walls (may be two skins of stretchers and snapped headers butted together). Bonding timbers and supporting timber struts in internal leaf. Brickwork faced with stucco, finished to resemble stone. Neoclassical and Greek revival influences. Absence of physical damp-proof courses – good ventilation and breathable construction materials ensured minimum internal moisture levels. Parapet gutter blockages with resultant damp ingress common. Alterations resulting in dampness, inadequate ventilation, wood-boring insects, fungal timber rot.
Victorian	1837–1901	Industrial age. Terraced/back-to-back streets. 225 mm "solid" brick walls. Bonding timbers and timber internal lintels – at risk due to interstitial condensation. Dark interiors. External façade decorative features. Arrival of factory-made bricks, uniform building materials, construction practices and designs. High-density, cheap housing clustered around centres of production (collieries, mills, factories) with lack of original sanitation. Iron and steel frame construction for commercial buildings.
		Late Victorian: emergence of semi-detached villas. Terraced properties structurally sound, standardised designs, with ventilation, drainage, lighting, and water supply. Narrow irregular cavity walls introduced. Slate dpcs introduced. Air-bricks introduced. Alterations resulting in dampness, inadequate ventilation, wood-boring insects, fungal timber rot, failing rainwater goods.
Arts and Crafts	1860–1914	Display of simplistic hand-craft forms and details. Half-timbering and tile-hanging. Idealised cottage styles.
Art Nouveau	1890–1914	Asymmetrical shapes, arches, curves.
Edwardian	1901–1918	Site concrete and damp-proof course more normal (bitumen, slate, asphalt, hessian). Garden villages and cities, e.g. Bourneville, Letchworth, New Earswick, Port Sunlight. Alterations resulting in dampness, inadequate ventilation, wood-boring insects, fungal timber rot, failing rainwater goods.

(continued)

Table 4.1 *(continued)*

Named period	Approximate dates AD/CE	Typical characteristics/features of extant buildings/parts of buildings
Inter-war	1919–1944	Includes Art Deco. Low-density residential planning. Semi-detached and detached. Bay windows (half-brick thick on shallow foundations). Two rows of blue engineering bricks for damp proofing. From around 1930, cement mortar (strong, not breathable) became more popular than lime mortar (flexible, breathable). Wall-tie corrosion. Bridged dpcs and underfloor ventilation.
Brutalist	1940–	Exposed concrete. Functionalist blocks of concrete/steel. Lack of decorative facades.
Early post-war	1945–1964	Cavity walls. Semi-detached, three-bedroom estates of smaller size than inter-war. High-rise concrete buildings. Materials efficiently used to create functional, uniform living spaces. Two-course engineering bricks for damp proofing make way for bitumen-impregnated dpcs. System build mass residential estates, various types based on pre-fabs, no-fines, precast, and in-situ concrete, timber, steel, and cast iron construction. Some types regarded as intrinsically defective e.g.: Airey, Arcon steel, BISF, Boot, Boswell, Cornish Unit, Dorran, Dyke, Gregory, Hamish Cross, Myton, Newland, Orlit, Parkinson Frame, Reema Hollow panel, Schindler and Hawksley SGS, Stent, Stonecrete, Stour, Tarran, Underdown, Unity and Butterley, Waller, Wates, Wessex, Winget, Woolaway.
		Varying problems – thermal insulation, noise insulation, condensation, failure of external cladding and structural components. Deteriorated dpcs, floor slab settlement, lack of dpm, deteriorated plastic plumbing goods, brittle sarking felt, rotted softwood timbers, failing flat roofs, wall-tie corrosion, spalling concrete, corrosion of metal frames.
		Other defective materials include high-alumina cement, asbestos, woodwool slabs, concrete carbonation, calcium chloride accelerators, corrosion of reinforcement and metal structures (e.g. fire escapes), fixings failures (external wall panels, parapet walls). Intrinsic low performance for dampness and insulation.
Early post-war	1965–1980	Parker Morris standards (1961). Introduction of national Building Regulations.

(continued)

Table 4.1 *(continued)*

Named period	Approximate dates AD/CE	Typical characteristics/features of extant buildings/parts of buildings
Postmodern	1960–	Reaction against the austerity, formality, and lack of variety of modern architecture. Factory-formed, softwood, structural timber-framed, internal leafed houses, fronted with brickwork, panelling, or tile-hanging. Plastic-strip dpcs used from the 1970s. Airtightness and insulation progressively increasing. Ventilation issues. Retrofit injection damp-proof courses introduced. Bungalows become not cost-effective. Problems as early post-war. Compaction of thermal insulation materials in cavity walls and roof voids.
High-Tech	1970–	Lightweight materials, sheer surfaces, new engineering, etc. techniques, display of the construction and services. Moderr methods of construction. Factory-builds.
		Problems include poor-quality concrete, carbonation, claddin fixings corrosion, shrinkage, joint failings, mastic shrinkage and embrittlement, mismatch of dimensional tolerances, pattern staining, water ingress, differential movement, progressive collapse, means of escape failings, combustibility problems, direct paths for fire spread, omission of fire barriers, thermal bypass, thermal bridging, condensation.
Retrofit	1980–	Thermal retrofitting to existing residential properties to improve occupant well-being and energy use. Roof void insulation, cavity wall insulation (CWI), external wall insulation (EWI), internal wall insulation (IWI). Encouraged by government grants and incentives. Consequences include interstitial condensation, surface condensation, mould, poor air quality, bridged dpcs, thermal stratification, fire hazards (combustible materials used for internal and external wall linings).

Acceptable standards change over time. Society changes; demands mad of its buildings change; they need to adapt and change to accommodate new uses Services need to be introduced or updated. Standards of human comfort are nov much higher, and insulation and heating need to be upgraded. It used to be accept- able for buildings to have draughts, some degree of moisture, limited insulation continual heat sources, and low internal temperatures. Now expectations includ airtight performance, complete lack of damp in any form, high insulation, hea on demand, and high internal temperatures. Often old buildings were not dam

Table 4.2 Victorian, post-war, Modern construction

Typical features of Victorian, post-WW2, and Modern construction		
Victorian (1832–1901)	**Post-World War 2 (1945–1970s)**	**Modern (about 1976 onwards)**
• Period of mass construction • To house workers during industrial revolution • Solid brick walls • Sash windows • Vapour open • Flexible	• High construction rate period • Replacing houses destroyed in war and to house increased population • Brick cavity or system built (e.g. concrete panel) • New materials • Damp proofing	• Progressive insulation increases • Airtightness measures • Double glazing • Gas/electric central heating • Brick/block cavity walls filled with insulation • Timber- or steel-framed inner leaf with brick outer rain shield

Table 4.3 Treating one like the other leads to inappropriate works

Victorian vs Modern	
• Breathable, flexible • Solid walls (no cavity) • No dpc or dpm • Lime-based mortars: soft, flexible, sacrificial • Walls absorb moisture which then evaporates • Porous jacket • Probably no wall or floor insulation	• Hermetically sealed • Cavity walls • Damp-proof course and membrane • Cement-based mortars: hard, rigid • Rigid and impervious cement mortars and renders • Water-tight, airtight envelope • Probably wall and floor insulation

until flues were sealed, draughts were stopped, and impermeable interior finishes applied. Natural ventilation allowed the building to breathe and prevented unacceptably high levels of moisture in the interior. Following such modern sealing, the building can no longer breathe and can become damp, cold, and insanitary.

Surveyors are aware of previous standards, not regarding these as "defects", and apply any upgrades with caution to not introduce new problems by disrupting building performance. See Table 4.3 Treating one like the other leads to inappropriate works.

Growing concern for the environment and increasing awareness of the importance of conserving finite resources and of minimising pollution are stimulating attempts to find more sustainable building practices, which involve more determined attempts to rectify defects and adaptation rather than just demolition and landfill. Annual materials consumption in the building industry is hugely significant. Most of these materials are from non-renewable resources, such as clay for brick manufacture. Cement is a vital building material, but it also makes a significant contribution to global carbon dioxide emissions. It is now recognised that re-use, not disposal or mere re-cycling, of building materials is needed. Unnecessary demolition and reckless use and disposal of building materials are no longer acceptable if we are to achieve a sustainable society.

Buildings should be designed, repaired, renovated, and altered for durability, but also for flexibility to defer obsolescence in a rapidly changing society.

Perspectives on building pathology and surveying

A building survey of a property, as defined by the Royal Institution of Chartered Surveyors and the UK Construction Information Council, is a thorough and detailed evaluation of its construction and condition, involving an appropriately extensive on-site inspection. The focus of the evaluation depends on the client's requirements, although the extent of the inspection itself may not be so restricted.

DEFECT MANAGEMENT

See Figure 4.2 Need for reflection, testing, monitoring, reporting in defect management. Inspections of a building at regular intervals can ascertain the condition and performance of the various elements. Annual inspections of guttering, for example, can reduce rainwater damage to buildings. Services need to be tested at least annually and any building defects for which the cause is not immediately apparent monitored over time. Monitoring of cracks will reveal whether the cracks are historic or live, and whether they are affected by seasonal or diurnal changes in temperature or moisture. Ecclesiastical buildings have quinquennial inspections to note the development and speed of any defects. Stock condition surveys are similarly often conducted five-yearly.

Surveyors are required to gather all the data to enable assessment of the situation with the information available, analyse test results and all relevant data, and

Reflection, Testing, Monitoring, Reporting

- Not always possible or desirable to give instant answers
- Reflection required
- Need to gather data
- Need to analyse data – often over a period of time
- Need to provide clear, concise, unambiguous information – reports
- The same defects described differently can lead the client to different decisions

Figure 4.2 Need for reflection, testing, monitoring, reporting in defect management.

recommend a course of action based on that analysis. To gather the data effectively, surveyors need to use most of their senses: vision, hearing, touch, and sense of smell, together with knowledge of construction materials, construction defects, and a large portion of common sense.

In building pathology books, different authors take different approaches to the subject. Some will consider the technology of how buildings are put together, what materials are used and how they perform and connect, and what processes of degradation are at work to bring about the defects observed. Others will focus on the use and abuse of buildings by their owners or occupiers and will consider the flexibility of the accommodation and ease of maintenance of the structure or services. Still others will concentrate on the building elements and components. All these aspects are inter-related.

THE TROUBLE WITH WATER

In a temperate climate using porous building materials, either excess or lack of water is the primary cause of defects. This can result in:

- Hydration of building materials
- Deterioration of cementitious materials
- Salt crystallisation
- Sulphate attack
- Frost action
- Corrosion of metals

During construction, a great deal of water is used, including:

- Water for the hydration of building materials – concrete, plaster, paints, mortar, grout, etc.
- Timber has an inherent moisture content that needs to stabilise
- The building is open to the elements until it is roofed
- At occupation the building is only beginning to dry out
- Drying and moisture stabilisation can take many months.

Water added to building materials is done in a controlled way. The water:cement ratio of concrete and mortars is very important in terms of the final strength and quality of the material. Thus, added water in the form of rain is unwanted and newly laid concrete and mortars are protected. Protection from freezing is also provided because water expands on freezing causing voids in the concrete, seriously weakening it, and mortar to be friable and weak. To allow sufficient time for hydration to take place, new work is protected from too rapidly drying out in hot or windy weather.

Water is also destructive after the event. It can erode stone, bricks, and mortar. Mortar should be weaker than the bricks so that the mortar erodes preferentially to the bricks. Pointing is much cheaper and easier to replace than the bricks themselves. However, this should not happen too quickly because of bad detailing or as an escalation of another defect. For example, sills or capping stones not projecting far enough away from a wall so that water is forced down the face – green algal staining is a typical indication of this. Repointing with a stronger mortar will cause the stones/bricks to erode. The answer is to replace the sill or capping stones with adequately sized ones having a suitable overhang and to repoint with the original soft mix of mortar.

Algal growth shows that excess and unwanted water is being discharged, e.g. due to insufficiently sized overhangs, corroded metal gutters or downpipes, or dislodged junctions of rainwater goods. Such water can lead to damp on interior faces and decay of timbers such as floor joists built into the wall. Defects to downpipes are common because decorators often only paint the front face of the downpipe as the rear is difficult to get to and the lack of painting unnoticed.

White staining may be due to **free lime leaching** from the mortar, running down and staining the brickwork face. This white staining will not disappear and will need to be removed. This is different from the harmless white powdery deposition of salts on the surface known as **efflorescence**, which eventually stops and can be brushed off. More damaging is the formation of salts inside the surface pores of the material, known as **crypto-florescence**, as the repeated crystallisation and dissolution of salts in the surface pores lead to stresses on the walls of the pores of the material and the breakdown of the surface. The crumbly, friable nature of the face is distinctive diagnostically. Where the surface of the stone has been eroded this can continue until some stones require replacement.

Sulphate attack is a common defect in chimneys and another example of the damaging effects of combining water with salts. Flue gases from open fires and boilers can contain sulphates, and the cooling effect of the prevailing wind promotes condensation to be formed within the chimney, depositing the sulphates on the internal brickwork flue surface. The sulphates begin to affect the mortar, causing expansion of the mortar on the affected side. This causes the chimney to bend away from the prevailing wind. This is a serious defect and frequently requires re-building of the affected stack to be undertaken (using sulphate-resisting cement). It is worth remembering that masonry collapses happen suddenly and can be disastrous: stacks can fall into the street below or through the roof into a bedroom below.

Frost damage is generally problematic in exposed situations because it occurs in masonry that is very wet in freezing conditions, so more commonly is associated with chimney stacks, parapet walls, etc. It can also occur in conjunction with rising damp or strong winds carrying prolonged and heavy rain deep into the pores, thereby thoroughly wetting the masonry, followed by a drop in temperature. The problem is associated with repeated freezing and thawing and leads to a cracking away of the surface or of projections. Old bricks are particularly vulnerable as they were fired to lower temperatures and only the surface is hard. Once this has been breached the softer inner brick erodes rapidly.

Metals corrode. There are some key points here: metals corrode in different ways; some like aluminium form a tightly bound impenetrable layer of corrosion that prevents further deterioration, whilst in others the decay is progressive; e.g. iron forms a loosely bound layer of corrosion that breaks away, leaving fresh iron to be affected. Galvanic or bi-metallic corrosion is where two different metals are in contact, and one metal will corrode in preference to the other, e.g. a copper roof fixed with steel nails. See Figure 4.3 Galvanic or Bi-metallic corrosion risk between different metals in contact.

When looking at galvanic tables, a more anodic metal will corrode when in contact with a more cathodic metal. Corrosion is more likely when there is more than 0.15 V difference in the anodic index. Note: these are usually shown as negatives. See Table 4.4 Simplified anodic index of construction metals.

Corrosion can affect structural members, such as lintels, wall-ties, and reinforcement. Current wall-ties are generally corrosion resistant, e.g. stainless steel or polymeric. Older ties were iron (rusted rapidly) or galvanised, but the galvanising

Risk of corrosion in "corroding metal"?		Metal in contact										
		Gold, silver, graphite	Some stainless steels	Bronze	Copper	Brass	Lead, tin	Some stainless steels	Cast Iron	Alumin-ium	Galvan-ised steels	Zinc
Corroding metal	Gold, silver, graphite											
	Some stainless steels	Yes										
	Bronzes	Yes	Yes									
	Copper	Yes	Yes	Yes			Yes	Yes				
	Brasses	Yes	Yes	Yes	Yes		Yes	Yes	Yes			
	Lead, tin	Yes	Yes	Yes	Yes	Yes		Yes				
	Some stainless steels	Yes	Yes	Yes	Yes	Yes	Yes		Yes			
	Cast iron	Yes	Yes	Yes	Yes	Yes	Yes	Yes				
	Aluminium	Yes	Yes	Yes	Yes	Yes	Yes	Yes	Yes			
	Galvanised steels	Yes	Yes	Yes	Yes	Yes	Yes	Yes	Yes	Yes		
	Zinc	Yes	Yes	Yes	Yes	Yes	Yes	Yes	Yes	Yes	Yes	

Figure 4.3 Galvanic or bi-metallic corrosion risk between different metals in contact.

Table 4.4 Simplified anodic index of construction metals

Anodic index (V)	**More cathodic**
Gold	0.00
Silver	0.15
Stainless steels	
Copper	0.35
Brass	0.4
Bronzes	0.4 to 0.45
Stainless steel	0.5 to 0.6
Tin	0.65
Lead	0.7
Cast/wrought iron	0.85
Aluminium	0.7 to 0.95
Galvanised steel	1.2
Zinc	1.2
	More anodic

could be either too thin or worn through. The idea of ties is to hold the two leaves of a cavity wall together to ensure they act in unison. Once the ties have corroded, then parts of the walls can begin to bulge and become unstable. Where the tie corrodes, it causes localised expansion, which further destabilises the wall. Wall-tie failure is quite characteristic in that the bulging and joint expansion will be associated with the tie spacing and will be less at foundation level, i.e. where there is more restraint.

Reinforcement of concrete or brickwork is potentially liable to corrode. The reinforcement in concrete is protected by the high alkalinity until such time as the alkalinity is reduced by carbonation or the corrosion is stimulated by some other agent, e.g. chlorides in the mix. Bars expand as they corrode and cause cracking of the masonry. Corrosion of steel reinforcement in concrete can occur where there is little cover to the concrete, making it susceptible to corrosion. As the reinforcement corrodes, it will initially cause staining, and then as the metal expands it will force the concrete away at the surface (spalling).

Green stains down masonry can indicate copper corrosion. Copper can be corroded by sulphur dioxide, damp wood impregnated with some fire retardants, and run-off from red cedar cladding. Copper corrodes aluminium, zinc, and steel. Copper cladding and roofs gain a green patina after 5–20 years, formed by a thin

Table 4.5 Possible remedies for typical water issues

a. Inadequate hydration of building materials	a. Ensure not allowed to dry out in hot or windy weather (protect with damp-proof membrane)
b. Protection of new cementitious materials	b. Cover with suitable material to prevent water ingress – e.g. cover newly laid brickwork with damp-proof membrane
c. Salt crystallisation	c. Use low soluble salt bricks where this is likely to be a problem, ensure weathering details are well designed and maintained
d. Sulphate attack	d. Use sulphate-resisting cement in situations where this is likely to be problematic, line and insulate chimneys
e. Frost action	e. Prevent excessive wetting of masonry, e.g. by good design and maintenance of weathering details. Specify frost- resistant bricks in exposed situations
f. Corrosion of metals	f. Careful selection of metals for the situation in which they're going to be used, avoid bi-metallic corrosion, ensure adequate cover to reinforcement, good maintenance

protective layer of copper salts. Aluminium is corroded by contact with brass, copper, and lead. Aluminium corrodes zinc. Zinc is corroded by copper, western red cedar, oak, some fire retardants, and soluble salts.

See Table 4.5, Possible remedies for typical water issues.

DAMP

Generally, damp is water that is where it should not be. Roofs get wet when it rains, but we would not describe them as having a damp problem. Damp accounts for the majority of problems in buildings in the UK. The climate is wet, and in our buildings we use one of two approaches: we either use porous materials, which soak up the moisture and begin to degrade, or we have impermeable surfaces which force the water off and that puts pressure on the joints, which then fail and let water in.

Aesthetically, damp spoils finishes by bleaching colours, causing water-marks, encouraging mould growth and mildew, flaking paint off, lifting lift liners, etc.

Structurally, it causes materials to degrade: plasters become friable; salts are leached out to the surface (or worse behind the surface); wood becomes infested with insects or is attacked by fungus, some of which progress and cause total failure

Types of damp
Condensation
Penetration
Rising

Figure 4.4 Types of damp.

of the timber; metals begin to corrode, so fixings can fail. If left untreated, in some cases, damp can eventually cause structural collapse.

Health-wise, the damp surfaces cause fungal spores, which trigger asthma attacks, allergic reactions such as rhinitis, conjunctivitis, etc., and cause bronchitis, pneumonia, etc. The reduction in thermal properties of the building can also lead to hypothermia, particularly of babies and elderly people.

See Figure 4.4 Types of damp. Damp can broadly be classified into three groups: condensation, penetrating, and rising.

CONDENSATION

Condensation can have severe consequences. The incidence of condensation and the severity of it have increased with the move to draught-proof buildings. This allows moisture vapour to build up in the air, which then condenses onto cooler surfaces. This has been recognised by the Building Regulations, which stipulate adequate background ventilation. Provision of ventilation is required for cool enclosed spaces such as roof and sub-floor voids. Five key concepts:

1. Warm air can hold more moisture than cold air, so air in a room at 20°C is capable of harmlessly holding more moisture than the same room at 10°C.
2. When warm, moist air meets cold surfaces, such as windows and poorly insulated external walls, floors, and ceilings, its temperature drops at that point, the air near that surface becomes saturated with the water vapour, and this results in water condensing onto those surfaces.
3. Water vapour can move throughout a building, i.e. by circulation routes or by percolating through the pores in the building fabric. A plasterboard ceiling for example will not stop water vapour from passing through it into the roof void above.
4. Water vapour is driven by changes in vapour pressure, so if there is a high level of moisture in the kitchen, say, and a low level in the bedroom, the vapour will move out of the high-pressure kitchen into the low-pressure bedroom.

Mould growth often occurs in bedrooms as a result of not having an extract fan in the kitchen. The kitchen itself may be condensation free due to warmer surfaces.

5. Normal human behaviour can exacerbate condensation. Failure to open windows or use extract fans when bathing, washing, drying clothes, cooking, etc. leads to an increase in internal vapour pressure, which in turn can lead to condensation. This failure may not be an occupier "fault"; for example, no trickle vents may be provided or fans may not be fitted, may not work, or may not be accessible, and security may be a problem.

Resolved by a **BALANCE** between heating, insulation, and ventilation. Ventilation can lower the temperature, which means a decrease in the amount of water held in the air and reduced surface temperatures in the room. Ventilation is required to remove the moisture that builds up through normal daily activities. Insulation is of limited value unless heat is provided. If there is too much ventilation or insufficient insulation, the heat energy will be lost without providing much benefit. The benefits of providing lots of heating, good levels of insulation, and extractor fans flounder if the occupiers turn off the radiators, do not use the fans, dry washing indoors, and boil cabbage all day.

There are two types of condensation:
Surface – visible on the surface.
Interstitial – occurs within the structure at interfaces between materials and is hidden by the construction and surface layers.

Interstitial condensation can dampen insulation materials, reducing their thermal resistance. Dew point is the temperature at which water condenses out. If it condenses on the outer face of insulation, then that insulation could work less well and further increase condensation problems. Depending on the nature of the construction, if you get interstitial condensation forming on timber, there is a risk of fungal decay. The problem with interstitial condensation is that damage often becomes quite advanced before it is discovered.

The way to control this type of condensation is to calculate the thermal and vapour transmission across the structure to identify a risk and reduce its risk of occurrence by reducing moisture build-up and by use of vapour barriers.

Vapour barriers come in many forms – from plastic sheets to foil backing on plasterboard, etc. Barriers are required to be imperforate – but will be punctured during construction by fixings, services, etc. If such holes are not totally sealed, a significant amount of vapour can be driven through due to the potential differences in pressure between the two areas. Vapour barriers should always be on the warm side of the insulation. If a barrier is placed at the cool side, there is the possibility that water could condense in the insulation itself.

PENETRATING DAMP

- Water that penetrates into the structure
- Joints/junctions are weak points
- Driven under pressure (e.g. heavy rainfall, wind-driven rain, snow, internal/external pressure differences)
- Pumping action (wind pumping through tiny gaps, e.g. in sealants around windows).
- Copious amounts in a short time
- Basements – special case: hydrostatic pressure

Penetrating damp can occur in walls because of missing or inadequate pointing in solid walls, mortar snots on cavity ties, failed cavity foam insulation, gaps in sealants/barriers, external ground high relative to internal spaces (e.g. at or above ground floor levels and/or damp-proof courses). Penetrating damp also results from roof leaks allowing water into the structure, especially flat roofs (should use sufficiently flexible materials and be carefully detailed).

A major problem with penetrating damp is that it often emerges a long way from where it comes in, as it tracks along structural members, services, along interfaces, etc.

Common problems include:
Leaking rainwater goods which can lead to prolonged saturation and consequent damage to masonry through frost damage, erosion of mortars, spalling associated with repeated wetting and drying cycles, moss and algal growth, etc. If sarking felts are too short (and they do rot from the bottom up) then the roof discharge water can run under the gutter and into the roof space, causing serious problems.

Defective (missing) cavity trays, inadequate sealant, warping of timbers, etc. The incidence of missing cavity trays is high in, for example, barn conversions, where it is difficult to dress them into random walls, and in conservatories added to houses. The upshot is that water runs down the face of the external wall, which then effectively becomes the inner wall of the conservatory.

Steel structures rusting. Steel is a popular modern method of construction (MMC) material for the structure of new buildings, residential and non-residential. The vastly reduced tolerances of factory/off-site manufacturing will impact the future of what can be changed in these buildings. The strength of such structures can come from the whole frame rather than individual parts, meaning that increasing opening sizes, e.g. for window or partitions, is not easily achievable because removing one part can make the whole structure too weak. This is important for surveyors to look out for. Because dampness, leaks, and excess humidity will not just be unpleasant, these could adversely affect the structural integrity of light gauge steel structures. The durability testing of these structures has often been done in warm, dry environments, which produce very good results. However, the UK conditions, after some time in use, could easily not be warm and dry. The NHBC and other housing warranty providers require a design life in excess of 60 years and accept the use of light gauge steel construction. The surveyor should thus look for evidence that the structures remain protected.

Basements are frequently damp because the water is forced in by hydrostatic pressure. This forces the water through the construction and into the internal space. This can be coupled with rising damp in basement floors. Wall and floor damp proofing must be continuous. The junction between the wall and floor is particularly vulnerable.

To remedy, basements can be tanked internally or externally. Internal cementitious tanking will eventually fail but will probably take quite a while to fail, and it is often the only practical solution. Extreme care should be taken with the floor edge detail as this is the weak spot, and the wall tanking must lap with the floor dpm. Drained cavity construction is a good method. Sometimes the only real option is to drain or pump the water out continuously.

RISING DAMP

- Water rises by capillary action up from the ground
- Only as far as it takes for gravity to start pulling it back (about 1200 mm)
- Salts – hygroscopic
- Not as common as used to be thought
- Solutions: damp-proof courses/membranes, drainage, ventilation, and heating

Rising damp does not occur instantly; it can take years for it to become a problem. Rising damp does not get any higher than 1500 mm max, usually about 1200 mm above ground level. Above this height, the capillary action drawing up the water through the pores of the masonry is overcome by gravity and by evaporation from the surface.

Properties may not have a dpc; the dpc may no longer be effective, be or become bridged, or be compromised. NOTE: no dpc does not mean rising damp.

External and internal walls may be affected. Similarly chimney breast walls.

Solid ground floors are often damp in older buildings because damp-proof membranes were not installed. Even in newer properties, the membrane may have been punctured or inadequately lapped. Where dpcs have not been included in sleeper walls, timber floors can become damp and rot. Note that including a dpm in the floor of an old property might make the damp in the walls worse.

Table 4.6 Remedies for rising damp

Remedies for rising damp problems: depends on cause
Removing whatever is bridging the dpc, e.g. mortar snots, earth, render, etc.
Reduce external ground level
Inserting a physical dpc
Drying out, e.g. thick stone walls with rubble infill, basements, etc.
Provision of chemical system, e.g. injection dpcs
Provision of patent systems, e.g. electro-osmosis
Internal covering or lining
Increase internal ventilation, improve heating, remove impermeable internal finishes (e.g. vinyl wallpapers)

See Table 4.6 Remedies for rising damp.

Dry lining may make the wall damper by preventing evaporation to the inside therefore it should be storey height, any battens should be pressure impregnated rust-resisting fixings used, and a dpm behind the battens.

Reoccurring dampness will continue following remedy unless the hygroscopic salts in the plaster are removed and the wall is dried and re-plastered with a renovating plaster.

MOULD, DECAY FUNGI, AND BACTERIA

There are different species of mould fungi causing problems and damage in buildings.

Typical mould fungi in damage cases are, e.g. *Alternaria* sp., *Aspergillus* sp. (e.g *A. fumigatus* and *A. versicolor*), *Aureobasidium* sp., *Cladosporium* sp., *Fusarium* sp. *Chaetomium globosum* (this is also a soft-rot fungus), *Penicillium* spp., *Stachybotrys atra*, *Trichoderma viride*. Bacteria such as *Actinomycetes*, *Streptomycetes*, *Bacillus*, and different types of yeasts often exist in moisture damage cases.

In the moisture stress situation, there are different microbial species reflecting the moisture or humidity level. The first colonisers of building materials belong often to the genera *Penicillium* and *Aspergillus*. Some of these fungi can grow at relative humidity (RH) 80–90% at room temperature. In addition, these fungi produce large amounts of air-mediated spores that can settle on material surfaces.

Most fungi need very high humidity conditions, e.g. RH above 95%, to grow Typical representatives of these fungi are *Chaetomium*, *Stachybotrys*, and *Trichoderma* As tertiary colonisers, decay fungi, actinobacteria, and other aerobic bacteria need a high RH to be able to grow. In addition, the microbial succession in the development of moisture damage follows the nutritional demands of different types of microbes.

Typical decay fungi causing brown rot in moisture-damaged buildings are *Serpula lacrymans*, other *Serpula* species, *Coniophora puteana*, *Leucogyrophana* species, *Poria/Antrodia* species, *Gloeophyllum trabeum*, and *G. sepiarium*. Humidity conditions must be above RH 95% (around the fibre saturation point of wood, wood moisture content about 25 to 30%) for the spore germination and mycelium activation of brown rot fungi.

Under unsuitable conditions, fungi are inactive but they can quickly become active under suitable conditions. The degree of decay in wood depends on the total active time of the decay organism.

The dry rot fungus *S. lacrymans* is a typical mesophilic fungus growing best at lower temperatures, while some strains of *Poria placenta*, *Coniophora puteana*, and *Gloeophyllum trabeum* are typical thermos-tolerant fungi growing best at higher temperatures (+25 to +46°C). The lethal temperature dose for these fungi is also higher. The lethal (killing) temperature for dry rot fungus *S. lacrymans* is about +35 to +50°C, but some fungi, e.g. *G. trabeum* and *P. placenta*, can resist some hours at temperatures of +60 to +80°C. The lethal temperatures of decay fungi vary between 35°C and 80°C.

Health problems caused by fungi and bacteria

Up to 20% of the population in Europe suffers from allergic-related diseases, with the problem increasing. These health problems may be caused by lifestyles where people spend more time indoors and thereby are more exposed to the indoor climate. Lifestyles are not always voluntary, for example for care home and hospital residents, for workers or residential occupants without suitable natural ventilation, etc.

Prolonged dampness in building materials can cause the proliferation of a number of fungal and bacterial genera. These microbes, whether embedded behind materials or growing on visible surfaces, are associated with a wide array of adverse health effects, for example:

- Irritative and non-specific symptoms
- Respiratory infections
- Allergic diseases
- Alveolitis and organic dust toxic syndrome
- Other chronic pulmonary diseases, such as chronic bronchitis

The indoor climate contains partly the same **volatile organic compounds** (VOCs) as the outdoor air and partly VOCs emitted from building materials, paint, furniture, etc. The VOCs, especially some terpenes, might have an important role in indoor air quality. In very high concentrations in indoor air, they may cause sensory irritation.

Legionella can result in Legionnaires' disease in humans, a potentially fatal form of pneumonia. There is a risk where any stored water is used so that a fine airborne spray is produced, which can be inhaled. It can be found in water towers, cooling towers, evaporative condensers, showerheads, and Jacuzzis. There is particularly a risk in commercial buildings, higher rise flats, and student accommodation. Water systems require regular maintenance regimes of inspection, testing, and cleaning. HSE produces guidance.

Toxic moulds such as *Pseudomonas bacterium* can be the unintended result of airtightness and high insulation with inadequate ventilation because it can develop and thrive in persistent condensation and moisture. It is a potentially fatal hazard in susceptible people of all ages. It can also form in air-conditioning or heating system water pipes, making them less efficient, reducing their lifespan, and producing a hazard for occupants. Specialist advice is required.

EVERYTHING UNDER THE SUN

Sun
Wind
Water
Chemical attack
Unwelcome visitors: insects, fungi, etc.
All together now!

Sun

Solar radiation is a powerful weathering agent, especially in combination with water.

The photochemical effects – ultra violet (UV) light (radiation) - break down the bond of polymers present in many organic materials.

Solar radiation fades colours of fabrics, paintwork, timber stains, etc. These forces also break down the bonds in long-chain polymers in these and other organic materials, such as adhesives, glass, plastics, elastomeric felts, etc. Results include:

- Embrittlement of glass
- Embrittlement of plastics
- Breakdown of elastomers in roofing materials and sealants
- Breakdown of paints and emulsions

Degradation through solar radiation is the reason for the required provision of solar protection to roofing felts and for UPVC windows having a shorter life expectancy on south-facing elevations than on north-facing ones.

Heat results from prolonged exposure to solar radiation. Materials expand and contract with diurnal and seasonal changes in accordance with their co-efficient of thermal expansion. This can result in differential movements. Thermal movement can result in cracking.

The **orientation** of the building is important since south-facing elements will have more solar gain than north-facing elements. Use can be made of this, e.g. putting larger windows to the south and smaller ones to the north for increased solar gains, or the opposite to reduce solar warming and to more beneficially place solar panels.

Internally placed insulation in/on external walls ensures that the external wall faces will stay colder. This results in an increased risk of frost attack to the brickwork and more differential movement between the interior and the exterior. This increased differential movement extends to buildings with frame and panel construction and puts more stress on the sealants used in the construction.

Wind

Wind damage is associated with gusts exceeding the usual design force of wind.

Roofs are particularly vulnerable and should be adequately strapped down. Framed buildings need their wall and roof members strapping and tying to foundations. Catastrophic damage can occur with winds above 80 miles/hour. Guidance is given in the Building Regulations, Standards, and Approved Documents.

Wind can cause draughts at windows and doors even when draught-stripped, and wind-washing can lead to heat loss. The wind can also penetrate the building fabric, and higher heat losses will ensue, particularly when associated with rain.

Older properties with very thick walls are often effective at keeping the heat in or out (warm in winter, cool in summer) except in windy conditions.

Water

- Driving rain (penetrating damp)
- Wind pressure can drive rain through joints and the building fabric
- Effects are more significant with increased height and exposure of building
- Drives atmospheric gases: sulphur dioxide and hydrocarbon emissions form acid rain – erosion of limestones and brickwork, corrosion of metals

Driving rain will find any and every weakness in the structure. See Table 4.7 Penetration of moisture driven by wind.

Table 4.7 Penetration of moisture driven by wind

Pushed uphill under slates
Driven behind loose flashings
Driven in under sills
Driven though fine cracks or gaps (including through render)
Penetrate deeper into porous building materials (such as bricks)

Rain is always acidic because it picks up carbon dioxide from the air. Since industrialisation the acidic content has been problematic because of the pollutants associated with industrial processes. These acids attack materials which are essentially alkaline in nature, such as limestone which is basically calcium carbonate particles cemented together in a calcium carbonate matrix. Acidic rain also promotes faster corrosion of metals.

Chemical attack

- Corrosion; oxidation of metals
- Sulphate attack; sulphates (salts) present in industrial wastes, gypsum products, etc. In damp conditions, the reaction causes cement mortar or render to expand and disintegrate
- Crystallisation of salts: soluble salts crystallise on the surface (efflorescence) or worse, within the materials (crypto-florescence)

The corrosion process is an electrochemical reaction which takes place faster in the presence of water and faster still if the pH of the liquid is low, i.e. acid. It is all to do with the balance of hydroxyl ions in the electrolyte (liquid).

Sulphate attack is a problem in concrete and in masonry, where the mortars are affected by sulphates (as are concrete bricks). Sulphate-resisting cement tackles the chemical reactions that take place when sulphates attack the tricalcium aluminate in cement. This causes the formation of calcium sulphoaluminate, which occupies a larger volume than the tricalcium aluminate and causes expansion. This sets up stresses in the mortar or concrete and results in significant deterioration and damage.

In the early stages of sulphate attack to a wall, the only visible evidence may be horizontal cracks to the internal face of the wall in response to expansion of the outer face. The distribution of these horizontal cracks depends on whether it is a cavity wall or a solid wall. Cracks are visible near the roof in cavity walls because the wall-ties offer restraint lower down. They are visible all over the external face in solid walls. On rendered walls, sulphate attack can be distinguished from shrinkage cracking (random over the external face) by the horizontal distribution pattern, which reflects the mortar joints.

Cementitious materials do shrink on drying (a characteristic, not a defect). Mistakes made with the application or protection after application that lead to excessive shrinkage can create a problem, e.g. when rendering is carried out on a hot day and no attempt is made to protect the work from too rapid drying, leading to inadequate hydration of the cement.

Chimneys, particularly unlined ones as the gases contain sulphates and water vapour, are prone to sulphate attack, as condensation can occur in the flue. Be careful to distinguish sulphate attack from frost damage, as chimneys are in a very exposed location and can suffer both problems.

Salts in the masonry can be dissolved in rainwater, and on evaporation these are brought to the surface as efflorescence. This will eventually resolve itself but can be speeded up by gentle brushing with a bristle brush (wire brushes can leave iron stains and may be too harsh for old or soft bricks).

Of more concern is the problem of **crypto-florescence,** the process by which these dissolved salts re-crystallise within the pores of the material. This recurring dissolution and re-crystallisation sets up stresses within the material as the salts occupy a larger volume and press outwards onto the sides of the pores of the stone or brick. This causes the breakdown of the material. A crumbly, friable quality is suggestive of this.

TIMBER INFESTATION: FUNGAL ATTACK AND INSECTS

Timber decay tends to be one of the more expensive problems in property maintenance. This is because it often occurs in places that are hidden or only rarely inspected; consequently the problem can be well advanced before being discovered.

Dry rot is a critical fungus to find in buildings. It thrives in damp wood, but not wet wood. It has an indicative odour – a fusty mushroomy smell. It is not localised but spreads beyond the immediate area, searching for its own moisture and feeding on available timbers. It is rarely found externally.

Wet rot is more common than dry rot. Wet rots need a higher moisture content. Wet rot is localised and not as destructive as dry rot. It does not spread beyond the immediate area and grows on wetter timbers. Wet rot likes very wet (not saturated) timber and can happily live outdoors.

Infestations

The worst insects in the UK are the common furniture beetle (woodworm), house longhorn beetle, and deathwatch beetle. Climate change may result in an increasing problem with termites. Termites are very destructive and difficult to control.

All infestations can be treated with chemicals and the replacement of badly damaged timber. The extent of the outbreak must be established. If there is no frass, it is likely to be a historic outbreak, no longer active, or not active in the larval stage. Important diagnostically are the size of the holes and the pattern of the tunnels in respect to the grain of the wood, the characteristics of the frass, the location of the outbreak, and the type of timber attacked.

Not all insects are harmful, e.g. woodlice do no harm, but can be indicative of a damp problem.

Common furniture beetles prefers softwood, especially plywood, leaving 1–2 mm exit holes. They generally prefer damp, rather than dry, timber. Softened timbers, for example in floorboards surrounding a w.c., are particular favourites along with damp floorboards and damp loft timbers. It lays its eggs on the timber and the larvae do the damage.

Deathwatch beetles are attracted to timbers that already have some fungal decay present, and they do not attack sound wood. They leave large exit holes, but the extent of the outbreak is usually limited in area because the beetles are reluctant to fly (although they can do), so spread by crawling over the surface.

House longhorn outbreaks are particularly destructive. They can tolerate dry timber and prefer a warmer climate (e.g. Surrey) than northern England. Roof spaces are a good place to look for them.

Rodents such as mice, rats, and squirrels cause problems because they chew through wires, nest in insulation such as mineral wool, and carry diseases. Chewing through wires in roof spaces has been known to cause fires.

Other infestations, e.g. of cockroaches, occur in warm, dry buildings. They are a serious health hazard and breed rapidly. They can spread through service ducts and along wiring runs, etc. They are a particular problem in multi-occupancy buildings as co-ordination of extermination attempts can be hampered by accessibility and co-operation problems.

Algae, moss, and lichen growing on walls and roofs signify poor drainage of roofs, sills, etc. They excrete some acids which can scar alkaline building materials, but they are indicative of other problems which may need to be addressed, such as a lack of drip grooves to undersides of cills, inadequate falls to flat roofs, etc. Moss is often seen on north-facing slopes of roofs as these are colder and wetter for longer. One associated problem is that chunks of moss get washed off in heavy rainfall and block gutters, which can then cause other problems such as water backing into the roof space.

ALL TOGETHER NOW! COMPLEX INTERACTIONS

Sometimes building defects are caused by one problem – it is more usual to experience a combination of factors. Care is needed in diagnoses of the problem and in prognosis and recommendation.

The biggest problem facing the surveyor is diagnosis. Diagnosis is difficult at the best of times but is made worse by the fact that problems rarely occur in isolation.

Wet rot and dry rot can occur in the same room; the problem can be caused by solar degradation of a flat roof covering or frame sealant that has resulted in water ingress. This can be exacerbated by the exposed location, leading to penetrating wind-driven rain, which dampens insulation and leads to condensation, etc.

CONCRETE

Concrete is a very old technology, used extensively by the Romans, although without steel reinforcement. Concrete on its own is strong in compression (strength depending on mix) but weak in tension. The earliest reinforced concrete structures date from around 1850. Some of these early structures have survived corrosion-free because they used a very high cement content and coarsely ground cements that were slow to gain strength. The manual compaction techniques were basic, leaving a good clearance between the steel and the formwork, thus achieving good cover to the reinforcement. Finely ground cements are more commonly used currently to allow for rapid strength gain and reduction in cement content for the same strength after 28 days.

Concrete was used extensively in the early part of the 20th century by the early proponents of the Modern movement – often to striking effects. The beauty of concrete is that it enables wonderfully organic and fluid lines to be achieved when poured in-situ. The beauty of concrete is this versatility.

Concrete was also used extensively as a post-war material (1945–1955), especially in Brutalist design, and often with poor results in low-rise houses and flats. The poor performance of these buildings, often attributable to poor design and workmanship, has contributed to a negative perception of concrete as a building material. The often poor functional design created social problems coupled with a lack of appreciation of how the concrete would look after weathering. The system building of enormous concrete towers demanded work to relatively fine tolerances and new skills from the workforce, who were largely unaware of the importance of removing packing pieces and ensuring all fixings were in place. A failure was the potential for progressive catastrophic collapse.

Concrete is a mixture of cement, aggregate, and water. It is mixed and will harden. It is important to get the right proportions of each. The water to cement ratio is critical to achieving the strength and durability required.

The nature of the aggregate is also important. Porous aggregates result in less durable concrete. It is also important that there is a range of particle sizes (for full compaction) and they are mixed well before adding to the mix. Too much fine material in the aggregate adversely affects workability, requiring more water, but then adding too much water reduces strength.

Where damaging sulphates are present, for example in ground water, sulphate-resisting cement should be specified. This has reduced tricalcium aluminate – one of the main constituents of cement – as it is the hydration products of this compound that are most susceptible to sulphate attack.

Common modifications

- Finely grinding the cement particles speeds up the rate of hardening.
- Reducing the tricalcium aluminate.
- Partially replacing cement with ground granulated blast furnace slag and pulverised fuel ash to slow hydration (also slows strength gain) and reduce heat generation – useful in massive constructions.
- Modifying aggregates for specific purposes.
- Adding admixtures for site-specific reasons, e.g. to improve workability, permit air entrainment, accelerate or retard hydration, etc.

The mixing method affects results, and quality control is key. Mixing by hand has relatively poor quality control. Site mixing can be variable. A batching plant operated by an experienced technician has excellent quality control. Factory-produced components such as prefabricated beams, etc., enable excellent quality control. The ingredients need to be fully mixed, and then the main aim is to get concrete to the site or location quickly and in a condition that enables it to be compacted properly. The main reason for this speed is that if there is a delay in getting it into position then it could have become less workable, which would make the process of compaction very difficult or impossible. In high-rise buildings the concrete is often pumped up to where it is needed and the material design, mixing, placement, and compaction can be carefully controlled. There has been a wealth of research and development in large-scale construction and expensive equipment is used – it is a specialised field.

Placing is important, as if air becomes trapped then voids will result, which will significantly weaken the finished concrete. Water should never be added to concrete that is setting. There is a distinction between air voids and deliberate air entraining, where a special air-entraining agent is included. The latter results in many very small holes that help to improve frost resistance – air-entrained concrete still needs to be compacted to get rid of air voids.

Adequate compaction is required to avoid air voids which weaken concrete and to overcome the difficulty in compacting around steel work. Spacers are used to hold the reinforcement in the correct place. Broadly speaking, 1% entrapped air = 5–7% loss of strength, which means even when achieving 95% compaction, the concrete can be 1/3 weaker than if it was fully compacted; 5% not compacted = 5 × 5–7% loss of strength, which is 25–35% weaker than if fully compacted. Concrete must therefore be well compacted during pouring. Pouring from height risks the various constituents segregating or separating. Similarly, the pouring concrete must not be allowed to move the carefully designed and placed reinforcement cages on site.

Curing – concrete hardens, it does not just dry out. The hardening process is exothermic: it gives off heat. As it hardens, it is sensitive to temperature:

- At 17°C it reaches 64% strength at 7 days, full at 28 days.
- At 30°C it reaches 44% strength at 7 days, 85% at 28 days.

Concrete should not be poured in freezing conditions as water expands as it freezes and air voids result, seriously weakening it. Generally, it should be 4°C and rising. There are ways of pouring concrete in less than ideal conditions by means of additives and managing the environment; this is a specialised field.

Thermogenesis (heat generation) has to be taken into account. The concrete should be protected from too rapidly drying out or from being rained on. The heat generated can result in cracking if the core heat is too different from the surface heat. This is combated by the cement specification (some produce less heat of hydration), the formwork material (for example, glass reinforced plastic (grp) allows better heat dissipation than timber), etc. If concrete is allowed to dry out too quickly as might happen on a hot, windy day, then inadequate hydration results and a weaker, dusty surface will result. If it is subjected to rainfall too soon, then additional water is added to the surface, which reduces strength.

Note: over-working a surface, e.g. by trowelling too much, can bring the water to the surface and result in a powdery weakened surface finish. This is often seen on floor screeds.

Long-term defects

See Table 4.8 Long-term concrete defects.

Table 4.8 Long-term concrete defects

Carbonation	Corrosion of the steel
	Surface shrinkage
Chloride attack	Corrosion of the steel
	Shrinkage
Aggregate reaction	Including AAR and ASR
Sulphate attack	Sulphated ground water

Alkali aggregate reaction (AAR) occurs when some aggregates contain constituents (carbonates and silicates) that react with the cement. In alkali silica reaction (ASR) silica reacts with the cement to form a gel that absorbs water and causes stresses within the concrete to disrupt it. This gel carbonates as it gets to the surface to form a whitish gel, which can be seen oozing from the concrete.

Sulphate attack destroys the concrete, and the severity of a sulphate attack is dependent on (these are the main ones, there are others):

- The type of sulphate (calcium and particularly magnesium sulphate are particularly aggressive)
- The concentration of the sulphates in solution and whether or not the supply is continually replenished (e.g. as with ground water)
- The type of cement. Sulphate resistance is increased by using sulphate-resisting cement, and by using pulverised fuel ash and ground granulated blast furnace slag, because they help to fix the calcium hydroxide in the concrete, thereby reducing its availability to react with the sulphates
- The permeability of the concrete (impermeable is good)
- Number and location of construction joints (these allow water to penetrate deeper).

Carbonation and chloride attack – note lack of cover is also a major cause of corrosion. As long as the concrete surrounding the reinforcing bar has a pH of 12.5 or more, the steel will not corrode. In new concrete, the effect of pH delays corrosion – when the alkalinity is above 12.5 a passive film forms on the steel, which prevents further corrosion.

Main factors affecting alkalinity are leaching, carbonation, and chlorides. It is the hydration of the cement that causes the concrete to be highly alkaline (>pH 13). It liberates alkalis as a by-product of this hydration process, and the pore solution rapidly becomes saturated with calcium hydroxide (lime). A reserve of calcium hydroxide from unhydrated cement in the concrete serves as a buffer, preventing the pH from dropping, but even neutral pH water passing over the surface will slowly reduce the calcium hydroxide levels. This is called **leaching** and is often exacerbated around construction joints. A homogenous, impermeable concrete offers better resistance to leaching, and incorporation of additives such as pulverised fuel ash (pfa) can help to reduces permeability and fixes the calcium hydroxide, thereby reducing the soluble lime content of the hardened concrete.

The speed at which this normally slow process works can be increased when atmospheric carbon dioxide penetrates into the capillary pores formed during the hydration process. The carbon dioxide removes the calcium hydroxide by converting it to calcium carbonate, thereby reducing the alkalinity of the concrete. This process is called **carbonation** and is most commonly the cause of corrosion of steel. Over a period of many years, this process works through the concrete from the outside inwards in what is termed a carbonation front. This is useful diagnostically, as by measuring the depth of carbonation (by using an indicator solution such as phenolphthalein on freshly fractured concrete: pink = alkalinity >9) and knowing the date of construction, estimations can be made for how long it will be before the steel starts to corrode. The depth of penetration of the carbonation front is generally considered to be proportional to the square root of the time of exposure. The rate of carbonation will not be uniform throughout the building however, as it is affected by:

- The concentration of carbon dioxide (this may be higher inside the building than outside)
- The saturation of the concrete. Saturation hinders the passage of CO_2 through the pores, thus slowing down the carbonation. For that reason, exposed walls may suffer less than sheltered walls
- The quality of the concrete
- The presence of cracks.

Chlorides were historically added to the mix to reduce setting times but have been banned in the UK since 1977. The presence of chloride ions in the capillary pore water of hardened concrete affects the protective passive film on the steel and thereby enables it to corrode. Chloride ions can be from various sources such as sea water, or de-icing salts, etc. The corrosive effect is largely dependent on the ratio between free chloride ions and the level of hydroxyl present, which is dependent on the amount of cement in the concrete and its alkalinity. Carbonation exacerbates the corrosive effect at levels of free chloride ions that might otherwise have been relatively harmless. Corrosion due to chlorides tends to be more localised than the widespread deterioration due to carbonation. The corrosion process is an electrochemical one, and it is this that makes it difficult to remedy. For example, patch repairs simply spur an adjacent area to start deteriorating.

Electrochemical methods of stopping corrosion include **desalination** and **re-alkalisation**, which are two sides of the same coin. An electric field is applied between the reinforcement in the steel in the concrete and an external electrode net, which is in contact with the concrete surface through a sprayed-on electrolyte. In re-alkalisation, alkalis are carried into the concrete, thereby increasing the pH level to stop the corrosion. In desalination, chloride ions are carried out of the concrete. The long-term efficacy of these methods is still under research.

Repair

- Predicting when to repair
- Testing strength and alkalinity
- Monitoring

Key issues are knowing the rate of deterioration, to enable predictions of when repairs are likely to be needed, ensuring the strength of the structure is maintained and there is no risk of injury from spalling and monitoring the structures. Concrete structures need regular inspection every three to five years to detect problems early and minimise repair costs. The age of the building might suggest the kind of difficulties that may be experienced, e.g. buildings constructed in the 1950s and 1960s might have had chlorides added to the mix as at this time the associated problems were not appreciated and chloride accelerators had not yet been banned.

Repairing concrete structures is time-consuming, often a permanent ongoing task, and is generally expensive when undertaken on any appreciable scale. Scaffolding will increase costs of remedial work. The defect can be quite well advanced before it comes to attention (hence the importance of regular monitoring of condition), and additionally the processes at work are often progressive. Maintenance is essential because concrete spalling and dropping from a building can pose a significant hazard. It is important is to ensure the building or structure remains able to carry the necessary loads safely; concrete is weak in tension, it is the steel that takes that load.

MOVEMENT AND STABILITY

Movement and stability: other than foundation movement

Why buildings move
Types of movement
Causes of movement
Inspecting, recording, and assessing
Monitoring
Repairs

All buildings move. Buildings move because of:

- Conditions above ground
- Conditions below ground

Movement and stability: other than foundation movement – Above ground

See Table 4.9 Reasons for building movement other than foundations. Types of movement (other than foundations):

Moisture and thermal movement. Materials need to be conditioned for use in buildings and changing environmental conditions when in use can cause

Table 4.9 Reasons for building movement other than foundations

Why do buildings/parts of buildings move? (Other than foundations)		
Expansion	Wetting/absorbing moisture	Some materials expand on wetting and shrink on drying
	Freezing water	About 10% increase in volume when water freezes
	Sub-surface crystallisation of salts (crypto-florescence)	Salts occupy more volume
	Sulphate attack	Permanent expansion, can be over months or years. Particularly problematic when cementitious materials are wet for prolonged periods
	Corrosion/oxidisation	Oxides often occupy greater volume, e.g. rusting of ferrous metals
	Alkali silica reaction (ASR) permanent expansion	Problem with concrete in wet conditions and containing reactive aggregates and silica
	Hydration of oxides permanent expansion over months or years	Problem with clay bricks that contain unhydrated lime or magnesia. Causes spalling of the face of the brickwork
Shrinkage	Temperature decrease	Causes contraction
	Drying/losing moisture	Causes shrinkage (particularly timber). May get noticeable initial drying shrinkage of cementitious materials
	Loss of volatiles	Noticeably with mastics, sealants, plastics, failure of curtain wall sealant
Failure	The effects of excessive loads	Loads may exceed the capacity of the structure because: • The structure was never adequate • The loads have increased (e.g. increased imposed floor loads, new heavier roof covering) • Or both

problems. Timber needs to be seasoned and clay bricks allowed to adjust to normal temperatures and moisture. Bricks begin a permanent expansion on cooling from the kiln, and this can continue for the life of the building. The greatest expansion is shortly after removal from the kiln. By contrast, concrete and calcium silicate bricks shrink after manufacture, so there is potential for movement in a brick/block wall.

Timber used to be seasoned over several seasons (hence the name), but currently this process of gradual reduction in movement is achieved by faster kiln drying. Central heating and air-conditioning dry the atmosphere and materials, exacerbate shrinkage after construction of timbers and plaster, etc. This can affect old buildings provided with such services for the first time (e.g. an obsolete factory converted to flats).

Materials also move in response to water absorption and drying out. These sorts of cracks are generally minor and have been minimised by the reduction of wet trades and the increased use of dry finishes such as plasterboard instead of plaster.

With moisture and thermal movements there will be an initial irreversible change as the material is built into the construction followed by further reversible movements in use in response to climatic changes. The main difference between thermal and moisture movement is that thermal movements are reversible, whilst moisture movement can be either reversible or irreversible; for example, fibrous ceiling tiles never recover from being wet.

The likely positions for this type of movement are at openings, chases, changes in level, buttresses, and at joints. Openings are the weakest point of a wall, so it is unsurprising that cracks open up there. It boils down to wherever there is a change in the otherwise uniform section of the wall the cracks will be seen. They are a response to tension in the wall. Masonry is good in compression, but not in tension. The heat or moisture causes the wall to expand and on cooling or drying, it tries to move back, but at that point tension develops in the wall and small vertical cracks appear.

So the **key points** to note are: response to tension, vertical, at changes in wall thickness or openings. They are generally fine (1–2 mm) and often step around the bricks, following the line of the mortar, sometimes in a diagonal fashion. The cracks are generally uniform in width and do not go below the damp-proof course (greater restraint and more even temperature, moisture balance).

Materials expand in response to heat, but they do not all respond in equal measures. This can cause differential movement and consequent stresses that can lead to distortion and cracking.

Figure 4.5 shows the typical distribution for shrinkage cracking in a brick cavity wall – note the cut off at the dpc. Quite common in long rows of terraced properties – the whole row expands but then cannot shrink back. The houses in the middle suffer quite noticeable cracking.

Overloading of structures can cause cracking that is often misdiagnosed as subsidence because the cracks extend below the dpc. See Figure 4.6 Overloading cracking.

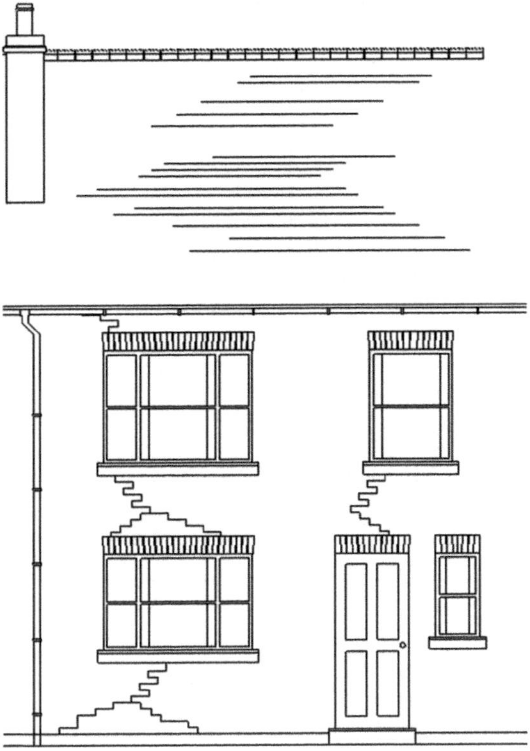

Figure 4.5 Shrinkage cracking.

Accommodating movement:

- Within their elastic range all materials will bend
- Ductile materials deform and will eventually yield.
- Brittle materials will crack without deforming, sometimes catastrophically
- Metals creep; timber will (sometimes) creep.

Figure 4.6 Overloading cracking.

Movement and stability: foundation movement – Below ground

Cases of cracking in buildings seem to be on the increase and there are several reasons for that, not least the fact that subsidence and landslip are now covered by the householder's insurance, which makes insurers more likely to investigate them.

Climate change has also had an effect, particularly with buildings constructed on shrinkable clay soils. The fact that people are now very willing to sue tends to encourage surveyors to suggest over-the-top measures to ensure that future movement is prevented.

Not all buildings have foundations, and early structures were raised straight off the ground, but they tended to be either lightweight or took a while to build and used lime mortars or clay, which allowed considerable settlement after construction without serious harm to the structure. See Figure 4.7 Foundations and Table 4.10 Below ground movement causes.

Foundations
• Foundations – buildings have not always had them • Purpose - they carry the building load down to adequate bearing strata, or they spread the load so that it can be built over weaker soils • Foundation design depends on the superstructure (loading) and the ground conditions

Figure 4.7 Foundations.

Table 4.10 Below-ground movement causes

Below-ground movement causes (including Financial Conduct Authority (FCA) insurance definitions)		
Settlement	Downward movement as a result of soil being compressed by the weight of a building within ten years of construction	Shortly after construction Usually small movements and non-recurring Differential settlement can cause serious problems Not usually covered by insurance
Subsidence	The ground beneath a building sinks, pulling the property's foundations down with it. The ground loses moisture and shrinks	Change in ground conditions, caused by: Prolonged dry spells especially when coupled with shrinkable clay soils Leaking drains, broken culverts; effects of trees and shrubs: which can absorb significant volumes of water from the soil Collapsed mine workings Decomposing organic fill, etc. Differential subsidence is problematic
Heave	Upward movement of the ground beneath a building as a result of the soil expanding	May be caused by: Tree removal Frost (frost heave) Demolition of a nearby building Adjacent excessive development, etc.
Inadequate consolidation	Of fill below foundations or floors	Type of settlement. May be caused by: Inappropriate materials in fill Lack of adequate compaction of fill Inadequate ground preparation before start
Chemical attack	e.g. sulphates	Deterioration of concrete causing it to become soft and spongy Symptoms are similar to, and may be mistaken for, heave Not usually covered by insurance Sulphate-resisting cement or a mechanical barrier/dpm can be used for protection
Landslip	Downward movement of sloping ground	

Below-ground failure crack patterns

See Figure 4.8 Differential settlement cracking; Figure 4.9 Subsidence cracking Figure 4.9 Subsidence cracking, and Figure 4.10 Uneven ground movement, e.g from mining.

Leaking drains can cause subsidence as the water gradually washes away the soi from the foundations and undermines them. See Figure 4.11 Subsidence cracks du to drainage leaks.

If trees are felled, there is a sudden ground swell known as heave where th soil has to reach a new equilibrium in terms of water content. Figure 4.12 shows typical crack pattern associated with heave; note how the cracks are widest at th bottom, the opposite of what is normally found with subsidence, although this i not always the case and heave cracks can appear similar to subsidence cracks. Se Figure 4.13 Movement due to tree removal.

Trees

See Figure 4.14 Effects of trees. Trees are wonderful things: in addition to thei obvious beauty, they provide shelter from winds and weather, but they do need t be a safe distance from houses and other buildings unless special precautions hav been taken to prevent root damage (e.g. sheet steel piling), in which case they ca be planted very close – as often happens in town centres. In fact, trees and building are quite a happy mix except where there are clay soils – this combination of tree and clay is challenging and needs consideration.

Figure 4.8 Differential settlement cracking.

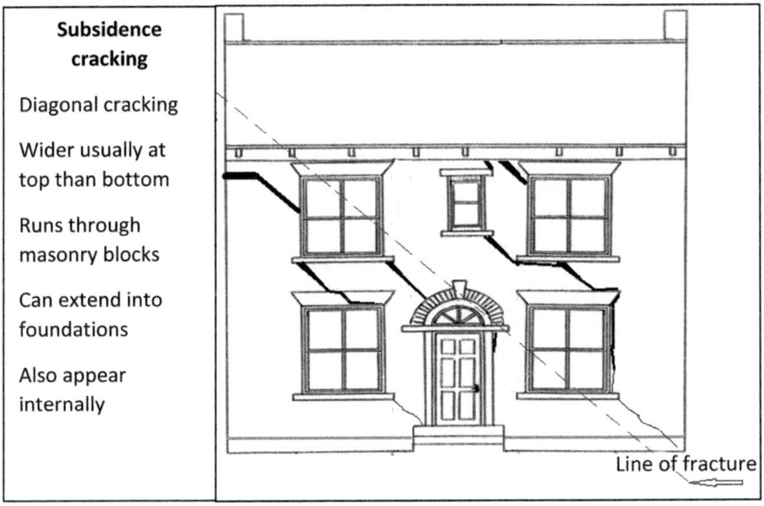

Figure 4.9 Subsidence cracking.

Figure 4.10 Uneven ground movement, e.g. from mining (exaggerated).

See Table 4.11 Typical heights of trees and safe building distances and Figure 4.15 Reasonable tree distances to buildings.

Tree roots rarely damage the building itself but can often damage drains. It is the effect on moisture levels of soils which is damaging and can cause serious problems. Trees are huge pumps that draw water from the soil and lose it by

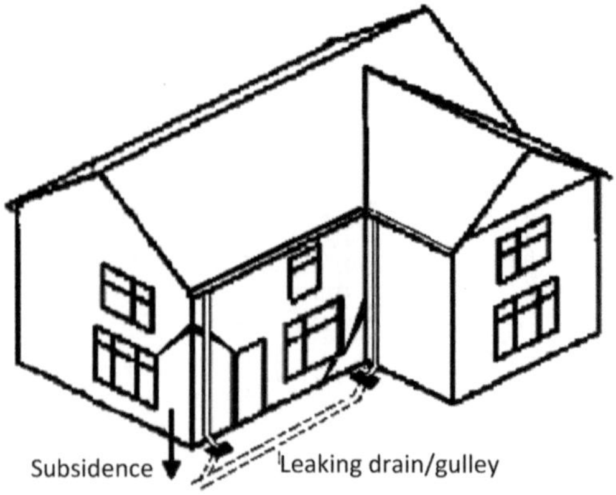

Figure 4.11 Subsidence cracks due to drainage leaks.

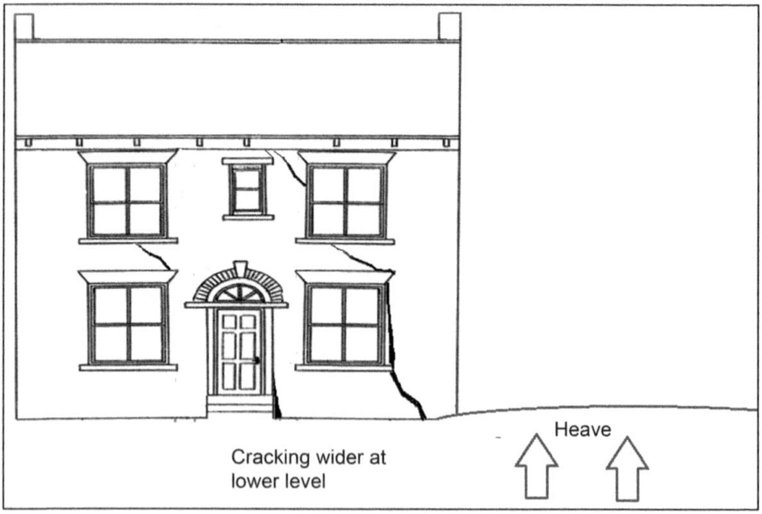

Figure 4.12 Cracking due to heave.

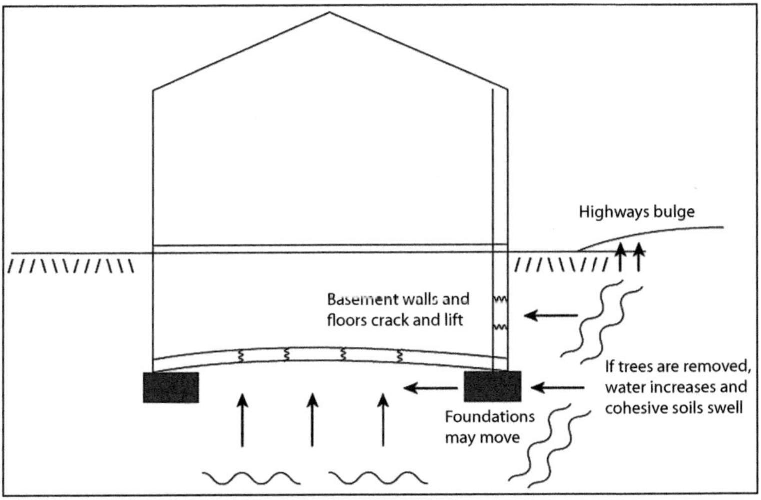

Figure 4.13 Movement due to tree removal.

Effects of trees

- Caused by moisture changes in soils, particularly shrinkable clay soils
- Excessive drying (shrinkage)
- Heave (expansion)
- Can be progressive and can cause complete collapse if not addressed – significant problem

Note: Trees make a massive contribution to local environmental conditions, and damage is rare

To determine a safe distance from a building, knowledge of the following is needed:

- the species of tree
- its water demand
- its mature height
- the soil type

Figure 4.14 Effects of trees.

transpiration through their leaves. The roots bind the soil, which can be useful on sloping sites to avoid soil erosion, but also extend in search of moisture. Some trees are deep rooted, and others are shallow rooted.

In order to determine a safe distance from a building, knowledge is needed of the foundation type and depth, the species of tree, its water demand, mature height, and the soil type. It is not enough to consider just the tree; account must also be taken of the number, type, and location of all trees and shrubs in the immediate area as they will all affect the water content of the surrounding soil.

- Most trees cause no damage
- Tree roots spread up to three times the height of the tree
- Modern buildings are seldom affected
- Shrinkable clay soils are most at risk
- Subsidence is worst in dry years.

It is a complex assessment, and the advice of an arboriculturalist is useful. See also Table 4.12 Action if trees are a problem.

It is often not practically feasible to take root barriers to an adequate depth. In addition, roots may grow under a barrier since provision introduces air and water, making an attractive growing environment.

Table 4.11 Typical heights of trees and safe building distances

Type	Typical mature height in m	Typical safe distance from buildings in m
Apple, cherry, damson, pear, plum	12–17	11
Ash	23	21
Beech, walnut	20	15
Birch, laburnum	12–14	10
Cypress	25	3.5–20
Elm, oak	25	25–30
Hawthorn, rowan	10–12	12
Holly, laurel, magnolia	9–14	5
Horse chestnut	20	23
Lime	25	8–20
Maple, sycamore	20–25	20
Pine	30	8
Plane	30	7.5–22
Poplar	30	30–35
Spruce	18	7
Willow	24	40
Yew	12	5
All these depend on cultivar, ground/soil/environmental conditions. Take specialist advice.		

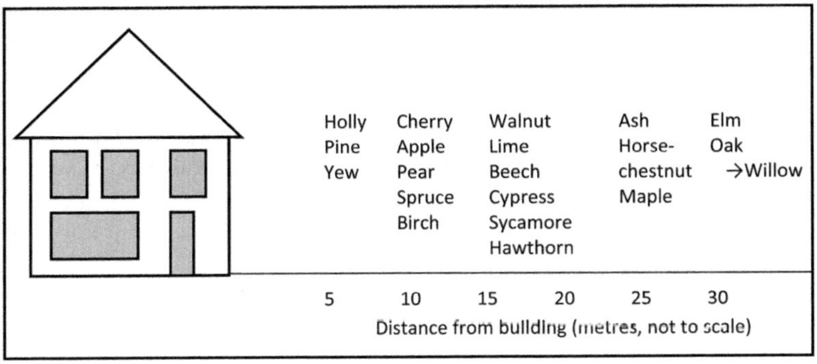

Figure 4.15 Reasonable tree distances to buildings.

Table 4.12 Action if trees are a problem

Options	Positive	Negative
Prune	Inexpensive Trees preserved	Temporary solution
Root barrier		Uncertain effect
Underpin	Permanent solution Trees preserved	Expensive Disruptive
Fell	Permanent solution Inexpensive	Possible heave Loss of the tree
After winter recovery, repair damage	Little disruption Inexpensive Trees preserved	Damage may recur in future dry summers Probable further damage if the tree is young
Arboriculture advice should be taken before pruning		

Inspecting, recording and monitoring, and specifying repairs

Assessing movement

- All buildings move
- Often it is not a problem
- Some deflection is inevitable in floors.

If the structure remains sound, services continue to work, the enclosure remains weather tight, finishes are not damaged, and aesthetics are not impaired, then there is little cause for concern. Freestanding walls may be significantly out o plumb and still stand, and if offered restraint can take additional load too. Wall in buildings often bulge at some mid-point, and there are various causes fo this. What needs to be assessed is the degree of movement out of the plumb When restraint is offered, as is normally the case at roof and foundation level movement out of vertical can be surprisingly large. The **stability** of walls i influenced by:

- Materials used
- Restraint offered
- Openings
- Piers
- Loading

A common cause of wall bulging is a lack or inadequacy of **lateral restraint** This might be due to failure of cavity wall-ties. Floors and party walls should b tied into the structure; where they are not – and this is common in older proper-ties – the external face can move away from the floors or party walls. Other cause include rotting of in-built timbers and snapped header brickwork. Sometimes add-ing restraint back in, in the form of restraint straps or resin grouted ties, etc., is suf-ficient to restrain further movement. However, when the bulging is too severe, th walls may need to be rebuilt incorporating restraint. Old stone walls are thick an comprise two leaves with a rubble infill. Very often the "mortar" in the interior o the walls was simply clay or earth. Over time this can be washed away along wit the quarry waste or rubble infill. The through-stones that tie the leaves togethe can also be inadequate, so the two leaves begin to separate and the wall starts t bulge. Lateral restraint is included as a matter of course these days, but it is impor-tant to check site practices to ensure proper fixing of these very important compo-nents. Sometimes it is the effect of additional lateral loads that causes walls to lean for example, where roof spread occurs. The reason tie beams came to be include in roof design was that the earlier roof designs revealed an inherent failure. Th load of the roof covering and of the timbers themselves causes the inverted V shap to collapse downwards, placing rotational pressure on the wall plates and forcin the upper parts of the walls outwards. The greater restraint at the floor and/o ground/foundation level accentuates the differential movement.

Cracks in masonry. The severity of cracks in traditional loadbearing masonr should be considered in the light of its:

- Width
- Depth
- Extent
- Oversailing.

Table 4.13 Remedial action related to crack width (adapted from BRE Digest 251)

Crack	Crack width	Action
Hairline cracks	Up to 0.1 mm	No action
Fine cracks	0.1–1 mm	Decorate
Small cracks	1–5 mm	Fill
Wide cracks	5–15 mm	Repair/repoint
Extensive cracks	15–25 mm	Replace
Severe cracks	25 mm +	Major repair

Assessing movement: Inspecting, monitoring, recording

See Table 4.13 Remedial action related to crack width. An inspection only provides a snapshot in time. Assessment should build up a picture of the history of the defect using evidence. Monitoring can provide information that is far more detailed. Tell tales allow you to measure changes in crack width over time and can also gauge vertical movement by measuring the cross-hair movement against the grid on the lower plastic plate. This is difficult when the wall is not flat and is of limited accuracy (about 1 mm). The first question to consider – is it imminently dangerous?

Recording:

- Record movement at regular intervals
- Take records over the likely movement cycle
- Record weather conditions, temperature, and any other factors that may be relevant

Repairing cracks:

- Has it stopped?
- Do nothing?
- Cure the cause
- Fill, patch, or replace.

Deleterious materials

See Table 4.14 Deleterious materials risking health and building failure.

Table 4.14 Deleterious materials risking health and building failure

Entity	Risk	Comment
Alkali aggregate reactions	Building	Chemical process producing a calcium silicate gel at the surface, which absorbs moisture, swells, resulting in concrete cracks. In unrestrained concrete, manifests as random network of fine cracks bounded by fewer large cracks (map cracking). When restrained, cracking tends to follow lines of restraint.
Asbestos	Health	Boarding, cladding, insulation, tiles, fire protection, pipes, etc. Risk is from breathing in airborne fibres. No type or colour of asbestos is considered safe. Results in cancers, asbestosis, mesothelioma. Building materials required to be positively identified as not containing asbestos. Those that do contain asbestos (ACMs) to be recorded, assessed, labelled, and/or disposed of by registered means, with a "duty to manage".
Calcium chloride additive and atmospheric/ applied	Building	Used as an accelerating admixture in concrete. Applied as de-icer. Atmospheric in marine or coastal positions. Reduces passivity of concrete. Results in corrosion of pre-stressing tendons, reinforcement, and embedded metal.
Calcium silicate brickwork, concrete brickwork	Building	Shrinkage. Thermal movement is approximately 1.5 times that of clay bricks. General construction detailing important, e.g. sufficient flexibility in wall-ties to permit the differential movements, allowing for discontinuity around cavity closers, flexible expansion joints. Risk of shrinkage or expansion cracks; therefore the mortar needs to be flexible with brickwork.
Combustible cores	Building; Life	Cladding in high-rise buildings and larger industrial/ commercial framed buildings. Fire can spread unobserved inside core. Fire can spread rapidly with toxic fumes. Ignition not usually instantaneous, varying core materials used with differing ignition properties. Polymeric core materials, e.g. EPS, PUR, ignite at relatively low temperatures, contributing to fire spread. Cores become exposed to ignition risk through damage, fixings, or penetrations. See Chapter 7.
Formaldehyde	Health	Adhesives, furnishings. Risk is from vapours, dispersed in good ventilation. Carcinogenic.
High alumina cement (HAC)	Building	Used in precast roof and floor beams, lintels, etc. Decreased strength of concrete over time, especially in high humidity or temperature, leading to structural weakness or vulnerability to chemical attack.

(continued)

Table 4.14 *(continued)*

Entity	Risk	Comment
Lead	Health	Risk of lead in older water pipes, joint solder, paint, and brass fittings. Risk is from ingestion. Headaches, stomach pains, anaemia, kidney damage, nerve/brain damage, infertility.
Machine-made mineral fibre (MMMF)	Health	Mineral wools (glass wool, rock wool): thermal/acoustic insulation and fire protection. Ceramic fibres: insulation boards, furnace /incinerator insulation, e.g. refractory ceramic fibre (RCF). Continuous glass filament fibres: reinforced cements and plastics. Risk is from contact and breathing in. Local skin/eye/throat irritations, dermatitis, development of asthma/bronchitis, decreases in lung function, possibly carcinogenic.
Mundic	Building	Deterioration in concrete/concrete blocks due to the decomposition of mineral constituents within the aggregate. Particularly in south-west England. Loss of integrity in damp conditions, mechanical weakening of the building.
Polychlorinated biphenyls in building materials (PCBs)	Health	Synthetic organic chemical used in the manufacture of transformers, thermal insulation, paints, sealants, adhesives, etc., but progressively banned by year 2001 in many countries. Persistent organic pollutant. Released into the environment during demolition and refurbishments. Do not readily break down once in the environment, remaining for long periods cycling between air, water, and soil. Poisonous. Found in food chain of fish and sea mammals, accumulate in the fatty tissues of animals and humans. In UK and other countries, holding of PCBs requires registration, and disposal is regulated.
Reinforced autoclaved aerated concrete (RAAC)	Building	Used in UK construction, mid-1950s to 1980, possibly beyond, in planks and panels for roofing, walls, floors, and internal partitions. Various building types, especially schools. Excessive and progressive deflections in service, widespread hairline cracking of soffits. Association with excessive damp. Need regular inspections.
Toughened glass containing nickel sulphide impurities	Building	Spontaneous fracturing after time (e.g. up to 20 years). Avoid toughened glass in overhead positions.
Urea formaldehyde foam (UF)	Health	Cavity wall insulation. Risk is from vapours, dispersed in good ventilation. Linked with respiratory irritation, nausea, sore eyes, skin rash, headaches, general lethargy, muscular pain, disturbed sleep.
See also biological entities above: "Mould, decay fungi, and bacteria".		

5

Retrofitting and refurbishment

Chris Gorse, Melanie Smith, B. N. West, Cormac Flood and Lloyd M. Scott

INTRODUCTION

A city without old buildings is like a man without a memory

Konrad Smigielski

No building is ever finished or completed; periods of activity are merely interspersed with periods of quiescence, but maintenance will continue for as long as the building's economic life. Throughout their own history, all buildings undergo a replacement process of fittings, members, and major parts, coupled with periodical cleaning and redecorating. In some cases, this replacement process takes a step further into the region of restoration or renovation. Periodic changes to the structure often enhance the appeal it has in the present.

Building retrofit, refurbishment, and upgrading inevitably include aspects such as maintenance, repair, restoration, and extension. These major components of construction activity regularly account for only slightly less than half of the UK construction industry's total output. The activities can be generally described as "repair and maintenance" by governments for statistical purposes.

Aspects such as finance, sustainability, location, logistics, etc., mean that stocks of old, redundant, and obsolete buildings can provide high-quality, desirable, ultra-modern accommodation through their refurbishment and re-use, and at a lower cost and more quickly than new construction.

There are, nonetheless, considerable technical issues in upgrading old buildings, for example controlling performance.

Perceived problems:

- Heat loss
- Damp

- Human comfort
- Draughts
- Incompatible materials and technologies
- Unknowns

Requirements:

- Human comfort
- Durability
- Low carbon
- Reduced energy use
- Fashion

A major factor in favour of refurbishment is that many existing buildings are still in existence because they are well-built and structurally sound. They may be currently run-down, neglected, and unfit for modern usage, but their traditional methods of construction have left an abundant legacy of sound, durable structures which provide an ideal basis for refurbishment and re-use. Notwithstanding this, it should not be assumed that they are always of high structural quality. It is essential that any building being considered for refurbishment, even though it may appear to be sound, is subjected to a detailed survey in order to confirm its quality and condition, and to ascertain the likely cost of any repairs deemed necessary and their effect on the feasibility of going ahead with a refurbishment scheme.

Truism claims that the older the building and the longer it has been empty, the greater the repair and refurbishment costs for the existing structure and fabric. Dampness is the principal agent in most forms of building deterioration. If the building has remained empty and neglected for a prolonged period, dampness is likely to have penetrated, leading to timber decay and damaged finishes. Empty, neglected buildings are frequent foci for vandalism, which can be very expensive to rectify. Additionally, ordinary entropy of materials due to age will mean that certain items will need attention. Items that may still be in good condition may need replacing just because they are obsolete or outdated in their design, common examples being sanitary, heating, and elevator fixtures, fittings, and installations.

Generally, overall costs of refurbishment schemes can be directly proportional to the age of the building and its degree of neglect, so that the proposed refurbishment of any old, neglected building should thus be given very careful assessment before proceeding. See Table 5.1 Opportunities and consequences of common upgrading measures.

Table 5.1 Opportunities and consequences of common upgrading measures

Opportunities	Consequences
Reduced heat loss	Rebound comfort-taking
Seal draughts	Reduced ventilation
Fit air barriers	Loss of moisture equilibrium
Add insulation	Potential condensation risk
Double-glaze	Displace condensation
Microgeneration of energy	Offset costs, but is there net carbon reduction?
Removal of damp	Is this necessary?
Retrofit dpc	Increase energy usage
Increase heat	Increase energy usage
Improve ventilation	Reduce energy usage, but is there increase in damp?

CONSIDERATIONS FOR REFURBISHMENT

Structure

For a number of reasons it may be necessary to strengthen existing walls, floors and roofs when undertaking refurbishment works. Cavity ties can be a problem in buildings constructed before 1981. Galvanised and mild steel ties used before this date did not have good resistance to corrosion, resulting in a regular horizontal cracking pattern visible on the external surface; stainless steel and plastic are more resistant. Where the structure is sound but the existing wall ties have failed, remedial wall ties can be installed, fitted into holes drilled into the wall. Otherwise rebuilding the walls may be required. Walls and floors are required to be strapped together; similarly roofs are required to be strapped to walls. This was not done in the past; therefore additional retrofit strapping may be required. A structural appraisal of the property may be advised.

Fire protection

Generally, properties are built originally to the appropriate standards of fire protection required at the time for their size, shape, occupation, and use. If any alterations have been made since those alterations should have been constructed and provided with appropriate safety measures. However, it is vital that it is understood that:

1. Past standards may not now be sufficient
2. Past uses might be different from current and/or proposed uses
3. The building and any alterations may not have complied with standards of the time – especially where hidden

4. Any measures providing safety may now have deteriorated or been compromised by acts of amendment, maintenance, alteration, decay, distortion, shrinkage, or other entropy.

How do old buildings work?

See Figure 5.1. which shows the complex range of moisture control, and pathways of airflow and moisture movement in a traditionally constructed building. Sufficient ventilation is necessary to prevent moisture build-up and subsequent damage to the building.

Questions to ask prior to upgrading choices

The original building:

- How has the building been altered over time?
- Was it built badly in the first instance?

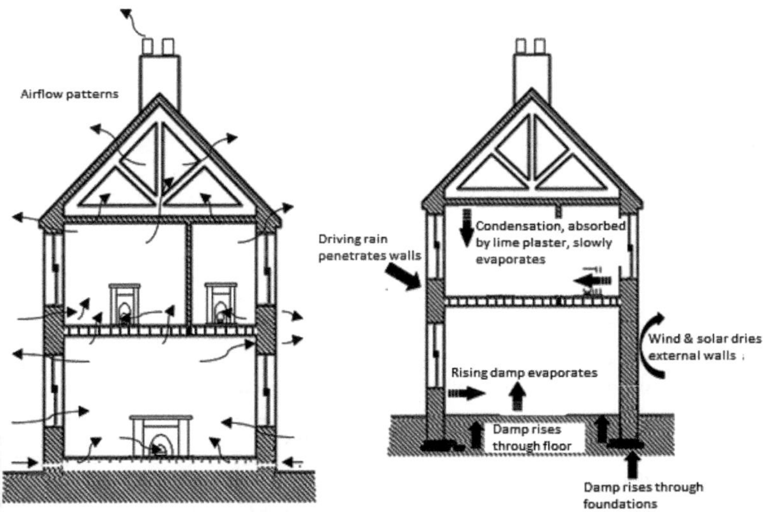

Figure 5.1 Section through a traditionally constructed building showing the complex range of moisture control and pathways of airflow and moisture movement. Sufficient ventilation is necessary to prevent moisture build-up and subsequent damage to the building (adapted from Jenkins (nd), "Improving Energy Efficiency in Traditional Buildings". Historic Scotland, and Picketts Historic Building Conservation).

- Was there enough knowledge available at the time of construction?
- What has been done recently?
- Have incompatible materials been used? e.g. iron or timber in potentially damp situations.
- Have technologies been used without the correct materials? e.g. flat roofs with insufficient allowance for movement.
- Have recent alterations made things worse? e.g. vinyl wallpapers.

What can be done?

Refer to Table 5.2, which gives opportunities and challenges in thermal upgrades.

What is the latest research telling us? Example desktop research:

- SPAB – hygrothermal modelling
- STBA
- Historic Scotland
- Historic England, e.g. Energy Efficiency in Historic Buildings
- University outputs
- Journals, e.g. International Journal of Building Pathology and Adaptation

What are the unknowns?

- The moisture content of the walls
- The presence of micro cavities
- The continuity of mortar
- The properties of materials within the wall
- What else don't we know?

Table 5.2 Some opportunities and challenges in thermal upgrades

Opportunities	Challenges
Control ventilation paths	Increased condensation/indoor pollution
Control air movement through the fabric	Moisture trapped at air barriers
Insulate main elements	Condensation risk at thermal gaps
More efficient heating systems	Balance with requirements (fast response? thermal mass?)

What do we need?

- A process for surveying and understanding the risks of unintended consequences
- Understanding of systemic design
- Understanding of human behaviour
- Understanding of environmental conditions
- Understanding of building physics
- A method of measuring a U-value of a building in situ

Is the building or part of the building historic, or just old, or is it defunct?

- "If it ain't broke – don't fix it" is valid here
- Don't need overkill – if there's no worthwhile character you don't need to preserve it
- BUT you do need to maintain durability
- Biggest issue is control of moisture
- Consider the ramifications of every alteration
  - Airtightness – is it appropriate, is it possible?
  - Removal of ventilation routes (e.g. flues) – effect?
  - Increased insulation – where, how much, what type, effect?
  - Condensation risk – increased, reduced, displaced, or removed?

No one right answer – but there are plenty of wrong ones!

- You need to aim for a balanced approach.
- Durability can easily be compromised by wrong decisions and insufficient joined-up thinking.
- What is achievable?
- What is practical?
- What is desirable?
- **Does it all work together?** Remember construction of old buildings represents a SYSTEM – it all needs to work together.

UPGRADING THERMAL PERFORMANCE

There is much to consider prior to upgrading and enhancing a building's thermal performance. Primarily, the focus is on improving the whole building, whilst ensuring that the changes made to one part of a building do not adversely impact another part, or result in latent defects, or result in unexpected consequences. Simply adding insulation onto a part of the building can serve to trap in problems that are already present or bring about changes that result in moisture forming behind the insulation or in another part of the structure. A number of points to consider when upgrading are listed below.

What can be learned from what has happened so far?

- Effect of inappropriate insulation
- Insufficiently assessed buildings being chosen for insulation
- Designs involving cut-outs for eaves, services, abutments, etc., leading to large areas of thermal bridging
- Low standards of detailing on site – even if appropriate design
- Lack of understanding of the risk of insulating older properties, with many unknown factors being assumed
- Lack of understanding of the effect of wind-driven rain on buildings
- Lack of basic maintenance prior to retrofit
- Insufficient emphasis on good-quality finishing materials and workmanship

Thermal upgrades of external fabric

Refer to Table 5.3, which discusses external wall insulation (EWI), internal wall insulation (IWI), and no wall insulation.

Table 5.3 External wall insulation, internal wall insulation, or no wall insulation?

External wall insulation	Internal wall insulation	No wall insulation
Thermal mass	Fast response	Thermal mass and slow response
Affects exterior aesthetics – Planning Permission required	May not require Planning Permission	High energy usage – missed opportunity?
Problems at roof level, openings, dpc level, etc.	Loss of internal space	Still need to control hygrothermal performance
Affects the wall's ability to deal with moisture	Associated works: electrical sockets, switches, etc., and door frames, windowsills and reveals, etc.	What can be done elsewhere in the building to compensate for loss of heat?
	Affects the wall's ability to deal with moisture	How comfortable will it be to live in?
		Has fear of adverse effects held the designer back or are these justified?

Before starting on any fabric upgrading, need to consider:

- Space heating – boilers, radiators, etc.
- New sanitary and plumbing services requirements
- Research has shown that effective use of these can be more useful in reducing carbon emissions and fuel consumption than doing anything to the fabric
- Think about ensuring that the occupier knows how to use the controls – need to be user-friendly. Over-complicated systems will invariably be misused.

What about the other elements?

- Ceilings and roofs – up to 25% of heat is lost through the roof
- Floors – essential to consider moisture movement and ventilation
- Doors – draught proofing and consider the panels, not just the frame
- Windows – draught proofing can be very effective; curtains were invented for heat retention, not just privacy
- Flues – very important function in an old house. Soot is hygroscopic.

Ceilings and roofs

- Lath and plaster ceilings – may be fragile and liable to collapse
- Insulation materials – choice is important – natural materials control moisture better than others
- How much and where? Historic Scotland advise that you need at least 270 mm of natural insulation to be fully effective
- Need to maintain a gap of about 50 mm between the insulation and the sarking felt and a robust airway at the eaves.

Floors

- Strength in relation to any new purpose
- Timber floors – may be able to insulate from below if there is access
- Consider materials: impermeable insulation may lead to timber decay
- Solid floors – what sort? Rammed earth, flags? What are the benefits of insulating weighed against leaving it in situ?
- Stone flags will be damaged in lifting; an insulated limecrete floor can be a good choice in some circumstances
- Existing concrete floors can be insulated in the usual way.

Doors and windows

- The least effective part of the door is usually the panels.
- Draughts - draught stripping around frame, letterboxes, keyhole covers, etc. (cat flaps?)

- Windows – why double/triple glaze? What effect will that have on other elements? What other alternatives are there? See Hitroic Scotland for shutters and secondary glazing.
- For UK climates, Historic Scotland and Historic England have published useful documents for thermal upgrade.
- Considerations: repair, draughtproof, shutter, secondary-glaze, provide insulated glass units (IGUs), re-glaze, replace.
- Position glazing in line with the plane of the external wall insulation (to reduce thermal bridging).

Flues

Very useful things:

- Structural buttressing
- Passive stack ventilation

But,

- Soot is hygroscopic, so will absorb moisture from the air
- Heat is lost up the flue
- Down draughts can be created
- Chimneys might not be lined, is the integrity of the feathers compromised?
- Sealed flues result in a lack of ventilation in the flue, leading to moisture retention.
- Moisture movement and evaporation from chimney breasts can result in aesthetically poor patches to be visible in rooms.

So,

- Do you remove the flue? – easy, difficult, stability?
- Do you line the flue?
- Do you seal the flue – completely, partially, not at all?

External wall insulation

Depending on the construction, and historic relevance, of the property, any improvement to the external fabric of a property could be:

- Cavity wall insulation (CWI)
- Internal wall insulation (IWI)

- Thin internal wall insulation (TIWI)
- External wall insulation (EWI)

There are benefits to each system, but it is important to remember that there are disadvantages to each system as well, not least in the way each system changes the fundamental ways the building works as a moisture moving process system.

Before considering external wall insulation, a **pre-design thermal retrofit survey** is necessary. This is an on-site anticipatory inspection and assessment of a building. It is undertaken to inform the design for a retrofitted energy efficiency measure, and must include:

a. The nature and characteristics of the building,
b. The risks to that building and its occupants posed by the measure,
c. The risks to the efficiency of that measure posed by the building, and
d. The challenges required to mitigate these risks through and by the design.

See Tables 5.4, 5.5, and 5.6 which show typical defects found in properties having, respectively, retrofitted IWI, EWI, and CWI.

Windows thermal upgrade

There are obvious benefits to alternate solutions for windows rather than replacement. According to Historic Scotland Technical Paper 1, "Thermal performance of traditional windows", timber shutters reduce heat loss by 51%; ill-fitting existing shutters can be restored or new shutters made. Heavy, lined curtains can reduce heat loss by 14%. This publication also gives the information shown in Table 5.2, as well as further measures. Historic England's publication "Traditional windows: their care, repair and upgrading" gives more advice. Also refer to Table 5.7 which gives the test performance results of different improvement methods to sash and casement windows.

DISCUSSION SURROUNDING THERMAL UPGRADES

Cormac Flood and Lloyd M. Scott

Policy drivers and strategies for improving the energy efficiency of existing dwellings

Energy efficiency of buildings has become an important issue. Improvement in the performance of buildings is strongly linked to energy savings and a reduction in CO_2 emissions, as identified by many reports and protocols internationally. The

Table 5.4 Typical defects found in properties having retrofit IWI

Typical defects in retrofit IWI
Room in roof thermal bridge. Lack of effective seal at wall and ceiling junctions
Lack of thermal separation between (unheated) basement and upper storeys
Missing IWI at floor voids
Pointing/render concerns
Penetrating damp
Brick deterioration
Rising damp
Thermal bridge at eaves
Flashings deterioration
Thermal bridge at door/window reveals/frames
Lack of sealing for unnecessary warm air loss or draughts
Additional ground floor/ground level concern
Thermal bridge at ground floor level
Service leaks
Condensation
Roof ventilation blocked
Designed ventilation concerns
Sloping roof soffits not insulated

basic principle of improving the energy efficiency of a building is to use less heating, cooling, and lighting without affecting the health and comfort of its occupants.

Retrofitting of existing dwellings offers significant opportunities for reducing global energy consumption and greenhouse gas emissions. Although there is a wide range of retrofit technologies readily available, methods to identify the most energy-effective retrofit measures for particular projects remain a major technical challenge. The generic building retrofit problems and key issues involved in building retrofit investment decisions are presented below.

The EU Directive on the Energy Performance of Buildings (EPBD) contains strict provisions for further energy reduction in buildings, stating "Measures to improve further the energy performance of buildings should take into account climatic and local conditions as well as indoor climate environment and cost-effectiveness". Individual countries have therefore set up various schemes to achieve the requirements.

Table 5.5 Typical defects found in properties having retrofit EWI

Typical defects in retrofit EWI
Thermal bridge at ground floor level
Significant cut-outs (missing EWI) at pipes, services, abutments, etc.
Lack of thermal separation between (unheated) basement and upper storeys
Thermal bridges at eaves
Lack of sealing for unnecessary warm air loss or draughts
Condensation
Room in roof thermal bridge: Lack of effective seal at wall and ceiling junctions
Thermal bridge at door/window reveals/frames
Rising damp
Sloping roof soffits not insulated
Penetrating damp
Roof ventilation blocked
Additional ground floor/ground level concern
Flashings deterioration
Service leaks
Designed ventilation concerns
Pointing/render
Brick deterioration behind EWI

Challenges for improving the energy efficiency of existing dwellings

The complexity of energy efficiency improvement of existing dwellings cannot be underestimated. See also the section in this chapter, "Upgrading thermal performance", the section in chapter 6, "Emerging themes: Gaps and needs in building performance simulation for building retrofit", and Chapter 9, "Environmental considerations". Investments in homes are particularly labour and cost intensive; thus it is difficult to convince homeowners to implement retrofit strategies. Therefore, the central issues are discussed in this section to provide context for which challenges to retrofit implementation are relevant. This discussion surrounding thermal upgrades concludes with a further overview of the challenges that are encompassed by the technical solutions that exist when improving the energy efficiency of existing dwellings, specifically external wall fabric upgrades.

Table 5.6 Typical defects found in properties having retrofit injected or blown-in cavity wall insulation

Typical defects in retrofit cavity wall insulation CWI
Thermal bridge at ground floor level
Lack of thermal separation between (unheated) basement and upper storeys
Thermal bridges at eaves
Lack of sealing for unnecessary warm air loss or draughts
Condensation
Room in roof thermal bridge: Lack of effective seal at wall and ceiling junctions
Thermal bridge at door/window reveals/frames
Sloping roof soffits not insulated
Penetrating damp
Designed ventilation concerns
Lack of insulation at cavity trays
Lack of insulation at window/door heads and sills

Table 5.7 Test performances of improvement methods to sash and casement windows (Baker, 2008)

Improvement method	Possible reduction in heat loss	Possible approximate resulting U-value (W/m^2K)
Unimproved single glazing	–	5.5
Lined curtains (closed)	14%	3.2
Shutters (closed)	51%	2.2
Insulated shutters	60%	1.6
Roller blind (closed)	22%	3.0
Closed blind and shutters	58%	1.8
Closed blind, shutters, and curtains	62%	1.6
Secondary glazing	63%	1.7
Double-glazed panes fitted in original sash	79%	1.3

Source: Historic Scotland Technical Paper 1, "Thermal performance of traditional windows".

Research (Clinch & Healy, 2000; Curtin, 2009, 2013) has established a number of barriers to action regarding domestic energy conservation and the implementation of retrofit strategies in Ireland as restated from Chapter 6, including:

- High upfront costs
- A lack of understanding amongst homeowners about energy use and potential savings from the retrofit of their homes
- A lack of disposable income for retrofit works
- Uncertainty surrounding energy prices, particularly with payback periods over five to seven years
- A lack of incentive
- Inconvenience
- A lack of skill in the construction industry to carry out retrofit works
- A lack of a concerted marketing strategy targeting homeowners in getting its message delivered to consumers

In dwellings where fuel poverty is a factor in restricting energy use, the effect of retrofit measures may be to enhance thermal comfort rather than reducing energy consumption. The purported "Jevons/take-back" effect (Greening et al., 2000) proposes that much of the improvement in thermal performance of the building fabric would be used to achieve higher internal temperatures and comfort levels than pre-retrofit rather than achieving a reduction in energy usage. A social housing study (Lowery, 2012) found that retrofitted homes consumed more energy as a result of heating the dwelling more regularly attributable to a perception that less energy and money is required due to improved efficiency.

Increasing awareness and encouraging changes in occupant behaviour appear to be fundamental in reducing energy use and subsequent carbon emissions (thus alleviating fuel poverty). However, this must be coupled with accurate and efficient technical solutions that reduce heat loss through the fabric of dwellings to a level that can be designed, measured, verified, and quantified.

Technical solutions

There are two essential methods of employing technical solutions to improve the energy efficiency of existing dwellings – the measures-based approach and the whole-house approach. The former, which involves installation of individual measures, is the most common approach adopted, usually due to the limitations imposed by cost constraints and acceptable levels of disruption for occupants. In addition, the opportunities that arise as a result of being triggered by "other" work that may be required within the dwelling, for example replacing an irreparable boiler, are usually more frequent and thus lead to individual improvement measures being installed. Whereas the whole-house approach is generally only undertaken where a deep retrofit project is being undertaken. The easy to install, relatively low-cost and thus common improvement measures installed in existing dwellings primarily consist of cavity wall insulation, loft insulation, draught proofing, hot water tank insulation, replacement windows and doors, replacement boilers, and low-energy lighting. More intrusive, costly, and thus less common interventions include solid wall insulation, ground floor insulation, mechanical ventilation with heat recovery, and micro renewable energy technologies (Killip, 2008). Whilst all of these improvement measures are applicable to enhancing the energy efficiency of existing dwellings, this sub-section focuses upon external wall fabric upgrades due to their relevance to this doctoral research project.

Cavity wall retrofit with full fill insulation is considered quite simple, yet there are a number of issues to be considered prior to execution. Retrofitting cavity walls with full fill insulation is claimed to only be suitable in dwellings that were built between the 1930s and 1980s, where the cavity is at least 50 mm deep and uninsulated, as the thickness and thus overall thermal performance of the insulation that can be retrofitted is determined by the depth of the cavity. Prior to the 1930s, most buildings were constructed with solid walls and consequently cannot receive retrofitted cavity insulation. Additionally, many early cavity walls built during the 1920s have a depth of less than 50 mm, which results in unsuitability for retrofitted cavity wall insulation as the insulation will not align with certification of the full fill system, nor would the thermal benefit be substantiated. The move towards the introduction of the Building Regulations during the 1980s brought about the installation of cavity insulation in new dwellings, rendering these dwellings typically unsuitable for the application of retrofitted cavity wall insulation (Immendoerfer et al., 2008; UWE, 2008; CPA, 2010).

Contradictory to the BER/DEAP assessment methods, a determination on the suitability of a wall assembly for cavity wall insulation retrofit requires a survey to be undertaken by a suitably qualified professional in accordance with the National Insulation Association code of practice; this is to verify the presence of a cavity in the wall, in addition to the depth and quality of the cavity and that there is no debris or mortar obstructing the cavity. Where retrofitted cavity wall insulation is deemed suitable, injection of the insulation product from the outside is the best method (SEAI, 2019). Once cavities are full, the holes are then filled and made good so that the holes can no longer be seen. The two most common insulation materials injected into existing cavity walls are mineral or glass wool and expanded polystyrene beads (Immendoerfer et al., 2008).

A solution to retrofit cavity walls may be full fill insulation, which is ratified and promoted by NSAI and SEAI with fully certified systems carrying NSAI Agrément certification. Nevertheless, there are concerns about filling cavity walls with insulation retrospectively. Cavity walls were introduced to prevent moisture from being able to pass from outside to inside the dwelling, particularly in geographical locations that are classified as being at high risk of exposure to wind-driven rain. If moisture can bridge the cavity through the thermal insulation, this has the potential to cause dampness, surface condensation, and mould growth on the internal surface of the external walls (Immendoerfer et al., 2008; CPA, 2010). Moreover, if these cavity wall insulation materials get wet, the thermal performance is reduced (Stirling, 2001; Greenspec, 2018). Where cavity wall insulation is determined not to be suitable for retrofitting for reasons of cavity depth or wind-driven rain exposure, these walls should be reviewed in line with retrofit solutions for solid walls using internal or external wall insulation, which is discussed in the next section of this chapter. These are also the recommended methods for dwellings with existing cavity wall insulation, and there is a requirement for these to be further improved.

Solid wall insulation (SWI) can be installed either internally or externally, the decision of which is dependent upon a number of factors including but not limited to conservation architecture, wind-driven rain exposure, cost, and floor area, as well as unintended consequences such as the risk of overheating and changes to the distribution of moisture in a building resulting in severe effects on occupants' health and the building (Doran et al., 2014). Solid wall construction was the primary method used to build dwellings up until the 1920s, a form of construction classified as hard to treat for retrofit due to the intrusive nature of insulation retrofit measures applicable for thermal upgrade.

SEAI (2019a) promotes EWI as the best solution for retrofitting external walls. The compliance or benchmark U-value for retrofit funding schemes is presented as 0.27 $W/m^2 \cdot K$ (SEAI, 2019b). To achieve this, a broad spectrum of insulation materials such as phenolic, polyisocyanurate (PIR), expanded polystyrene (EPS), and mineral wool can be used for both EWI and external wall insulation (IWI) solutions (EST, 2006). Historic England (2016), however, argues that mineral wool does not have the necessary hygroscopic properties required for externally insulating solid walls. Conversely, Historic England (2017) suggests that insulation materials based on natural fibres should be used due to their ability to absorb, disperse, and evaporate moisture. They further maintain that the critical factor for IWI is that there is a continuous vapour check/control layer (VCL) incorporated on the warm side of the insulation so long as it is sealed into window reveals and imperforate, as even small holes will allow water vapour through. The use of VCLs has, however, been questioned, and current research suggests that a VCL in a building fabric may cause the accumulation of toxic mould growth at the interface of the VPC and existing wall, creating the potential for backfill of moisture in the wall structure leading to an inflated risk of freeze-thaw frost damage (Doran et al., 2014). However, Little (2010) argues that vapour open VCL materials, in lieu of foil facings, could facilitate drying out of the wall rather than contributing to increasing moisture.

While EWI and IWI both present some degree of condensation risk, it has been purported that IWI carries the higher risk of the two systems. Thermal bridges contribute to this risk together with additional heat losses, particularly at common details such as window and door reveals, party walls, internal walls, intermediate floors and eaves junctions. The effects of these can be lessened and temperature gradients exploited by returning insulation into reveals of openings and along abutting internal walls (Doran et al., 2014).

Solid walls are considered to be a breathable construction, as they allow the absorption and evaporation of moisture through the fabric (May, 2005; Historic England, 2016). Kingspan Ltd (2013), however, argues that 1.7% airborne moisture passes through building elements as a result of vapour diffusion ("breathability") and 95% through bulk air exchange (intentional ventilation plus air leakage), weighting the importance of adequate ventilation in buildings over breathability of the wall fabric. May (2009), however, disputes these findings, highlighting that breathability was established to describe vapour permeability in conjunction with two other key components of hygrothermal performance: hygroscopicity (how a material absorbs and desorbs water vapour in relation to relative humidity) and capillarity (how a material absorbs liquid water). In any case, the jury is out on how external walls cope with moisture management, and to what extent external sources of moisture deriving from internal and external climate conditions influence the performance, hygrothermally, and how this may impact the U-value of the building element. The CIBSE Guide A3 specifies a typical moisture content of 5% for "exposed" brickwork and a typical moisture content of 1% for "protected" brickwork and recommends that the brickwork in solid walls be considered "exposed", even where they have been rendered. As noted (Doran et al., 2014), moisture content can strongly influence thermal conductivity of a material. Consequently, it is crucial to establish the intended and expected moisture load for each assembly within the existing fabric, noting any further potential post-retrofit. Understanding a U-value, then, relies on the consideration of this probable moisture load in the detail design for the building fabric.

It can therefore be concluded that moisture within external walls is an inevitability, and with it present, the impact on U-values needs to be quantified, in line with the moisture content itself and highlighting key contributors to it. Design intentions then need to be executed on site to ensure they are not undermined by poor workmanship or material failure. In existing dwellings, the existing fabric needs to be adequately known to understand the issues at hand to be resolved and accommodated within retrofit design systems. However, to determine if design intentions have been achieved post-retrofit, the collection and analysis of empirical data are fundamental.

UNDERPINNING

Chris Gorse

Underpinning is taking the level of an existing foundation to a deeper, firmer stratum by adding a new foundation construction beneath the existing foundation. It can be necessary:

- Due to settlement of the existing foundation, causing structural damage to the building
- To enable vertical extension below the building for new basement storey(s)
- To enable the level of adjacent ground to be lowered, e.g. to construct an adjacent new building with foundations deeper than the existing building's foundations
- To increase the load-bearing capacity of the existing foundation, e.g. required for increased floor loadings or to construct additional storeys
- To prevent settlement, where adjacent proposed developments have potential to undermine the strata below existing foundations.

Methods

Mass concrete underpinning is one of the most common methods. It is conducted in successive sections. This leaves the greater proportion of the existing foundation fully supported throughout the operations. The maximum length of each section depends on the stability of the existing structure, its loadings, and the subsoil conditions. Each section is generally between 0.9 and 1.5 m in length.

Refer to Figure 5.2 which gives example of sequenced sections of mass concrete underpinning.

Beam and pier underpinning comprises a reinforced concrete beam, inserted either directly above or below the existing foundation, supported by mass concrete piers constructed at 2.5–3 m centres.

Pile and needle underpinning comprises reinforced concrete needles inserted horizontally through the existing wall, above foundation level, and supported at each end by small-diameter piles that transmit the building's loads to a

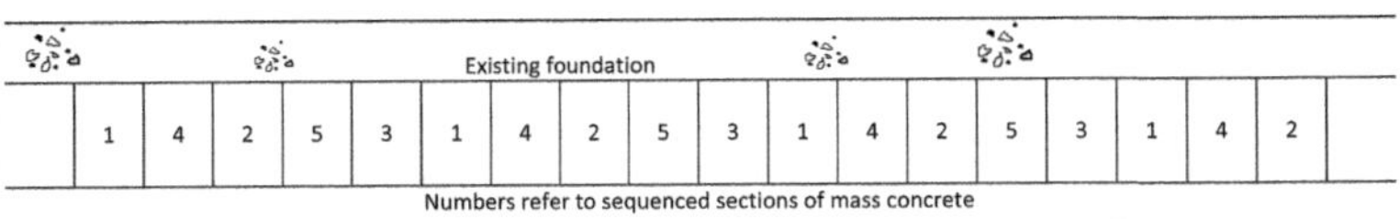

Figure 5.2 Example of sequenced sections of mass concrete underpinning (adapted from Highfield & Gorse, (2009) "Refurbishment and Upgrading of Buildings").

deeper, firmer stratum. The needles are inserted at approximately 1.5 m centres along the length of the wall being underpinned. The piles are taken down to a suitable load-bearing strata.

Cantilever ring beam underpinning comprises horizontal steel I-section cantilever needles inserted into the wall and supported on a reinforced concrete ring beam and mini-piles. The needles transmit the loads from the wall to the beam and thence piles to a deeper bearing stratum.

Double angle mini-pile underpinning involves the installation of small-diameter piles in pairs formed at an angle through the existing foundation at between 1.0 and 1.5 m intervals.

UPGRADING ACOUSTIC PERFORMANCE

Unwanted and intrusive sound is a common environmental problem and can lead to health and well-being issues. This can be improved by attention to acoustic design and construction of buildings and can be improved in existing buildings when refurbishment and alteration work are carried out. Building regulation and standards often apply statutory requirements for new build. In England and Wales, this is held in Part E of the Building Regulations. For properties that are not new build or, for example, undergoing change of use, the statutory requirements can often be limited. Designers and responsible persons for buildings in use therefore may have to consider measures which are not mandatory for the well-being of occupants. Refer to Table 5.8 for places which need good acoustic separation.

It is therefore essential that refurbishment/conversion schemes consider incorporating acoustic upgrading of walls and floors to ensure satisfactory environmental conditions for their occupants.

Table 5.8 Places which need good acoustic separation

Examples of places requiring good acoustic separation (but not necessarily mandatory)
Dwelling-houses, flats
Rooms for residential purposes
Hotels, hostels, boarding houses
Schools/university teaching spaces
Separation between workshops/factories and offices
Separation between medical consulting rooms and other areas
Separation between rooms for confidential use (e.g. solicitor office) and other areas

Airborne sound, e.g. from speech, musical instruments, and audio speakers, creates vibrations in the air which move out and subsequently cause vibrations in the enclosing walls and floors (elements). These vibrations spreading throughout the elements and into connecting elements cause air particles on the other side to vibrate. These new airborne vibrations can be heard as airborne sound.

Impact sound, e.g. from footsteps, knocking, and machinery vibrations, creates vibrations directly in the element they strike. These vibrations spreading throughout the elements and into connecting elements cause air particles on the other side to vibrate. These new vibrations can be heard as impact and airborne sound.

Reduction of impact and airborne sound transmission

- Sound-proofing, like water-proofing, is only as good as the weakest point.
- Resilient materials (e.g. carpeting, cushioning material) can reduce impact sound.
- Breaks in the structure reduce structure-borne sound. The sound energy must change back to airborne sound to pass through structural breaks; this uses energy and reduces the sound intensity.
- High-density materials (e.g. dense blockwork, concrete) can absorb and reduce sound energy.
- Low-density material (e.g. mineral wool with minimum density 10 kg/m^3) can absorb sound.
- Sound energy is measured in decibels (dB). This uses a logarithmic scale so, for example, a 10-dB increase equates to approximately doubling of the perceived noise. So the lower the decibel figure, or the bigger the reduction in decibels, the better.
- A normal decibel reading in a normal domestic situation may be about 30–40 dB.
- Guidance can be found in England and Wales' Approved Document E, even for works which do not need compliance.

Further advice:

- Avoid chasing walls for cables and sockets to be recessed; this creates weak spots in the wall
- If necessary to chase, do not place sockets back-to-back or create voids or holes in the walls
- Tape and seal all gaps, cracks, or holes in the masonry
- Do not fill the wall cavity with concrete, mortar, or other masonry materials (likely to create a bridge for the passage of sound); fill with resilient material

- Dense linings, such as dense acoustic plasterboard or dense mineral wool absorb sound energy
- Close cavities with flexible cavity closers to prevent sound travelling along the cavity.

Improving the acoustic performance of floors

In the methods described below, a perimeter gap may be required around all edges of new flooring, and a gap left between the skirting and the floor to prevent sound transmission into the walls. The ceiling perimeter must also be sealed along with all other airborne sound paths. Any incompatibility between this requirement and the floor's performance for fire protection and thermal insulation must be resolved.

Floating platform floors, of a new, dense, floor surface on a layer of resilient mineral wool on top of existing floors, are one of the most effective ways of upgrading the sound insulation of a separating floor. However, consider the increase in loading and the raising of the existing floor level affecting door thresholds, skirtings, services, sanitary fittings, etc.

Proprietary laminated acoustic flooring systems are similarly designed to be placed on top of the existing floor but can again raise the floor level by up to 50 mm, thus needing adjustments to doors, skirtings, services, etc.

Floating floor on resilient strips with heavy pugging includes inserting a layer of heavy "pugging", sound-deadening material (e.g. dry sand, limestone chips, or thin aggregate). If the ceiling is not capable of supporting the pugging, it should be laid onto plywood pugging boards supported by timber battens fixed to the sides of the joists. Replacement of the existing floor-boarding can be done using tongued and grooved floor-boarding floated on resilient (impact sound-reducing) strips laid along the top edges of the existing floor joists.

Proprietary resilient flooring systems usually involve filling the voids between joists with sound-absorbent material, replacing floor boards/surfaces with proprietary boarding on resilient strips, and replacing ceilings with proprietary ceiling boards. These systems are usually lighter and have less impact on room heights than other methods but involve replacement of floor and ceiling coverings, which may not be acceptable in historic properties.

Independent ceilings can be an alternative means of sound-proofing without affecting the existing construction. A new, independent ceiling is added beneath the existing ceiling, carried on its own set of joists, and spaced as far below the existing ceiling as possible, with an acoustically absorbent material between the new and existing ceilings. This method may not be feasible or acceptable if the existing headroom in the room below is limited, or if the existing ceiling has an ornate finish, or is an important feature in a listed building.

Improving the acoustic performance of walls

The most effective means of upgrading the acoustic performance of an external wall is to improve the windows. External noise usually enters a building via the windows, since single glazing is a very poor sound reducer and relatively insignificant gaps around ill-fitting casements adversely affect their acoustic performance. Improving the airtightness of window frames and increasing the thermal performance of the window by double-glazing or secondary glazing also increase the acoustic performance. All clear air-paths between casements and frames, and frames and reveals, should be properly sealed to prevent leakage of airborne sound. Flexible sealing gaskets can be incorporated between the casements and frames, and gaps between the new secondary frames and the existing window reveals sealed with acoustic sealant.

UPGRADING FIRE PROTECTION

Melanie Smith

Introduction

Any larger refurbishment project is likely to involve the necessity to review and possibly improve the fire safety provisions. Reference to Chapter 7, "Fire safety", will probably be useful, as well as the subsections relating to fire in Chapter 3 "Legal and regulatory frameworks". This section, however, aims to provide thoughts for consideration during a refurbishment project's initial design.

Walls

If the walls are masonry and plastered, they are probably providing sufficient fire protection because even 75 mm thick brick or block walls can provide one hour's fire resistance. The challenge will be to find where there is no masonry inside the walls – i.e. holes and gaps in the structure, and at the edges between walls/ceilings, wall/doors, walls/windows, walls/internal walls, walls/floors, etc.

For stud, lath and plaster, wattle and daub, and similar partitions, some exploration of the thickness of the layers will be required. The partition may provide 30 minutes fire resistance from both sides in terms of stability, but its resistance for insulation or integrity may require improvement. A simple way, provided this is acceptable, is to nail 9 mm plasterboard to each side, seal the joints, and skim over. It may not be acceptable from a historical or structural viewpoint.

Doors

If the door is not of historic importance, replacing the whole door set (door and frame) with a new one of the required fire resistance (integrity) will be the easiest option and possibly least expensive. In some instances, it has been acceptable for a historic door set to be replaced with new, provided that the original door is properly conserved, in a recorded suitable place, and labelled.

It is the door set (i.e. door and frame), not just the door, which is required to be fire-resisting. The door on its own will not provide integrity if fire smoke and products of combustion can easily bypass the door through the gaps between the door and frame and/or through gaps behind the door frame architraves. Improving the protection of the gap between the door and its frame with the proper use of intumescent smoke seals will provide a much better performing door set. Few door sets can be considered a fire door without this. Consideration may be required for any key holes, letterboxes, cat flaps, or gaps in the timber due to moved/damaged doorknobs, etc.

If the door and its frame cannot be replaced, then it may require upgrading. There are various materials, including intumescent papers, intumescent paints/varnishes, but these have limitations. Most modern "normal" internal doors cannot be upgraded to achieve 30 minutes' fire resistance because they have a hollow core, are very thin, or are lightweight and perhaps warped. Existing timber doors can only be improved to a maximum of 30 minutes. The following offer some observations.

- Covering only a single face of a door leaf with sheet fire-resisting material puts an uneven load on the door, causing it to distort out of its frame, affecting its potential fire resistance.
- Upgrading a door by applying an intumescent coating must be done by precisely following the manufacturer's instructions for applying the coating. Such coatings need careful maintenance following application.
- For most door types and construction, conversion to a fire door cannot be achieved using fire-protection paint or varnish. Advice should be sought. Intumescent coatings are generally designed to control surface spread of flame and fire propagation, and not to help upgrade internal doors to fire doors.
- Maintenance and redecoration after treatment may not be possible without negating the fire protection.
- Agrément or similar testing records and certificates should be studied for how, exactly, the door in its treated form was tested. It has been known that the test circumstances, and treatment method for the test, are nothing like how a manufacturer suggests the door is treated in its advertising material.
- Not all doors can be upgraded. Possible upgrading depends on the strength and construction of the door. The following doors are unsuitable for upgrading:
  - Unframed, hollow core, flush type
  - Ledged and braced type

- The door should be a good fit in its frame (gaps not exceeding 3 mm), and an intumescent seal should be fitted (with a flexible cold smoke seal if smoke control is required).
- Any architraves should be removed to check that there are no gaps between the door frame and the surrounding structure. Any gaps should be filled with fire-resisting material.
- Hardware used on fire doors should have a melting point in excess of 800°C. Letterboxes should be of a proprietary fire-resisting type.

Refer to Table 5.9 for methods of upgrading fire doors.

Floors

One of the most common elements to be improved is existing floors. Existing buildings can have timber floors, rather than cementitious. Whereas for a refurbishment, particularly a change of use, the floors may need to be compartment floors and/or to offer 60 or 90 minutes' fire resistance in terms of structure, integrity, and insulation. If the floors can be replaced, then that is merely a structural challenge. If the building is a heritage or listed building, floor replacement might not be permitted.

In such a project, the designer must consider whether it would be better, for protecting the heritage of the building, to go into the floor's structure from above or from below. The challenge might well be decorative ornate plaster ceilings and covings, versus the original oak tongued and grooved exposed floorboards. Which would it be better to put at risk of destruction?

A similar decision would need to be made for exposed floor joists. A simple decision to underdraw the floor joists with two layers of plasterboard (with staggered and intumescent sealed joints), fastened to the bottom of the joists, and packing the floor void with chicken-wire hung mineral wool, might not be acceptable if the exposed joists were an important or aesthetically pleasing feature. Setting the plasterboard and mineral wool sandwiches between the joists and, using the acceptability of the sacrificial timber of the oversized (to current standards) oak joists, might be an acceptable solution or might make the reduced extent of the remaining exposed joist look awkward, and therefore be unacceptable.

Placing the additional protection on the upper surface of the floor may work, or the raised floor might make room heights too low, door opening heights too low, or compromise movement around the building with effects on stairs, lifts, etc.

See also Cooke (2003) and Table 5.10 for methods of upgrading timber floors,

Table 5.9 Methods of upgrading fire doors

 Fire-resisting sheet/board	
	Over-panel both faces of door with fire-resisting boarding. This can be done in ways that are acceptably "reversible", i.e. the original door is not unacceptably damaged. The finished door does not have the aesthetics of the original. **Fire-resisting glass** of proven integrity (e.g. Pyran) can be used as an alternative to boarding. The glass will have little effect on the insulation performance of the door, but this may be unimportant. The fixing needs special consideration to allow for different thermal movement of timber and glass.
	Half and half with a panelled door, by retaining the (often thick, solid, oak) door stiles and rails, and replacing the (often, thin, split, pine) panels with fire-resisting boarding (covering panel with veneer and intumescent paste used in the joints). The new boarded panels can be much thicker than the original, which may not be acceptable aesthetically and for conservation purposes.
	Split in half, to produce two thin separate faces, sandwich these back together with fire-resisting board between the two faces. The door has to be thick enough to be able to do this. Panelled doors are unlikely to be suitable for this method.
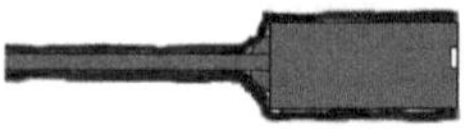	Intumescent paper. This has to be applied in thick layers, which can look rather clumsy when applied to the door. Sharp edges are rounded; the mouldings are reduced in clarity.
	Intumescent paint or varnish can improve fire resistance (and flame spread). The colour of the finished paint or varnish may not be acceptable. The intumescent must be compatible with the door's existing coating to perform adequately in a fire situation. It may need the existing coating to be removed for the coating; this might not be acceptable. Fire safety and maintenance management are important – **Notice** may be required that the intumescent coating should not be removed or improperly cleaned. Possibly only treat thin portions or minor parts of the door.

Table 5.10 Methods of upgrading timber floors

Methods of upgrading timber floors (not to scale and illustrative only)	
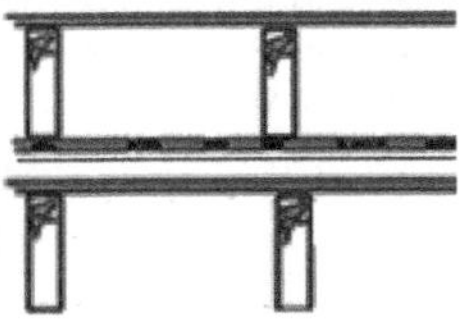	Two typical timber floors, not to scale. • Floor boards • Floor void • (With or without) Lath and plaster ceiling
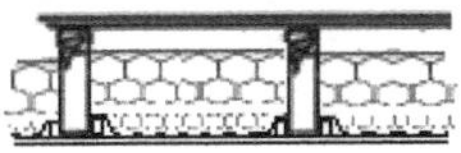	**Exposed floor joists**. Simple solution: underdraw the floor joists with two layers of plasterboard (with staggered and intumescent sealed joints), fastened to the bottom of the joists and pack the floor void with chicken-wire hung mineral wool. May not be acceptable if the exposed joists are an important or aesthetically pleasing feature.
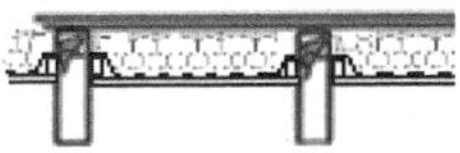	**Exposed joists**. Setting the plasterboard and mineral wool sandwiched between the joists, and using the acceptability of sacrificial timber of the oversized (to current standards) oak joists. May make the reduced extent of the remaining exposed joists look awkward, and therefore may be unacceptable.
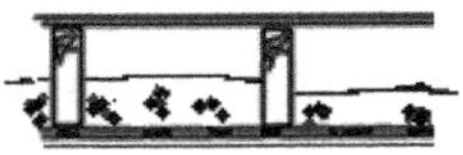	**Adding plaster from the top**. Some historic floors are found to have already been treated this way, either at the time of construction or after. Also useful when floorboards are in poor condition and/or where the historic moulded ceiling in poor condition, e.g. laths fixed with corroded nails. Work from upper face, remove floorboards, fix dpm and expanded metal to sides of joists, pour in lightweight plaster, replace floorboards. **Only possible** where floor and ceiling can carry additional weight. Whilst wet the plaster will be considerably heavier than when eventually dried out.
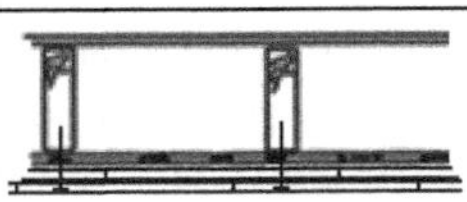	**Where the ceiling is intact but unimportant**. The ceiling can be left in place and underdrawn with one or two layers of fire-resistant boarding with staggered joints and joints taped, filled with intumescent paste, etc. Fixings are important.

(continued)

Table 5.10 *(continued)*

Methods of upgrading timber floors (not to scale and illustrative only)	
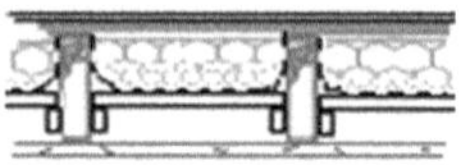	**Decorative ceiling**. Floorboards carefully lifted. Noggins carefully fixed to joists above ceiling. Fire-resisting boards laid onto noggins. Void filled with mineral wool. Floor could be overboarded for further integrity. All dependent on: • Sensitivity of original decorative ceiling, etc., to vibration • Condition of joists to carry extra weight • Importance of floorboards

Case study: University of Leicester Engineering Building, UK

This case study was originally published by *Context*, the magazine of the Institute of Historic Building Conservation (Smith, 1999), author Melanie Smith FRICS GIFireE. See also the University of Leicester (n.d.).

Current aspects required of the built environment such as accessibility, thermal comfort, reduced energy use, fire safety, and their associated regulation can be at odds with the construction, layouts, and facilities of existing, historic, and listed buildings. The demands for current standards can be in direct conflict with the conservation conditions or character. A collaborative, creative, holistic, and innovative approach is needed by the designers of interventions and the appropriate authorities, to reconcile vital discrepancies.

One of the problems with listed buildings is how to upgrade them to meet modern safety standards without compromising their architectural qualities. The case study presented involved the collaboration of a Building Surveyor as overall designer and fire engineer, the Fire Authority, the Planning Department Listed Building Section, Historic England (then English Heritage), a Building Services Engineer for design of the air-handling systems and services engineering works, the University, and excellent contractors.

The case study is about the retrofit fire safety works incorporated into the internationally renowned Sir James Stirling Building at the University of Leicester. The home of the School of Engineering is a structure designed by Stirling and James Gowan, built in 1963. A seminal building, unlike any post-war architecture seen before, it broke the hold of Le Corbusier upon British architects. It was the first high-rise structure in the region, and its striking red tiled and glazed 11 storey Tower with the diamond-shaped north-light roofs on the workshops and southern laboratory block was the inspiration for similar buildings around the world. Its lecture halls projecting out at the lower levels of the Tower are seemingly supported on impossibly slender columns. Internally, the red tile and glass features continue with open atria providing a sensation of colour and space. The building is listed Grade II*.

Despite its architectural merits, the building had not been without its problems. The curtain wall glazing on the Tower was replaced to improve conditions for the occupants, and the glass workshop roof, affected by leaks, has been replaced. There were no accessible facilities, and the School found itself periodically severely spatially restricted within its walls. This was the case when they decided to increase floor space by inserting two new floors over the boilers in South Block, where the boiler rooms had previously extended to the third floor.

On discussing these proposed works with the Fire Authority and querying whether the existing means of escape would be adequate with the increased number of occupants, the fire safety of the whole building was called into question. There were two main problems: the existing stairs in South Block were not adequate for the proposed new floors, but more significantly, the unprotected single stair in the Tower serving ten floors needed upgrading work as a matter of urgency to bring the building in line with current minimum safety standards. The occupants of the upper floors were at severe risk if there were to be a fire.

The Fire Authority indicated to the University what would be needed in relation to protected lobbies and new fire doors. Because of the listing, the University contacted the local authority Planning Department, and as it is Grade II*, English Heritage would also have to approve any proposed alterations. Enclosure of the Tower stairway, as required by the Fire Authority, was not acceptable to English Heritage because it would so drastically change the essential aspect of this important building.

Faced with the possible closure of the building, the University appointed a Building Surveyor and Fire Specialist as Lead Designer to find a solution acceptable to the Planners, Conservation Officers, Building Control Officers, English Heritage, the Fire Authority, and of course the University and the School of Engineering.

The first challenge for the Lead Designer was to decide what to comply with. The guidance at the time was more limited than currently available, but those documents available were all reviewed, discrepancies between them noted, pragmatic decisions made about best-fit, and comparables were used where guidance was not necessarily directly written for this circumstance. Where possible, the "best practice" method was used. Where possible, compliance with UK Approved Document B was proposed for alignment with fire safety regulation in place for new buildings. However, in a (then) 35-year-old building, compliance with a document intended for new design is not easy. In addition, every proposal had to be acceptable to English Heritage.

Compliance is often a serious problem with listed buildings. The existing design is often not conducive to meeting current standards but it is the existing design which is the reason for the listing. Sensitivity, a knowledge of architectural and construction history, and a solid grounding in the principles of fire safety, as well as knowledge of the regulations, are all vital ingredients for a successful solution to these problems.

A fundamental assessment and review of the entire building was undertaken by the Lead Designer. The way the building was used by all occupants, and all safety features incorporated were considered: structural protection, detection and alarm, travel distances, final exits, security, health and safety, protection of travel routes, and so on. For much of the building, simple upgrading methods to improve the passive fire resistance were possible, including determining compartment boundaries, upgrading doors, fire stopping, and intumescent or automatic pipe closers. The Tower, with its single stair and atria, was the greatest challenge. The radical decision was finally taken to insert into the Tower (probably) the biggest retrofit pressurisation system in England.

The building was originally designed for the School of Engineering, and the architects, Stirling and Gowan, ensured that internally it has an essential functional feel, visually and in its spatial organisation, to reflect the engineers using it. Most of the air-handling specialists asked to tender for the design and installation of the pressurisation system were unable to appreciate the absolute necessity of not affecting the look of the external envelope and wanted to insert large fans in the external façade, all the way up the Tower. The Lead Designer (as well as English Heritage) insisted that this was not an option and another way had to be found.

This was an example of consultants who are experts in their field, and highly skilled, but whose responsibilities may not extend to appreciating all the elements involved in sensitive works involving listed buildings. By using the procurement appointment system they did, the University was able to appoint separate specialists knowing that all proposals and work would be assessed for impact on the building's intrinsic merit, etc., as well as for the fire strategy.

The final design was produced for the Tower. Fire safety requirements were met by active measures that would, in the case of a fire, apply positive pressure to the building's central well using fans on the Tower roof and use internal pipes to pump air down large-diameter pipes carried on the internal faces of the stairway enclosure. This was a solution acceptable to the Fire Authorities, English Heritage, the City Conservation Officers, and Building Control. The pipework did not have to be hidden because it enhanced the functional aesthetics of the central shaft.

The next problem was – would it work? The challenge was to design a system that would be capable of pressurising the open stairway with its atria. The building was arranged to be "blown-up" (like a balloon) by the Building Research Establishment to check the air-leakage calculations. Rearrangement of the ground floor WC accommodation had to be made to house one pressurisation plant room, incorporating a fan and uninterruptible power supply. The opportunity was taken during this rearrangement to incorporate accessible sanitary accommodation previously not incorporated into the building.

A second plant room went in at the top of the Tower stairway. A new floor was constructed in the open space under the roof to house the second fan, and two apertures had to be cut in the roof of the Tower for air inlet.

The essence of pressurisation is that (in this case), in the event of a fire, the stairway and atria are pumped with air to put them at a positive pressure in relation to the adjoining accommodation. Smoke and toxic gases from a fire in the adjoining rooms will therefore be unlikely to find their way into the escape route. Occupants of the building should then be able to make their way down the only route of escape to a place of safety without having their movement inhibited by the presence of smoke and gases. In many cases, it is these toxic gases that harm and kill occupants in a fire rather than heat from the flames.

The pressurisation system was used in place of providing protected lobbies to the stairway. However, where possible, by simply upgrading or replacing existing construction on the upper floors to provide protected lobbies, this was done. It is vital to use passive protection wherever possible and avoid reliance solely on active measures.

In the South Block, more traditional methods of fire upgrade were utilised to provide protected stairways and escape routes in accordance with Approved Document B: Fire Safety of the Building Regulations.

One of the "turning points" in English architecture identified by the Council's Conservation Officers was the use of brick over-cladding to conceal external doors to a store. Two new fire exit doors were necessary to the single-storey workshops. These had to be placed along an originally unbroken single-storey brick elevation facing across the adjacent park. English Heritage and the Listed Building Planners agreed that the brick-clad door design could be copied along this wall. Thus, the "quirkiness" of this architecture was celebrated.

A new automatic fire detection and alarm system was installed throughout the whole building with detectors suitable for the different types of engineering research being undertaken and the different materials being used in the different areas of the building, especially the workshops. The system was an L2 system to BS5839 and designed to the performance standard required for new build. The AFD system also had to contend with planning restrictions. It was intended that orange pyro cable be surface mounted onto the red tiles of the Tower. The red and orange would clash unacceptably and so the plastic coating was stripped off, leaving the copper casing exposed. The copper sat acceptably with the red tiles and met the approval of both fire and conservation authorities.

The aluminium and glass doors off the Tower stairway at the upper levels were an essential part of the design but offered very little fire resistance. Steel-framed fire doors replicating the original aluminium were installed in both the main Tower and South Block.

The project is an example of all the authorities and consultants working together in conjunction with the University Estates and Buildings Office and the School of Engineering to make a unique building as safe as possible whilst not destroying its essential features. A spirit of cooperation existed between all the parties, ensuring success.

FASHION

For a building to survive, it needs to overcome obsolescence. It will not be feasible to retrofit or refurbish a building that does not lend itself to adaption for current requirements. The current and the old need to be able to co-exist. Fashion can be described as a custom, look, or way of doing things that is considered acceptable, particularly that which prevails at that time among those whose lead is accepted. The challenges associated with this include fleeting fashions, established views, views of the establishments (e.g. societies of architects, planners, government ministers, customers/clients, global trends/issues, suppliers, accepted science/knowledge transfer), and those influenced by these.

Fashion has been around in buildings since medieval times – inserting bay and then bow windows in existing houses, recladding timber-framed Elizabethan with brick in Georgian times, etc. Obsolescence, redundancy, retrofitting, and refurbishment result from rising living standards, changing social aims, and technological advancement. As a society gets richer, its views on what constitutes tolerable living and working conditions change radically. Sanitation, light, and ventilation are typically currently insisted on. Reduction in energy use is a relatively new requirement, and its lasting fashion relevance will depend on society, security, adverse effects of climate change, and sufficient wealth to make improvements that will result in its reduction.

Preservation, which is the static maintenance of an object in an unaltered condition, can only be of limited usefulness in a building. Financial viability is key. Conversely, conservation consists of cherishing the building so that it does not merely continue but also receives more abundant life from the changes and fresh associations that occur (Harvey, 1972). Successful conservation regards preservation as an essential element – but only one element. The way to conserve old buildings is to make them a living, useful, enjoyable part of the environment. To do this at any point in time involves the consideration of fashion, including the making of a new fashion or revival of an older one, but rarely harking back to a recently declining or discarded fashion. Notwithstanding, of course, the fashion to preserve as much as possible that which is worth preserving from the past, e.g. as shown in the listing requirements of listed, etc., buildings.

Finishes are an essential tool in a building achieving fashionableness, whatever its age or type. For example, current house fashions for refurbishment projects include:

- External render painted in white or other light, bright, fresh colour
- Maximising open space
- Stone, timber, seamless, or laminate floor finishes
- New windows in styles that complement the age, style, and proportions of the original building

- Replacing conventional windows with bi-fold doors or full-height fenestration
- Integrated open kitchen, living, and dining areas
- Large glazed façade giving access to rear gardens
- Spacious bathrooms
- Uncomplicated circulation routes
- Limited use of corridors and passageways
- Whatever the size of the property, increasing room area, light, and ventilation

The drive for opening up spaces is due to the lack of space, especially in UK housing, which is the smallest of all western developed nations, according to the interesting material in http://shrinkthatfootprint.com/how-big-is-a-house. In the UK, the "Technical housing standards" set out the UK government's nationally described space standard.

Current office fit-outs are similarly showing clean lines, mainly white or light colours, with contrasting steel, glass, and/or pale timber furniture incorporating bright accent colours. Open plan, break-out spaces, and relaxation spaces with an eye to well-being are all also currently fashionable. Corridors and passageways have always arguably been a waste of space, but it has been fashionable to have "cellular" individual rooms, especially in offices where there may have been a hierarchy of seniority levels or "pay-grades". This has been out of fashion, and the more open and "all-together" we can be, the better. This may revert or change again in the future, of course. The experience of covid may result in different views.

Opening up spaces, removing walls and communication areas, increases the efficiency of a layout, giving a feeling of space and the chance to use the space for dual purposes. In the case of new build, it means the square footage can be driven down and unit numbers increased, whilst still selling for the same price. In the case of refurbishments (barring extensions), an open plan enables a chance to be more efficient with the profitable space.

Sustainable design is increasingly desirable, using more recyclable and sustainable materials. "Sustainable" is a jaded term, a buzz-word that very few people fully understand, often used where not relevant or apt, for example, using imported stone or slate from the other side of the world, rather than that locally available albeit at a much greater cost. What appears to have been a success is the removal of hardwoods from materials' menus, e.g. replacing mahogany and other tropical hardwoods with oak and other temperate hardwoods, although de-forestation is still a major problem.

The use of outdoor space, for working as well as relaxing, shows increases in productivity, as does biophilia. Biophilia, internally and externally, has become an accepted, if still exotic, part of the building palette. It has proven itself to be good for the environment and for the well-being of occupants and does not need to be expensive to maintain (green roofs), although it is always quite expensive to install. It needs to be used where it can be enjoyed by as many people as possible, i.e. in public or communal spaces rather than private.

Of predominate importance is IT, cloud, and electronic communication provision, which must be easily adapted and upgraded due to the speed of technology advancement. Smart, remote, and flexible working, in the home as well as in traditional workplaces, is arguably the norm and of increasing importance. Video call from home became the norm for businesses during the 2020 covid pandemic. In the home, smart speakers can integrate with and control several smart devices, acting as a home automation hub, including thermostats, lights, door locks, security and media. Smart commercial buildings with sensors and integrated technologies help with energy management, comfort, security, and space-use optimisation which can be reflected in productivity. Residential and commercial buildings that are completely run by computer installations may become the way that things are done, simply because that is the only way forward, rather than a choice. However technology is in its infancy for automation, with competing standards and incompatibility of those standards. The good automation packages may be currently too expensive or too complicated for most projects.

It is not all about looks, however, e.g. the fashion for hard floor finishes is allied to underfloor heating and the fact that it works better with stone or tiled finishes rather than wood or carpet. It is also about cost, e.g. stripped back, rough, and elemental finishes, with little effort at dressing up surfaces or concealing services, and clashing with highlights of very well-finished elements such as doors, windows and kitchen installations. These can currently achieve an "up to the minute" look without the price tag. Although technology is driving finishes in flooring, it can be expensive to retrofit these into an existing structure; the depth of floor build-up can be a barrier. Extensions do not suffer this problem as much. Conversely for fenestration, improvements in glazing allow for bigger windows without as much of a penalty in energy bills.

An interesting point for the future of refurbishments are modern methods of construction (MMC). Many house builders are keen for factory-made off-site houses to become mainstream. These have benefits for many reasons; however future retrofit or change will be challenging due to the very nature of their design and their method of construction (e.g. light gauge metal frame). Structurally, they are not as robust as traditional or timber-frame construction because safe tolerances have been reduced so much by design. The strength of these MMC structures comes from the whole frame rather than individual parts, meaning that increasing opening sizes, e.g. for windows or partitions, is not easily achievable because removing one part can make the whole structure too weak. Even minor amendments such as extra lighting, sockets, or altered plumbing can be problematic The risk of rusting in light gauge steel MMC structures has been discussed in Chapter 4, "Building pathology", and this might complicate any desires for refurbishing such structures. It may be that construction of this type will not be suitable for refurbishment after time, perhaps having to be replaced in 40 to 60 years' time and surveyors need to be aware of this. This has happened before when housing types (e.g. Airey Houses) constructed in the UK from the mid-1940s onwards, to meet the housing needs due to the war, needed to be amended or replaced in the 1970s and 1980s.

Attention should be paid to the acoustics of any working environment, although acoustic performance in an open office environment does not appear to be as important as it used to be. With open office spaces, acoustic treatments to cut reverberation have improved over time, with completely open spaces sounding as "dead" as small rooms. Without this, in a room full of hard surfaces and low ceilings, noise may be an issue, but this is not raising many concerns except in private houses (floors and bedrooms) and between adjacent residential flats. People are rightly intolerant of intrusive noise, and building regulation usually reflects this.

The current fashion for refurbishment methodologies in the UK, prevailing since the 1970s, usually does not include "restoring" the building back to some mythical past. Ruskin, Morris, and others complained that Victorian Restoration swept away all signs of the building's own story, making it a pastiche of itself. Another side to this is replacing parts of an unlisted building because those parts are old, rather than because they are failing. The old adage of "if it ain't broke, don't fix it" comes into play here – but of course factors such as safety, energy use, and indeed fashion are intrinsic parts of whether or not it is "broke".

REFERENCES

Baker, P. (2008). Technical Paper 1: Thermal performance of traditional windows. Technical Report. Historic Scotland, Technical Conservation Group

Clinch, J. P. & Healy, J. D. (2000). Domestic energy efficiency in Ireland: correcting market failure. *Energy Policy*, 28, 1–8.

Cooke, G. (2003). Upgrading the fire resistance of floors and doors in heritage buildings. International Symposium on Protection of Cultural Heritage Buildings from Fire, Kyoto, April 6–7, 2003.

CPA (2010). *An introduction to low carbon domestic refurbishment*. London, UK: Construction Products Association.

Curtin, J. (2009). *Jobs, growth and reduced energy costs: greenprint for a national energy efficiency retrofit programme*. Dublin, Ireland: Institute of International and European Affairs.

Curtin, J. (2013). *Unlocking investment in home retrofit*. Ireland: Passive House+.

Doran, S., Zapata, G., Tweed, C., Suffolk, C., Forman, T. & Gemmell, A. (2014). *Solid wall heat losses and the potential for energy saving: literature review*. Watford, UK: BRE.

EST (2006). *Practical refurbishment of solid-walled houses*. London, UK: Energy Saving Trust.

Highfield, D. & Gorse, C. (2009). *Refurbishment and upgrading of buildings* (2nd ed.). Spon Press. https://doi.org/10.4324/9780203879160

Greening, L., Greene, D. L. & Difiglio, C. (2000). Energy efficiency and consumption—the rebound effect—a survey. *Energy Policy*, 28, 389–401.

Greenspec (2018). *Insulation materials and their thermal properties* [Online]. UK: Greenspec. Available: http://www.greenspec.co.uk/building-design/insulation-materials-thermal-properties

Harvey, J. (1972). *Conservation of buildings*. London, UK: Mackay Ltd.

Historic England (2016). *Energy efficiency in historic buildings: insulating solid walls*. v1.1 ed. UK: Historic England.

Historic England (2017). *Energy efficiency and historic buildings: application of part l of the building regulations to historic and traditionally constructed buildings*. v1.2 ed. UK: Historic England.

Immendoerfer, A., Houh, C., Andrews, A. & Mathias, A. (2008). *Fit for the future: the green homes retrofit manual—technical supplement*. London, UK: Housing Corporation.

Killip, G. (2008). *Transforming the UK's existing housing stock: a report for the federation of master builders*. London, UK: Environmental Change Institute, University of Oxford.

Kingspan (2013). *Breathability—a white paper: a study into the impact of breathability on condensation, mould growth, dust mite populations and health*. 3 ed. UK: Kingspan Insulation Ltd.

Little, J. (2010) Breaking the Mould 5: Comparative simulation of internal insulation systems. *Construct Ireland Magazine, Issue 12 volume 4, April 2010.*

Lowery, D. (2012). Evaluation of a social housing retrofit and its impact on tenant energy use behaviour. Doctor of Philosophy Doctoral thesis, Northumbria University.

May, N. (2005). *Breathability: the key to building performance*. UK: Sustainable Traditional Buildings Alliance (STBA).

SEAI (2019a). *A homeowner's guide to wall insulation*. The sustainable energy authority of Ireland. Dublin, Ireland: SEAI.

SEAI (2019b). Energy savings credits. The sustainable energy authority of Ireland. Dublin, Ireland: SEAI. Available from: https://www.seai.ie/publications/Energy_Saving_Credits_Table.pdf

Smith, M. (1999). The Engineering Building, University of Leicester fire safety works. Context *62*. June 1999. Available: https://www.researchgate.net/publication/299389357_The_Engineering_Building_University_of_Leicester_fire_safety_works; https://le.ac.uk/engineering/about/building/articles

Stirling, C. (2001). *BRE good building guides and good repair guides: a library of information for all construction professionals*. Watford, UK: BRE.

University of Leicester (n.d.). School of Engineering Webpage. Available: https://le.ac.uk/engineering/about/building/articles

UWE (2008). *House ages: specification by period*. 2008 ed. Bristol, UK: University of the West of England, Bristol, UK.

6

Thermal performance

Chris Gorse, Melanie Smith, Matthew Brooke-Peat, Martin Fletcher, Cormac Flood, Lloyd Scott, and John Spillane

INTRODUCTION TO THERMAL PERFORMANCE

Chris Gorse and Melanie Smith

Thermal performance currently presents one of the major global energy challenges. The practice and profession of building surveying holds a respected position in construction, but it is not clear how successful general surveying practice is when used to assess issues and conditions associated with thermal performance. This chapter sets out to improve knowledge transfer in the understanding of building thermal performance. There is inevitably some crossover between this and other chapters in this book.

Performance assessment includes testing and measurement, often using new technology and equipment. This type of survey is a relatively new concept as shown in Chapter 1. It is also a practice that requires considerable skill and interpretation. Building Performance Evaluation methods (Figure 6.1) tend to be self-regulated by adherence to the methodologies stipulated in industry-accepted standards.

A certain level of temperature control is required inside buildings. Energy is used to provide and maintain this level in the conditioned space. This heat energy is gradually transferred and lost from the building, through conduction (through the building fabric), radiation (to the sky), and convection, associated with exfiltration and infiltration (air moving in and out of the building). The heat lost is greatest at the building's weak points, through thermal bridges and gaps, resulting in uncontrolled energy losses, as shown in Figure 6.2. Measurement of the losses can be translated into an evaluation of the building's thermal performance. Substantial losses can be attributed to thermal bridging and lack of airtightness.

Thermal bridge

Sometimes referred to as a cold bridge, a thermal bridge is an area of a building's fabric which has a higher thermal transmission (i.e. the area is a poorer thermal

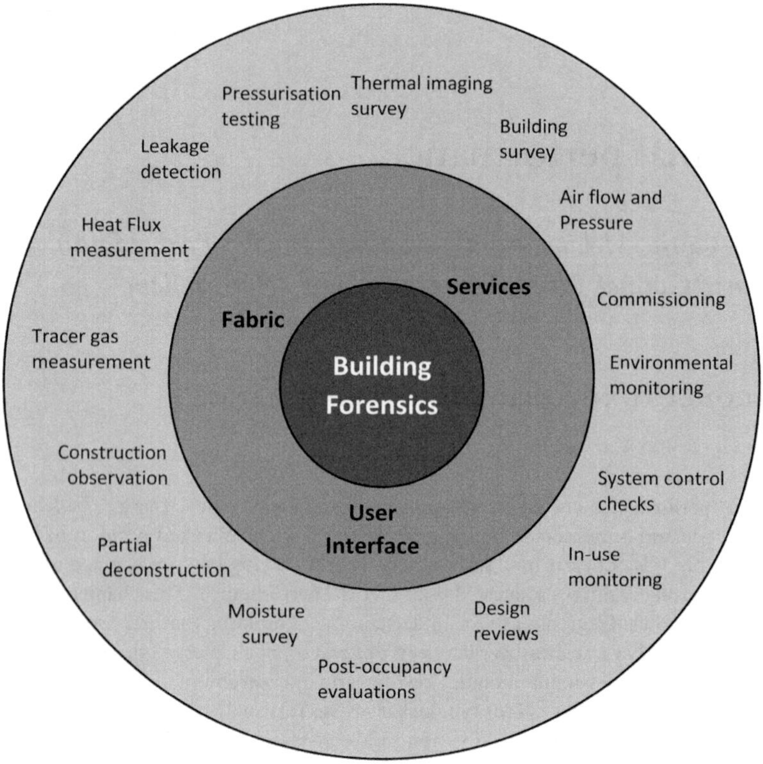

Figure 6.1 Forensic tests used in building pathology and building surveying (source: Leeds Sustainability Institute).

insulator) than the surrounding parts of the fabric. Thermal bridging occurs when materials that have a much higher thermal conductivity than the surrounding material penetrate the thermal envelope, or where there are discontinuities in the thermal envelope. Heat then flows through the path created – the path of least resistance – from the warm space (inside) to the cold space (outside). As heat flows through the fabric, the surfaces of the interior side of the bridge become cooler. The thermal bridge results in a reduction in the overall thermal insulation, and therefore performance, of the structure.

Thermal bridging can have a significant impact on the thermal and energy performance of the building envelope. If the building is relatively poorly insulated,

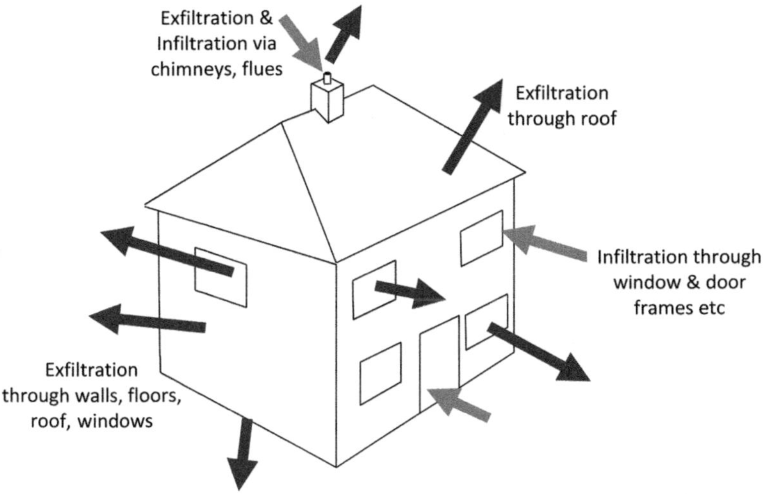

Figure 6.2 Building energy losses through a typical house and through a typical solid wall.

thermal bridging has little influence on its overall thermal performance. However, with improved insulation, the relative importance of thermal bridging increases. In very well insulated dwellings, the effect on the thermal performance of a dwelling can be significant. Even when measures are taken to reduce thermal bridging, remaining thermal bridging in very well insulated dwellings can still account for a significant proportion of the overall fabric conduction heat loss.

Thermal bridging can also result in:

- An increase in fabric conduction heat loss
- An increase in solar heat gains during the summer
- Reduction in internal surface temperatures
- Cold spots and stratification cold occurring within the building
- An increased risk of both surface and interstitial condensation, which may result in pattern staining and mould growth
- Reduction in indoor air quality due to condensation and mould growth
- Damage to building components

Efforts should be made to minimise thermal bridging as much as is possible. The aim should be to design and construct dwellings that are free from obvious and avoidable thermal bridges. However, it is not generally possible to eliminate all thermal bridging.

Thermal bridges occur due to:

- Geometrical effects, such as corners
- Penetrations through the thermal envelope, for instance windows and doors
- Junctions between different elements (wall and floor) and different component (around windows and doors)
- Poor construction practice (gaps or discontinuities in thermal insulation)

The occurrence of thermal bridges can be attributable to:

- Issues associated with the design of the thermal envelope
- Buildability and construction issues
- A combination of both of these

Airtightness

Taken from Low Carbon Housing Learning Zone, Leeds Beckett University

Airtightness is crucial to improving the energy performance of buildings. The converse of airtightness is air leakage, and airtightness is a measure of how minimal is the unwanted air leakage of the building. Desirable, controlled ventilation does not form part of measurement for airtightness. Ventilation is required for occupant and bacterial health, unwanted air leakage though the construction is not.

Uncontrolled Background Ventilation + Purpose Provided Ventilation = Total Ventilation Rate.

Airtightness can be expressed in terms of a whole building leakage rate at an artificially induced pressure, in the UK 50 Pa is used (n50), or in terms of an equivalent leakage area. Traditionally, airtightness was expressed in air changes per hour (with units conventionally used being ach or h^{-1}). Air permeability is more commonly used (with units $m^3/(h{\cdot}m^2)$ representing m^3 of air flow per hour, per m^2 of envelope area) as it takes into consideration the effects of shape and size. The lower the air permeability of a building, the greater the airtightness.

In many countries, the temperature of the outside air is lower than the temperature of the air inside the building; thus any air leakage from the inside to the outside of the building can result in:

- A significant reduction in the thermal resistance of the thermal insulation, due to air leakage past the insulation (thermal bypassing), leading to increases in realised fabric U-values
- An increase in the building's ventilation and fabric heat losses, resulting in an increase in space-heating requirements
- Increased energy costs

Where the fabric performance of dwellings improves, the proportion of total heat losses attributable to ventilation is likely to increase, unless air leakage is addressed. For instance, in reasonably well insulated but relatively leaky dwellings, ventilation heat losses can account for up to one-third of the dwellings' total heat loss.

Air leakage is the uncontrolled exchange of air both into (infiltration) and out of (exfiltration) a building through cracks, gaps, and other unintentional openings in the building envelope. It is driven by the same physical processes that drive natural ventilation:

- Wind effect
- Stack effect

Wind effect is where air moving around and over a building induces pressure differences across the building and generally results in a small negative pressure internally. Outdoor air will enter the building through cracks and gaps in the construction (infiltration) on the windward side of the building that is under positive pressure, and indoor air will exit the building (exfiltration) on the leeward side of the building that is under negative pressure.

Stack effect is where air inside the building is heated by solar gains, heating appliances, people, and equipment. As the internal air increases in temperature, it becomes less dense and rises due to convection. This results in higher pressure air at the top of the building, relative to the external air pressure, leading to exfiltration. As the warm air exits the dwelling, cooler denser air from outside infiltrates the building through cracks and gaps in the construction.

The rate of air leakage is therefore dependent upon the wind speed and direction, the temperature difference between the inside and outside of the building, and the air permeability of the construction.

Ventilation

All buildings require ventilation. This is necessary for:

- Human respiration
- The health and comfort of the occupants
- The control of condensation and humidity
- Fuel-burning appliances
- The dilution and disposal of pollutants, dust mites, etc.

Ventilation is the process of changing air in an enclosed space. In order to maintain optimum air quality in buildings, a proportion of the air contained within any enclosed space should be continuously removed and replaced with fresh outside air. If the incoming "fresh" air is contaminated by pollutants, then measures have to be taken to remove them. Buildings are ventilated by a combination of infiltration and purpose-provided ventilation.

Air infiltration: The uncontrollable air exchange between the inside and outside of a building through a wide range of air leakage paths in the building structure.

Purpose-provided ventilation: The controllable air exchange between the inside and outside of a building by means of a range of natural and/or mechanical devices.

Build tight, ventilate right

Ventilation requirements for UK dwellings are typically:

Recommended: Between 0.5 and 1.0 air changes per hour (ach)
Minimum: Between 0.3 and 0.5 ach

A dwelling cannot be too airtight, but it can be under-ventilated. The more airtight, the more consideration is required for ventilation. For example, reducing the airtightness from 0.5 to 0.4 ach, without increasing the ventilation, can increase the dust mite population by 100 times (Davies et al., 2004).

Further information can be found at:

Low Carbon Housing Learning Zone, Leeds Beckett University (2020). Available at https://virtualsite.leedsbeckett.ac.uk/low:carbon_housing/airtightness/introduction/index.htm

Davies, M., Ucci, M., McCarthy, M., Oreszczyn, T., Ridley, I., Mumovic, D., Singh, J. and Pretlove, S. (2004). A review of evidence linking ventilation rates in dwellings and respiratory health—a focus on house dust mites and mould. *International Journal of Ventilation*, 3(2), pp. 155–168.

THERMAL COMFORT

Martin Fletcher

This section covers:
Thermal comfort
Primary environmental variables
Personal factors
Thermal equilibrium
Thermoregulation
Indices of thermal comfort

Thermal comfort – an introduction

The environmental conditions within a building can have a significant impact on the satisfaction and wellbeing of occupants. One metric by which the quality of the thermal environment may be evaluated is thermal comfort; the primary purpose of space conditioning in buildings being the provision of a thermally comfortable environment. Thermal comfort is defined as "that condition of mind that expresses satisfaction with the thermal environment" (ASHRAE, 2017) and is the subjective response to aggregated environmental and personal factors (Figure 6.3).

Primary environmental variables

The primary environmental variables that affect thermal comfort are air temperature, radiant temperature, humidity, and air velocity.

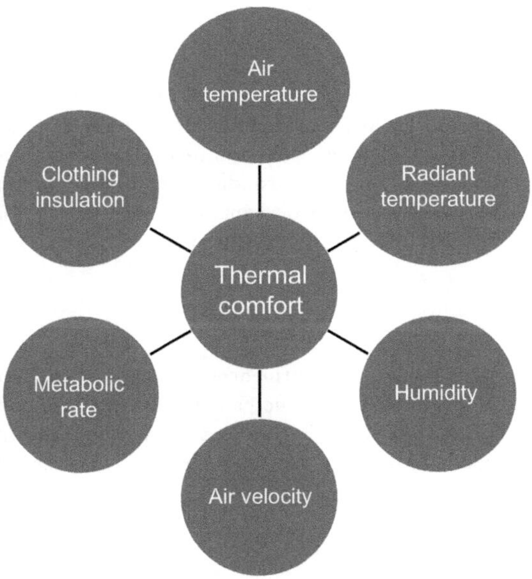

Figure 6.3 Factors affecting thermal comfort.

Air temperature

Air temperature is defined as the temperature of the air surrounding the body. Specifically, this refers to the temperature which governs heat exchange with the human body; air temperature at the body surface or at great distance from the body will not necessarily be representative of that which drives heat flow.

Radiant temperature

Radiant temperature refers to the heat exchange between two bodies via radiation, with net heat flow from a hot to a cold body. When considering thermal comfort, two radiant temperatures are relevant: mean radiant temperature (MRT) and plane radiant temperature (PRT). MRT is the average radiant exchange between an individual and their environment, aggregating the effects of heat gains and losses. PRT, on the other hand, considers radiant exchange in one direction and is important where a large radiant difference causing temperature asymmetry is present, such as proximity to a fireplace.

Humidity

The absolute humidity of the air affects the ability of the human body to control internal temperature via evaporative heat loss, i.e. sweating. In practice, humidity has little effect on thermal comfort in internal environments; however, it is increasingly relevant in warmer climates where air-conditioning is uncommon, or in sport and exercise facilities where sweating is of increased significance for thermoregulation.

Air velocity

The velocity of the air relative to the human body influences the rate at which heat is transferred to the environment. As air movement fluctuates in both direction and speed over time, thermal comfort analysis typically considers the average value in all directions over a given period, e.g. 1 minute. Air velocity is particularly relevant when effects are localised, as is the case with draughts.

Personal factors

In addition to environmental variables, thermal comfort is affected by personal factors of metabolic rate and clothing insulation.

Metabolic rate

The activity of an individual influences the amount of heat produced and released via metabolism. In thermal comfort study, the metabolic rate of an activity is expressed in MET values, where 1 MET is equal to 58.2 W/m^2. As the calculation of an accurate MET value is complex, thermal comfort researchers typically use reference values presented in standards such as BS EN ISO 8996:2004 (BSI, 2005).

Clothing insulation

Clothing provides a layer of insulation, reducing the exchange of heat between the human body and the surrounding environment. In thermal comfort the insulative property of clothing is expressed in Clo units, where 1 Clo is equal to 0.155 $m^2 \cdot {}^\circ C/W$. As the measurement of clothing insulation is complex, field studies of thermal comfort typically use reference values presented in standards such as BS EN ISO 9920:2009 (BSI, 2010) and ANSI/ASHRAE Standard-55 (ASHRAE, 2017).

Thermal equilibrium

The human body seeks to maintain an equilibrium between heat production, heat loss, and heat gain. The factors affecting thermal comfort all contribute to this thermal balance, which is conceptually described by the equation:

$$M - W = E + R + C + K + S$$

Where:

M is the metabolic rate, in watts per square metre (W/m^2)
W is the external mechanical work, in watts per square metre (W/m^2)
E is the evaporative heat transfer, in watts per square metre (W/m^2)
R is the radiative heat transfer, in watts per square metre (W/m^2)
C is the convective heat transfer, in watts per square metre (W/m^2)
K is the conductive heat transfer, in watts per square metre (W/m^2)
S is heat storage, in watts per square metre (W/m^2)

Thermoregulation

When combined, if the mechanisms of heat production and loss are balanced, storage is equal to zero. It follows, therefore, that if there is a net heat gain then storage will be positive and body core temperature will rise. Conversely, if there is a ne heat loss then storage will be negative and body core temperature will fall. Th processes that manage the thermal balance are referred to as **thermoregulation** and are grouped into two types, autonomic and behavioural.

Autonomic

Autonomic responses are involuntary responses to excessive heat or cold, and ar primarily driven by change in core temperature. Autonomic responses include:

- Sweating
- Shivering
- Vasodilation
- Vasoconstriction

Behavioural

Behavioural responses are primarily driven by skin temperature, and result in a individual making a conscious decision to adapt to their thermal state. Behavioura responses include:

- Adding or removing clothes
- Moving away from hot or cold surfaces
- Closing or opening windows
- Wetting the skin
- Seeking shade

Whilst the maintenance of a thermal balance is important from a health perspective, it is important to note that the perception of a thermal stimulus is not simply the result of a deterministic relationship. A warm or cold stimulus may be perceived positively or negatively depending on the thermal and psychological state o the individual at that time, even if the stimulus would be unpleasant in a differen context. For example, a cool breeze on a warm day may be pleasant if the individual is warm, but would be an uncomfortable draught if they were cold. This subjective context is important to remember when evaluating thermal comfort.

Indices of thermal comfort

Several indices are used to objectively evaluate thermal comfort. These indices aim to combine the factors affecting comfort into a single output to enable compariso between buildings and situations, and this is done in two ways. Firstly, rational indices use heat transfer equations and mathematical representations of the thermoregulatory system to predict the response to a given environment. This prediction of thermal state may be used to infer occupant comfort.

The second class of indices are empirically derived, based on observation of human subjects under a comprehensive range of thermal and personal conditions, with correlation between conditions and thermal strain used to develop multiple regression equations. Empirical indices assume that conditions of equal environmental stress result in equal physiological strain, and use this assumption to create models that accommodate all possible combinations of conditions.

Of these two index classes, there are two indices that are most frequently used and have been adopted into international standard. The first is the **predicted mean vote**, developed by Povl Ole Fanger in the 1970s, and the second is **adaptive thermal comfort**, developed to overcome the limitations inherent in rational indices.

Predicted mean vote

The predicted mean vote (PMV) index is a multi-nodal model describing the thermal exchange between the human body and the environment. This rational model encapsulates the absolute and relative influence of the six factors relevant to the human thermal balance: air temperature, mean radiant temperature, air velocity, humidity, clothing, thermal resistance, and metabolic rate. The conceptual leap introduced by Fanger was to introduce a thermal judgement scale, which is shown in Figure 6.4, to allow model calibration against subjective evaluation (Fanger, 1970).

By exposing subjects to a range of conditions in a climatic chamber and collecting their feedback on the thermal judgement scale, Fanger was able to elaborate an equation to describe the thermal sensation, incorporating the variables relevant to the heat balance. The output from this model is comparable to subjective feedback, and the precise calculation forms the basis of international standard BS EN ISO 7730:2005 (BSI, 2006).

-3	-2	-1	0	1	2	3
Cold	Cool	Slightly cool	Neutral	Slightly Warm	Warm	Hot

Figure 6.4 Thermal judgement scale.

Adaptive thermal comfort

The adaptive thermal comfort index was developed in response to criticism of rational indices, which have been observed to show poor predictive accuracy of thermal comfort in field studies. Specifically, the PMV model has been found to perform poorly in naturally ventilated buildings. Also, as rational indices consider only the temperature flows of an individual, they may neglect the hedonistic qualities of thermal sensation, and also overlook adaptive behaviours.

Adaptive thermal comfort seeks to incorporate behavioural and psychological adaptations in addition to those that are physiological. Humphreys (1978) and de Dear and Brager (1998) evaluated the empirical relationship between comfort and environment using field measurements, indicating three contextual variables influencing thermal adaption behaviour and thermal comfort: climate, the building, and time.

This analysis prompted the development of an index whereby a comfortable temperature is defined relative to the external temperature. The external temperature figure incorporates the influence of time on the thermal history of an individual by applying an exponentially weighted running mean value, such that the temperatures of the previous days are incorporated. The influence of the building is acknowledged through the application of thresholds around the comfort temperature, with buildings that promote a higher expectation of favourable conditions, such as new buildings, having tighter thresholds than older buildings.

The adaptive comfort index is illustrated by Figure 6.5 with description of threshold categories given. If the measured data is inside the pairs of lines, it satisfies comfort criteria I, Category I being the most strict (with lines closest together)

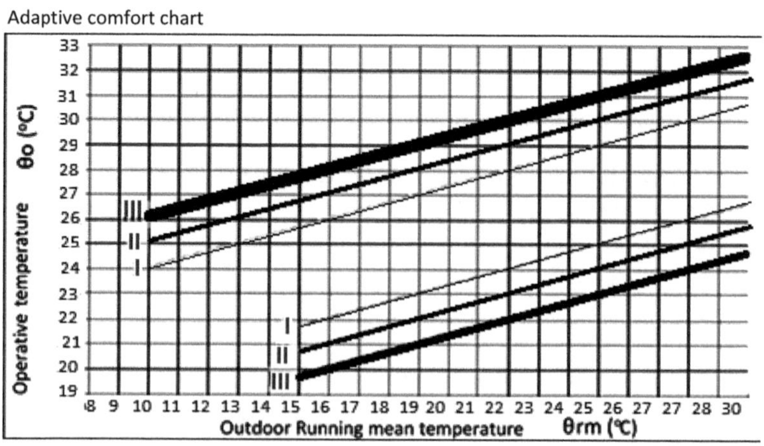

Figure 6.5 Adaptive comfort chart (adapted from BSI, 2019).

and Category III being the most forgiving. The adaptive method is contained in the standard BS EN 16798-1:2019 (BSI, 2019).

In addition to the objective evaluation of comfort, researchers also use subjective methods to establish actual comfort within an environment or context. As thermal comfort is an inherently subjective phenomenon, engaging with building occupants is important. International standard BE EN ISO:2001 (BSI, 2002) provides a range of scales which may be used to evaluate comfort, with correct semantic terms to avoid misinterpretation.

Developing research is applying established subjective methods using novel technology and user interfaces, to combine comfort evaluation with participatory feedback networks. This data may be used to inform, or even define, conditioning behaviour within buildings in the future, providing real-time occupant comfort to support energy managers.

References for "Thermal comfort"

ASHRAE (2017). Standard 55-2017—thermal environmental conditions for human occupancy. American Society of Heating, Refrigerating and Air-Conditioning Engineers. Atlanta, Georgia, USA, ASHRAE.

BSI (2002). BS EN ISO 10551: 2001 Ergonomics of the thermal environment—assessment of the influence of the thermal environment using subjective judgement scales. London, UK, British Standards Institution.

BSI (2005). BS EN ISO 8996:2004 Ergonomics of the thermal environment—determination of metabolic rate. London, UK, British Standards Institution.

BSI (2006). BS EN ISO 7730:2005 Ergonomics of the thermal environment—analytical determination and interpretation of thermal comfort using calculation of the PMV and PPD indices and local thermal comfort criteria. London, UK, British Standards Institution.

BSI (2010). BS EN ISO 9920:2009- Ergonomics of the thermal environment—estimation of thermal insulation and water vapour resistance of a clothing ensemble (ISO 9920:2009) (Vol. 3). London, UK, British Standards Institution.

BSI (2019). BS EN 16798-1:2019 Energy performance of buildings—ventilation for buildings. Part 1: Indoor environmental input parameters for design and assessment of energy performance of buildings addressing indoor air quality, thermal environment, lighting and acoustics—Module M1–6. London, UK, British Standards Institution.

de Dear, R. and Brager, G. S. (1998). Developing an adaptive model of thermal comfort and preference. *ASHRAE Transactions*, 104(1), pp. 1–19.

Fanger, P. O. (1970). *Thermal comfort.* Copenhagen, Danish Technical Press.

Humphreys, M. (1978). Outdoor temperatures and comfort indoors. *Batiment International, Building Research and Practice*, 6(2), pp. 92–92. doi: 10.1080/09613217808550656.

REFOCUSING SUSTAINABILITY ASSESSMENT METHODS: A COMPARATIVE INTRODUCTION TO THE BUILDING ENERGY RATING (BER) CERTIFICATE

John P. Spillane

Abstract to this paper

As international perception moves towards focusing on the need to address climate change, there is also a need to consider the impact of our buildings. It is widely acknowledged that the buildings in which we live and work have a significant and long-lasting impact on their surrounding environment. This then begs the question – how harmful, or more importantly, sustainable, is the operation of our buildings for the environment in which they are located?

To answer this, we need to not only manage, monitor, and control, but even more so to measure the impact our buildings have on the environment. To move towards addressing this, the aim of this section is to provide a brief introduction and overview of the Building Energy Rating (BER) certification scheme, as provided by the Sustainable Energy Authority of Ireland (SEAI) in 2009. Following its current ten-year history, there is now an apt opportunity to revisit, review, and provide an overview of the BER certification process, while also considering some of its counterparts. In doing so, an introduction to sustainability assessments in buildings is provided, considering BREEAM, LEED, and SAP ratings, before focusing on BER certification.

In the context of BER, focus is provided on the assessment of both residential and non-residential properties, followed by the BER rating system. To conclude, the discussion moves towards a comparative overview of the BER system in conjunction with those identified earlier, posing the question: Are current building energy rating methods fit for purpose?

This provides an apt opportunity to not only provide insight into a lesser-known measure of building sustainability, but also to draw comparisons with other more well-known and widely adopted methods of assessment. The results will be of benefit to those who not only measure and assess building performance and/or whole-life sustainability criteria, but also for individuals and organisations who are looking at reviewing either their existing or potential new properties, under the remit of energy efficiency, sustainability assessment, and the wider impact their buildings have on the environment.

Introduction to building energy assessment

With the continued global focus on sustainability and reducing one's carbon footprint, the built environment comes to the fore. Numerous approaches have been taken to measure and ascertain the contribution the built environment exerts on the environment. Attention has been directed towards buildings and their occupancy, due to their being one of the largest contributors to greenhouse gases and energy consumption over other sectors, particularly over their life cycle (Ortiz et al., 2009). With our buildings' insatiable appetite for energy, to fulfil and satisfy the needs of their occupants, it is essential to redress this approach to the design, manufacture, and ultimate usage of our buildings. In doing so, influence in the pre-construction phase comes to the fore, with the significant advantage in that design iterations and interjections can have significant impacts on reducing the carbon footprint of the infrastructure in use. **But how can you monitor what you don't measure?**

To address this, that is, not just the need to manage, monitor, and control, but to provide for objective and accurate measurement, there are numerous approaches one can adopt, depending on the focus, be it energy consumption, use of sustainable resources, waste minimisation, or reduction of the carbon footprint, among others. But with the need to provide measurement, there comes significant difficulties and challenges, mainly due to assessment methods varying to a great extent, primarily due to the tools being developed for different needs and purposes. The tools can be used to assess building products and buildings as a whole, including existing buildings, new buildings, and buildings under refurbishment, among others (Haapio and Viitaniemi, 2008); hence the need to select and implement the more appropriate assessment method, rather than the more popular one. In this context, Leadership in Energy and Environmental Design (LEED) and the Building Research Establishment Environmental Assessment Method (BREEAM) are two of the most prevalent and internationally recognised methods adopted (Zenial Hamedani and Huber, 2012). However, such methods focus on the overall sustainability of the respective projects under consideration, while focusing on new, sustainable projects globally.

But, with the focus on the measurement and assessment of new developments, there is a need to assess existing structures, to provide an objective overview of their sustainability and impact on the environment, and more to the point, the level of consumption of energy; this is where the Building Energy Rating Certificate or BER Cert comes in. The BER is based on the United Kingdom (UK) Standard Assessment Procedure, or SAP rating, which is a measure of the energy efficiency of a property considering the envelope, structure, heating and hot water systems, lighting, insulation, and other renewable sources within the property. Both the BER and SAP ratings consider and assess buildings at the home/building owner level, while providing an opportunity to consider how they can proactively manage and undertake positive change within their respective properties, to make them more energy efficient, and therefore, reduce their impact on the environment – and their pocket!

Sustainability assessments in buildings: An overview

With an increasing need to measure a building's sustainability, not just at projec inception, but throughout its life cycle, the need for more accurate measurement of actuality comes to the fore. Although seen by many as being voluntary, many clients in both the public and private sectors are now recognising the need to specify, adopt and/or adhere to prescribed assessment assessment methods. However Diaz-Lopez et al. (2019) rightfully argue that the area of sustainability assessmen methods is broad, complex, and fragmented, due to a multitude of factors, such a differing criteria, the diversity of disciplines involved, and most of all, the unenviable task of assessing what is a complex and difficult environment for classificatior in a very often limited array of outcomes. To begin to address the need to provide a more objective viewpoint, assessments of buildings came into practice. Diaz-Lopez et al. (2019) provide an overview of this introduction, highlighting that, since 2000, sustainable building assessment methods began to become more prominen in the literature. The aim of these assessment methods is to move away from the more subjective assessment, to one that is objective, based on pre-set criteria, thus allowing a comparative viewpoint with similar designs or projects.

To address this, numerous sustainability assessments have emerged, with Haapio and Viitaniemi (2008) highlighting 16 different assessment methods, while Kamaruzzaman et al. (2016) identify 10 methods, more suitable for building refurbishment. However, both studies note that a direct comparative analysis of each i futile and even all but impossible, due to the differing focus of each of the methods based on the economic, social, and environmental aspects under scrutiny and the region from which they originate. However, it must be noted that the number and array of different assessment methods continue to change, due to the disbandmen and emergence of new and diverse organisations, all with the goal of developing new knowledge and subsequent methods of assessing the built environmen and its environmental impact on its surroundings. Furthermore, Diaz-Lopez et al (2019) argue that the subject of sustainable building assessment is continuing to evolve, with more methods emerging, while existing approaches are revised and re-developed, based on, among other factors, the digitisation and evolution of the built environment. This has resulted in an ever-more broad, complex, and fragmented approach to identifying and selecting the preferred method of assessment primarily due to the significant diversity of disciplines and approaches involved.

However, there are a number of more prevalent assessment methods used widely, both at a national and international level, all of which have gained prominence in recent years. In the UK, focus is on BREEAM and in North America LEED. In the context of both BREEAM and LEED, both assessment methods incorporate the whole life cycle of the building, including its design, impact on the environment during and post-construction, and the selection and location o materials used, in addition to, and among other factors, energy consumption ir use. Given that buildings are one of the largest energy consumers, significant focus is provided on this aspect. This is where BER, in addition to the likes of SAP ratings, comes to the fore. Although not a whole-life assessment of a building, BER focuses on the energy consumption of both new and existing builds. Thus it only assesses a single aspect of what would be considered as just one factor in a plethora of variables under assessment in, for example, BREEAM and LEED, in addition to their other international counterparts. Ultimately, the likes of BREEAM and LEED are focused at developer level, while the BER Certificate and SAP ratings are more orientated towards single/one-off property owners, many of which are focused on existing stock, requiring assessment, to aid in better and informed decision-making on how to improve the energy efficiency of their homes/businesses.

Building Research Establishment Environmental Assessment Method (BREEAM)

The Building Research Establishment Environmental Assessment Method (BREEAM) was created in the UK in 1990 by the Building Research Establishment (BRE). It is an international certificate, currently in use in over 80 different countries. This is one of the more prominent environmental assessment methods, with different schemes of certification, depending on the country, the building type (office building, retail, etc.), and the construction type (new, refurbishment, etc.). Haapio and Viitaniemi (2008) note that this assessment method was the first to attempt to assess a broad range of factors when assessing a building's sustainability. The BREEAM method of assessment focuses on sustainability factors, with assessment under varying criteria such as energy, wellbeing, innovation, land use, materials, management, pollution, transport, waste, and water. Each category is weighted and rated to signify its overall importance in the criteria being assessed, as documented in Table 6.1.

The BREEAM ratings range from "Acceptable" (In-Use scheme only) to "Pass", "Good", "Very Good", "Excellent", to "Outstanding", with the award reflected in the number of stars denoted on the certificate, with one to six stars respectively. To achieve a certain BREEAM rating level, corresponding global scores must be reached, and corresponding minimum standards achieved, as documented in Table 6.2.

The BREEAM standard can be applied in five differing contexts: Communities (master planning), infrastructure (civil engineering and public projects), new construction (homes and commercial buildings), in-use projects (commercial buildings), and refurbishment and fit-out (homes and commercial buildings). However, like all assessment methods, it is not without criticism. Marks (2015) notes that BREEAM is favoured, not just in the UK but internationally, as evidenced by around 425,000 buildings certified at the time of writing and another two million

Table 6.1 BREEAM categories and weightings

BREEAM category	Weighting
Energy	19%
Health and wellbeing	15%
Materials	12.5%
Management	12%
Land use and ecology	10%
Pollution	10%
Transport	8%
Water	6%
Waste	7.5%
Total	**100%**
Innovation	Bonus 10%

Table 6.2 BREEAM rating and corresponding minimum global score required

BREEAM rating	Minimum global score required
Pass	35%
Good	45%
Very Good	55%
Excellent	70%
Outstanding	85%

registered for assessment. Nevertheless, Marks (2015) also argues that there are two core disadvantages to this scheme. Firstly, there is the potential for the BRE assessment to be influenced by following a more commercial standpoint. Secondly, Marks (2015) argues that far too much credit is given to minor, peripheral, environmental aspects of building design, such as the inclusion of bike racks, over more environmentally important factors such as heating, lighting, and water consumption. There can be much debate here, as transport by bike is only suitable for certain workers depending heavily on geographical placement, road safety provisions, social commitments (e.g. dropping off/picking up children), and ability/disability. The main point is that fabric first and on-site energy use can have much better returns for everyone.

Leadership in Energy and Environmental Design (LEED)

Following on from BREEAM, another popular assessment protocol is that of LEED. Originating in the US and developed by the US Green Building Council in 1993, it is fast becoming one of the most popular forms of sustainability assessment, not just in the US, but internationally. Again, similar to BREEAM, it works on a rating scheme, where projects accumulate points across various categories such as "Location and Transportation", "Sustainable Sites", "Water Efficiency", "Energy and Atmosphere", "Materials and Resources", "Indoor Environmental Quality", "Innovation", and more.

LEED focuses on rewarding sustainable and environmentally friendly decisions made throughout the inception, design, construction, operation, and maintenance of a building. The LEED rating system consists of five different areas addressing multiple projects: "Building Design and Construction", "Interior Design and Construction", "Building Operations and Maintenance", "Neighbourhood Development", and "Homes". Within each context, the respective projects are broken down into smaller components and assessed, based on varying criteria, with the points totalled and scored; see Table 6.3.

The process of applying for and gaining certification under LEED is based on four key steps: registration, application, review, and certification. In the context of registration, this step aids in the selection of the most appropriate LEED rating system that would be most suited to your project. In doing so, key project variables, such as floor area, are included for initial identification. The second step involves the application, where, based on the registration process undertaken earlier, the LEED credits required to be addressed are identified and assigned to a specific project stakeholder. Subsequently, information is collated and analysed, with supplementary information provided, demonstrating that the building meets the necessary minimum standards. The third step covers the review process. In this instance, there are three stages to the review process: The preliminary review, the final review, and the appeal review. The preliminary review is where the completeness and accuracy of the submission is considered. The final review is where a re-review is undertaken, to address any items noted in the preliminary review,

Table 6.3 LEED Certificate and corresponding points required

LEED Certificate	Points required
Certified	40–49 points
Silver	50–59 points
Gold	60–79 points
Platinum	80+ points

while re-evaluating the submission based on new information provided. An appeal review is held where the project owner is unsatisfied with the final review and appeals the decision. The final step is certification, where the building is then assigned one of the four certification levels, as outlined in Table 6.3.

Standard Assessment Procedure (SAP) rating

A Standard Assessment Procedure or SAP rating is the calculation that is required to produce a "Predicted Energy Assessment" and an "On Construction Energy Performance Certificate". The SAP methodology is used by governments to assess and compare the energy and environmental performance of dwellings is similar to that of the BER Certificate. It was developed to aid in providing accurate and reliable assessments of the energy performance of buildings, thus, aiding in the development and incentivisation of homeowners to make more sustainable decisions.

The SAP uses a scale of 1–100, to measure the overall energy efficiency of a dwelling, where a higher score indicates a property with lower running costs (Roberts, 2008). Similar to the BER Certificate in Ireland, the SAP rating also considers various factors, such as the building envelope, structure, windows, and sun shading, among other factors. As with a BER assessment, a SAP assessor visits the property and carries out an inspection, completing an online form with the necessary information using the prescribed SAP software.

There are two types of assessments undertaken: one for new buildings, which gives a SAP rating, and the other, for existing buildings, which gives an RdSAP rating. The "Reduced Data Standard Assessment Procedure" (RdSAP) considers less data in the accumulation of a SAP rating where a complete data set, that would be required, is not available. The results are given a rating, based on the energy efficiency of the property, and an Energy Performance Certificate (EPC) produced (Figure 6.6).

When developing and inputting the data into the respective modelling software, 12 sections are considered (see Table 6.4).

The Building Energy Rating (BER) Certificate

The Building Energy Rating, or BER as it is more commonly referred to, was first introduced in Ireland in 2009 by the Sustainable Energy Authority of Ireland (SEAI). In its simplest form, it is an energy rating for your home, similar to the SAP rating or earlier BRE Domestic Energy Model (BREDEM) in the UK. BREDEM is similar to SAP; however, it is not as restrictive as SAP in that it permits users to adjust inputs as they deem necessary.

The BER is devised around producing a rating, similar to those used for rating domestic electrical goods such as televisions, refrigerators, washing machines, etc.

Energy Performance Certificate (EPC)

SAP

© Crown copyright 2009

17 Any Street, District, Any Town, B5 5XX

Dwelling type: Detached house
Date of assessment: 15 August 2011
Date of certificate: 13 March 2012

Reference number: 0919-9628-8430-2785-5996
Type of assessment: RdSAP, existing dwelling
Total floor area: 165 m²

Use this document to:

- Compare current ratings of properties to see which properties are more energy efficient
- Find out how you can save energy and money by installing improvement measures

Estimated energy costs of dwelling for 3 years	£5,367
Over 3 years you could save	£2,865

Estimated energy costs of this home

	Current costs	Potential costs	Potential future savings
Lighting	£375 over 3 years	£207 over 3 years	You could save £2,865 over 3 years
Heating	£4,443 over 3 years	£2,073 over 3 years	
Hot water	£549 over 3 years	£222 over 3 years	
Totals:	£5,367	£2,502	

These figures show how much the average household would spend in this property for heating, lighting and hot water. This excludes energy use for running appliances like TVs, computers and cookers, and any electricity generated by microgeneration.

Energy Efficiency Rating

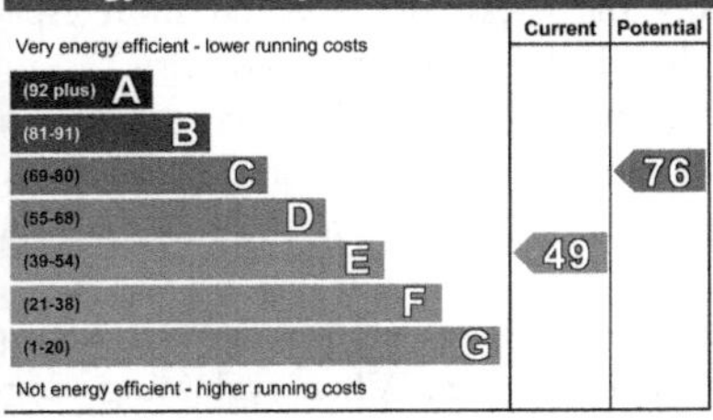

The graph shows the current energy efficiency of your home.

The higher the rating the lower your fuel bills are likely to be.

The potential rating shows the effect of undertaking the recommendations on page 3.

The average energy efficiency rating for a dwelling in England and Wales is band D (rating 60).

Top actions you can take to save money and make your home more efficient

Recommended measures	Indicative cost	Typical savings over 3 years	Available with Green Deal
1 Increase loft insulation to 270 mm	£100 - £350	£141	✔
2 Cavity wall insulation	£500 - £1,500	£537	✔
3 Draught proofing	£80 - £120	£78	✔

Figure 6.6 Sample Energy Performance Certificate (EPC) (UK Government, 2012).

Table 6.4 SAP section numbers and input information required

Section	SAP input
1	Job details
2	Dwelling
3	Heat loss floors
4	Heat loss walls
5	Heat loss roofs
6	Openings and summer overheating
7	Thermal bridging
8	Ventilation
9	Heating
10	Domestic hot water
11	Renewables
12	Other SAP input

It was introduced as a mechanism to measure and assess both new and, more especially, existing residential and commercial builds, throughout the country.

The BER was first introduced to aid in providing an objective measure o energy efficiency, primarily focusing on Ireland's ageing residential house stock but it also considers non-residential properties such as commercial, industrial, anc retail, among others.

In the context of new buildings, a desk-top assessment is undertaken, based or existing drawings and details available, pre-construction. When evaluating existing buildings, an assessor visits and inspects the property in person. Assessments in the majority, take between one and two hours to complete, with the certificate passed on to the home/building owner, usually within two to three days.

One of the key justifications for the introduction of the BER was to encourage homeowners to not only make more sustainable decisions regarding purchasing or renting a home, but also to provide information to aid in making more informec judgements on how best to upgrade their existing property. Furthermore, to aic in the development of the scheme and to establish compliance within the market anybody selling or letting property is required by law to have a BER Certificate for their property. From 2019, new building standards apply to all new residentia dwellings (houses or apartments) in Ireland, and new dwellings typically require a BER of A2.

In a residential context, to obtain a certificate, an assessor completes a Dwelling Energy Assessment Procedure (DEAP), which then provides the resultant BER rating for the property. The SEAI argue that having a BER Certificate, particularly one with a high rating, adds value to a home and make it more sought after when selling or renting. Furthermore, due to providing homeowners with more information and insight into how their property performs, it is also considered that this aids homeowners in improving and lowering their energy heating costs, thereby saving on energy and fuel bills, while also lowering CO_2 emissions.

For non-residential properties, BER is undertaken using the Non-Domestic Energy Assessment Procedure (NEAP), where Part L of the Building Regulations is considered. Under legislation, all new and existing commercial buildings are required to have an up-to-date BER Certificate when they are constructed, sold, or rented. Non-domestic or commercial buildings are divided up into three categories, namely 3, 4, and 5. This categorisation is based on the complexity of the commercial building under review. Category 3 covers relatively simple buildings, many of which would be similar to residential properties. Category 4 covers more complex buildings, such as offices, warehousing, and the like. The more complex buildings are considered in Category 5, which would involve factories, manufacturing, and other similar complex structures.

BER certification process for domestic properties

When conducting a survey, a BER Assessor, registered under the SEAI, either reviews pre-construction drawings for yet to be completed properties, or, in the case of existing properties, carries out a physical inspection. The inspection takes in the region of one to two hours to complete and does not involve any invasive inspection nor the opening of works. The assessor would be required to access the attic space, where appropriate, to complete the inspection. This method of assessment is observatory only, where they review and evaluate energy use for space and hot water heating, ventilation, and lighting, among other factors. On review, the assessor completes a DEAP, which is noted as the official Irish method for rating the energy performance of residential properties. This is an Excel-based calculation workbook programme, where the various particulars of the property are included for consideration. Items included range from the material used during construction, to ventilation, fuel used in the property, exposure of the property, and the methods used to provide lighting. All of this various information is considered and included by the assessor, under 12 worksheets, which correspond to tabs in Excel. The assessor works sequentially through each of the worksheets, as outlined in Table 6.5, such as sheets covering "Fuel", "Solar water heating", and "Summer". This leads to a display of marks, and the overall results in the 'Results' worksheet.

When undertaking the assessment, there are a number of assumptions made by the assessor. This includes the standard room temperature throughout the property, assumed to be 21°C in the family room, with all other rooms 18°C. The schedule for heating is assumed to be from 7am to 9am and from 5pm to 11pm,

Table 6.5 Overview of the DEAP calculation workbook

Worksheet name	Tab code	Details recorded
Cover	Cov	Cover page with SEAI logo, software version, etc.
Overview of workbook	Code	Guidance on how to complete the workbook
Project	Pro	Project, developer/owner, and assessor details
Overall dwelling dimensions	Dim	Ground and subsequent floor dimensions (where applicable)
Ventilation heat loss	Vent	Openings, structural airtightness, and ventilation method
Windows and other glazed elements	Win	Windows or rooflights on each façade or building element
Fabric heat loss	Fab	Recorded heat loss from the fabric of the building
Water heating	WH	Water-heating system recorded including associated details
Lighting	Light	Lighting and internal gains
Space heat use	HtUse	Required internal temperature, internal heat capacity, etc.
Space heating system	SH	Control and responsiveness, pumps/fans, emission efficiency
Energy requirements – individual heating systems	ER1	Space heating, water heating, electricity for pumps and fans, and electric keep-hot facility
Energy requirements – group/community heating scheme	ER2	Space heating – main and secondary systems, boilers, waste CHP, or heat from power stations
Results	Result	Results of space and water heating, pumps, fans, energy from lighting, central heating, etc., recorded and consolidated
Fuel data	Fuel	Gas, oil, solid fuel, electricity, group heating, and biofuel
Solar water heating	SWH	Calculation of solar water heating input, if any
Assessment of internal temperature in summer	Summer	Information sheet only, providing an overview of the threshold internal temperature during the summer months
DEAP report	Rpt	Consolidated overview of the inputs and results for the previous worksheets

seven days a week, from October to May. Once the BER Assessor has completed the DEAP, a BER Certificate is issued. This certificate is valid for ten years or until any material, significant change is made to the property which would affect its energy efficiency prior to this date, and includes, among other details, the following (BERIreland, 2019):

- The BER for the property
- The property's name and address
- The BER number
- The date of issue
- The date the BER is valid until
- The BER Assessor number
- The assessor company number
- The rating of the dwelling
- A carbon dioxide (CO_2) emissions indicator

BER certification process for non-domestic properties

The BER certification process for non-residential properties is a similar process to that described for residential properties. In this instance, the SEAI-certified assessor completes the NEAP, which is software used to provide the necessary results. There are numerous software packages available, including the default calculation tool, the Simplified Building Energy Model for Ireland (SBEMie), which was developed by the BRE. This tool provides monthly energy use calculations and CO_2 emissions, based on building geometry, construction, use, heating, ventilation and air-conditioning (HVAC), and lighting equipment used within the building. The key process behind the calculations is to ensure adherence to Technical Guidance Document – Part L, to work towards, in so far as is reasonably practicable, the achievement of a near-zero energy building (NZEB), particularly where projects commenced from 2019 onwards (SEAI, 2019).

When undertaking the assessment, a reference building is used as the baseline, on which the building under assessment is benchmarked against. This reference or benchmark building is a similar property of the same size, shape, and zoning arrangements as the property under evaluation. Each space within the reference building must also correspond with that of the building under assessment in the context of activities that are undertaken. Each activity is assigned based on prepopulated activities in the NEAP activity database. The subject of orientation and exposure is also considered, where the reference building must mimic that of the building under assessment, including such aspects as sun shading and other topographical features, thus, also providing an accurate baseline for measurement. Finally, whatever system type (heating, ventilation, cooling) is used in each of the respective areas within the assessed building must also be used in the reference building, again to provide an accurate representation for assessment. In undertaking the assessment, various elements based on the reference building are considered, as outlined in Table 6.6 (SEAI, 2019).

In addition to considering the reference building, there are also elements within the actual building under assessment to consider. Consideration of these elements is required, to complete the necessary information in the respective software, to facilitate the subsequent calculations, to determine the respective BER rating for the building. In addition to those listed in Table 6.7, aspects such as solar gain and overheating are also considered and assessed (SEAI, 2019).

Similar to residential BER assessments, commercial assessments consider the following (BERIreland, 2019):

- The geometry and age of the building
- Type and thickness of insulation materials
- The heating system in place
- The type of building ventilation used
- The type of lighting
- Any renewable technologies in use

Table 6.6 Specific requirements within the reference building

Requirements	Notes
Activity glazing class	The activity assigned to each zone determines whether it will have access to daylight through windows, roof-lights, or no glazing at all
Building fabric	U-values assigned to each of the building elements, including roofs, walls, floors, windows, etc.
Areas of windows, doors, and roof-lights	The areas of windows, doors, and roof-lights are calculated in the reference building including side-lit, top-lit, and unlit glazing classes
HVAC system	Each space in the reference building will have the same level of servicing as the equivalent space in the building under assessment
Auxiliary energy	The auxiliary energy is the product of the auxiliary power density and annual hours of operation of the heating system as taken from the NEAP Activity Database
Lighting power density	The general lighting in the reference building is based on lighting with efficacy of 65 luminaire lumens per circuit-watt, and the resulting power density (W/m^2) will vary as a function of the geometry of each zone modelled
Renewable energy	The reference building will have a renewable energy contribution from photovoltaics based on the equivalent of 20% of the reference building's primary energy use

Table 6.7 Specific requirements within the assessed building

Requirements	Notes
Building fabric	Building fabric is assessed, with thermal bridges and U-values documented, and ventilation openings disregarded
Lighting	Lighting is defined at zone level, considering LED, tungsten, halogen, etc.
Auxiliary energy	This considers how auxiliary energy, such as pumps and fans, is calculated
Demand control for ventilation	Ability to model demand control for ventilation for zones which accommodate mechanical ventilation

As with the domestic certification process, the NEAP is used to generate the BER rating and a supporting advisory report for both new and existing non-domestic properties. The process calculates the energy consumption and CO_2 emissions associated with the day-to-day use of the property. The energy consumption is expressed in terms of kilowatt hours per square meter floor area per year (kWh/m^2/yr) and the CO_2 emissions expressed in terms of kilograms of CO_2 per square meter floor area per year (kg CO_2/m^2/yr), which is again similar to the process for residential properties (BERIreland, 2019).

BER Ratings

Once the assessor has thoroughly inspected and assessed the property and completed the DEAP Calculation Workbook for residential or the NEAP for non-residential buildings, the results are produced in a report format (see Figure 6.7). This calculation considers the primary energy consumption, which is expressed in kWh per metre squared per annum (kWh/m^2/yr) and quantifies the CO_2 output in kilograms of CO_2 per metre squared per annum (kg CO_2/m^2/yr). This is then used as the basis for issuing a rating from "A" to "G", with a rating closer to "A" indicating a very energy-efficient property while those closer to "G" would be less efficient.

From the date of issuance, a BER Certificate is valid for ten years, as so long as there is no material change to the property that would affect its energy rating. Where a dwelling is purchased off-plan, a provisional certificate is provided, which is valid for two years. The builder/developer, in this instance, is responsible and must provide a full BER Certificate post-completion of the property, prior to selling/leasing the property.

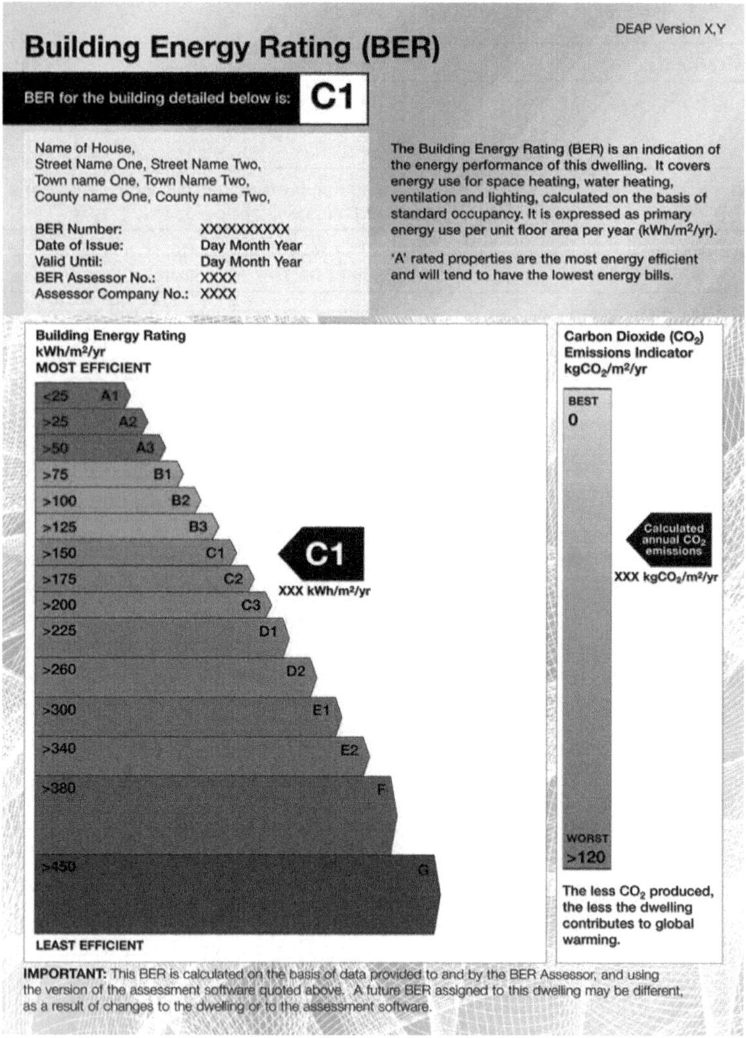

Figure 6.7 Sample BER Certificate illustrating kWh/m^2/yr and corresponding BER Rating (example source BerCert.com (2020), see also SEAI (2019)).

Conclusions to this comparative introduction to the BER Certificate

With the sustainability and energy efficiency of our buildings continuing to drive debate and investigation, there needs to be a mechanism by which to measure and assess a building's impact on its environment. On a large scale, there are numerous approaches one can choose, with BREEAM and LEED being two of the more globally recognised and widely adopted. However, such levels of investigation and certification need to also be considered at the home/building owner level, while also considering the extensive and ageing stock around the country. In doing this, SAP in the UK is used, while the BER Certificate is adopted in Ireland.

The relationship between BER or SAP and the large, multi-faceted assessment models, such as BREEAM and LEED, is that they are not in competition but complement each other. BREEAM, LEED, and other such certification models all set targets and provide recognition for sustainable buildings and their environments, whereas both the BER Certificate and SAP ratings provide a means by which developers, but more so, owners and occupiers, can make conscious sustainable decisions, not just during design and construction, but also, post-occupancy. Furthermore, the BER and SAP can also complement and aid in the betterment of BREEAM and LEED applications, as by obtaining a better BER and/or SAP rating, this can then be incorporated in the relevant material, thus, aiding in the justification for a higher result being awarded.

Interestingly, a study by Kelly et al. (2012) brings SAP back into focus and asks the question: "Is SAP fit for purpose?" SAP came to the UK in 1992, significantly prior to the BER Certificate in Ireland, with both following a similar methodology and underlying method in their assessment processes; hence there may be an argument to make that the findings of SAP may also accurately reflect those of the BER certification process in Ireland. Kelly et al. (2012) conclude that, although SAP has brought improvements to our understanding and appreciation of the energy that our buildings consume, it lacks real and quantifiable opportunities for homeowners to relate the 1–100 scale to energy consumption or emissions generated, an aspect that the BER Certificate provides. Furthermore, and similar to other studies which have used computational modelling, there are anomalies and/or inconsistencies between the actual consumption figures and those figures given using the various models. The differences, in some cases, can be substantial. (Stundon et al., 2015).

In defence of the BER certification process, and more importantly, the resultant report produced, it does not rely on a scale of 1–100 for the purposes of informing the reader of the energy efficiency of the property under scrutiny. Instead, it uses a similar system of letters from A to G – but more importantly, it clearly relates what each of these results mean in the form of annual energy usage (kWh/m^2/yr) and, also, quantifies the carbon dioxide output in metres squared per annum (kg $CO_2/m^2/yr$). This aids in the appreciation but also quantification of the results and their impact on the environment in annual energy usage and also CO_2 emitted. But more importantly, the report then, by using the above, provides an estimation of the costs associated with the results achieved, based on current factors, but also, what could be done to increase the rating, thus reducing the environmental impact while increasing the sustainability of the building in the environment. Unfortunately, BREEAM, LEED, and their counterparts do not provide such information, primarily due to the fact they are not designed for this purpose and is essentially not due to an oversight in their design.

In concluding, and to address the question – "Are current building energy rating methods fit for purpose?" – the answer may be more equivocal, querying whether any mechanism, for objectively assessing the subjective topic of sustainability, can be without fault. Based on this premise, both SAP, but more so, the BER, are positive solutions rather than failing to assess in the first place. After all, what you don't measure, you don't manage, you don't control, you can't improve, and ultimately, it costs you and the environment. Thus they may not be perfect, may have room for improvement, but are making a difference.

References for "Refocusing sustainability assessment methods"

BERIreland (2019). Commercial BER assessments. BER Ireland, Castletroy, Limerick, Ireland.

BerCert.com (2020). BER website. Available at https://www.bercert.com/index.cfm?page=about/index&q=sample-ber-certificate

Diaz-Lopez, C., Carpio, M., Martin-Morales, M. and Zamorano, M. (2019). Analysis of the scientific evolution of sustainable building assessment methods. *Sustainable Cities and Societies*, 49, pp. 101610.

Haapio, A. and Viitaniemi, P. (2008). A critical review of building environmental assessment tools. *Environmental Impact Assessment Review*, 28(7), pp. 469–482.

Kamaruzzaman, S. N., Lou, E. C. W., Zainon, E., Suzaini, N. M. Z. and Wong, P. F. (2016). Environmental assessment schemes for non-domestic building refurbishment in the Malaysian context. *Ecological Indicators*, 69, pp. 548–558.

Kelly, S., Crawford-Brown, D. and Pollitt, M. (2012). Building performance evaluation and certification in the UK: Is SAP fit for purpose? *Renewable and Sustainable Energy Reviews*, 16, 6861–6878.

Marks, C. (2015). Weighing up the pros and cons of the BREEAM environmental standard. *Fresh Workspace*, September 15th, Monmouthshire, UK.

Ortiz, O., Castells, F. and Sonnemann, G. (2009). Sustainability in the construction industry: a review of recent developments based on LCA. *Construction and Building Materials*, 23(1), pp. 28–39.

Roberts, S. (2008). Altering existing buildings in the UK. *Energy Policy*, 36, pp. 4482–4486.

SEAI (2019). Non-domestic energy assessment procedure—modelling guide (Q2–2019). Dublin, Ireland, Sustainable Energy Authority of Ireland.

Stundon, D., Spillane, J. P., Lim, J. P. B., Tansey, P. and Tracey, M. (2015). Building information modelling energy performance assessment on domestic dwellings: a comparative study. In Raidén, A. B. and Aboagye-Nimo, E. (eds.) *Proceedings of the 31st Annual ARCOM Conference*, pp. 671–679.

UK Government (2012). Sample EPC certificate. Available at https://assets.publishing.service.gov.uk/government/uploads/system/uploads/attachment_data/file/5996/2116821.pdf

Zeinal Hamedani, A. and Huber, F. (2012). A comparative study of DGNB, LEED and BREEAM certificate systems in urban sustainability. In Pacetti, M., Passerini, G., Brebbia, C. A. and Latini, G. (ed.) *The Sustainable City VII, Vol. 1. WIT Transactions on Ecology and the Environment*, Vol. 155, WIT Press, pp. 121–132. Southampton, UK, WIT Press.

THERMAL PERFORMANCE

Chris Gorse and Melanie Smith with Matthew Brook-Peat and Felix Thomas

An insight into Building Performance Evaluation

"Sustainable" and "sustainability" mean different things to different people and sectors. The concept and philosophy behind these terms require definition before they can make sense, be understood, or be evaluated. Therefore, when considering the whole-life sustainability of a building, this should be underpinned with demonstrable measurements and explained evaluation of direct environmental impact. A scattering of buzzwords creating greenwash is disingenuous. Performance is only valid if it is measurable. This chapter looks at different aspects of measuring building thermal performance.

A great deal of energy and resources are consumed in the construction of a building, but much more is expended in the building's operation, repair, maintenance, and energy use. These are less easy to predict than for construction, and can vary widely depending on the use, its occupant behaviour, and the building's energy efficiency. The first two can be only marginally influenced, but the latter can be largely controlled with directed attention.

The scientific community has been able to measure building performance and provide a perspective on a building's ability to meet sustainable factors for many years, but the resources involved for detailed measurement of its energy efficiency are too great for the mainstream construction industry. Notwithstanding this, energy-monitoring technology and data-capture methods have advanced to enable basic data, including live data, to be more financially feasible and more important to occupants through to energy providers.

The variance between theoretical and actual performance remains a major barrier to achieving both zero-energy targets and investors using pay-back forecasts. Research has shown that this performance gap can be reduced to acceptable levels, mainly by detailed design and thence construction, but such detailed design and on-site control occur only in isolated places and are not generally the norm. To meet the challenges of energy use, and ultimately climate change, more widespread improvement is required.

Building Performance Evaluation (BPE) is the assessment of a building's performance throughout its entire life cycle. It takes a holistic approach towards building operation by evaluating four key elements:

- Fabric performance
- Services performance
- Occupant satisfaction and comfort
- The changeable external environment

Building Performance Evaluation can be carried out at every stage of a building project, e.g.:

- Desk-based assessment at the design stage, to improve design practice and ensure the continuous improvement of design
- Observation of construction methods, to capture evidence of actual assembly and build
- Appraisal and analysis of in-use performance, including collation and appraisal of reflections and reviews
- End-of-life assessment of materials and energy cost

Figure 6.8 shows different phases of a building's life cycle, which can be subject to BPE. There are a number of different approaches. Some focus on pre-occupation tests and operational performance, others focus solely on occupation or operational phases, where most energy use is likely. To be effective, a whole-life assessment which follows the building through the construction process, its operation, and refurbishment and extends to demolition may be essential.

BPE helps understand operational energy use, embodied energy, and improving energy performance. It can quantify the energy performance of buildings, i.e. to determine if the performance achieved reaches the levels of performance intended at the design stage or if there is a shortfall – the "**performance gap**".

One vital strength of BPE lies in the **constant feedback at each stage of assessment** – essential if the evaluation process is to have a positive impact on building performance. Collecting qualitative feedback from designers, the construction

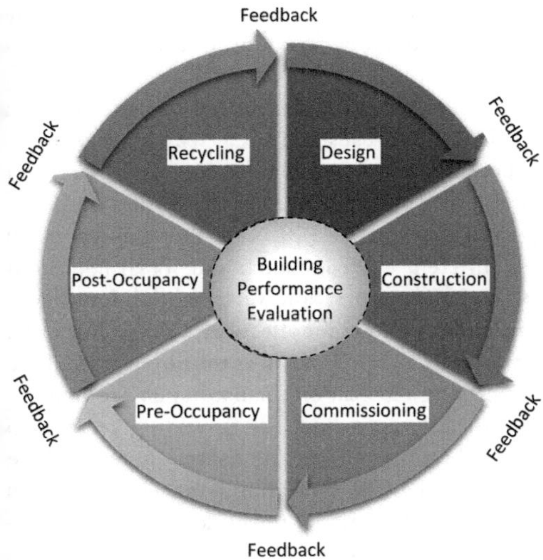

Figure 6.8 A whole-systems approach to Building Performance Evaluation.

team, building owners and managers, and importantly the occupants, throughout the building's life, is essential if the building is to achieve optimal performance, provide the comfort required, and reduce its impact on the environment.

Data is collected through:

- Design evaluations
- Testing
- Physical monitoring techniques
- Post-occupancy evaluation

It can be used to improve our understanding of efficiency, environmental conditions, building fabric quality, building services, control systems operation, and commissioning, all of which are important for an effective handover and operation of a building.

BPE undertaken during the construction phase of new-build developments can provide feedback to designers on the effectiveness of their design choices, providing the opportunity for stakeholders to remedy features that result in underperformance and to make improvements upstream in the build process.

Post-occupancy BPE is collecting information from the users and on in-use operation. It can give designers insights into building use and effectiveness when users interact with the building systems. This kind of post-occupancy evaluation

(**POE**) is invaluable when making future decisions on system improvements. The results of BPEs can identify opportunities for reducing running costs, maintenance requirements, and repair work.

Because the BPE process links design, construction, operation, and disposal it helps identify the key stages and key factors causing gaps at which performance discrepancies can arise, to enable potential solutions, Table 6.8.

Understanding the experience of using the building, and the comfort and control achieved as well as the energy used is essential to BPE. The interaction between the users and the building under different environmental and social condition needs to be **understood** to achieve effective operation. Optimal building operation may require the users to be **influenced** to use it effectively. The information on the building and the way it is **managed** is as important as the physical attribute of the building when balancing a building's impact on the natural environment.

Sometimes performance gaps are easily corrected, e.g. providing ventilation sealing unwanted air paths, and recalibrating services and control systems. Othe deficiencies are so embedded in the building that the feasibility of correcting them becomes difficult and they become an inherent part of the building, affecting performance throughout the building's life. The aim of BPE is to better understand buildings and the way that they are used, so that effective performance and operation are achieved throughout its lifetime.

Thermal modelling

Using thermal simulation software, two- and three-dimensional models of a building element can be created, complete with specified material properties and boundary conditions, such as surface resistance to the passage of heat and air temperatures. Thermal modelling, using thermal simulation software, enables user to predict the magnitude of thermal bridging occurring at junctions in a building fabric, and calculate the U-values of complex plane elements that may be difficul to calculate manually with formulae or with relatively simple calculation tools.

Table 6.8 Some causes of performance discrepancies

Factors resulting in performance gaps	
Design	Incorrect or incomplete design information
Design	Incorrect or ill-fitting components
Workmanship	Issues of workmanship
Workmanship	Commissioning
Workmanship	Inspection
In-use/design	Management
In-use/design	Operation

Model simulations are now a useful, necessary, integral, and ubiquitous tool for building design. They can be used to assess designs for moisture, and different scenarios can be modelled to finalise best compromise detail for new work.

- Model simulations can be used to assess designs for moisture predictions.
- Elemental modelling for thermal bridging and interstitial or surface condensation risk can be used in design to mitigate adverse consequences.
- Model simulation needs to be undertaken with skills in the use of specialist software, appreciation of limitations, and understanding of building pathology and materials science, for the competent use of hygrothermal modelling, predicting hygrothermal behaviour, and assessing associated risks.
- Requirements to use such a tool without providing the resources required are counterproductive.

Models can present an unrepresentative "perfect" illustration of, say, a wall, for example giving smooth surfaces, omitting air gaps, lacking an appreciation of extant damp. Thus, for example, any modelling used to assess a building's potential for retrofit needs to be undertaken with skills in the use of specialist software, appreciation of limitations, and understanding of building pathology and materials science. These are required for the competent use of hygrothermal modelling, predicting hygrothermal behaviour, and assessing associated risks. Using such a tool, without the understanding required, is counter-productive.

Whole-building models can be used to generate simulated energy buildings for both pre- and post-construction. These simulations can be used to predict energy and fuel consumption for buildings as well as to provide detailed information about the capability of building services, and to predict the risk of over- or under-heating in the internal environment.

Software, such as IESVE or DesignBuilder, can be used to generate 3D models of buildings. To set up a specific building, properties can be assigned to different building components, and building services (e.g. heating and ventilation systems) specified. Specific location and weather data can give context to performance predictions during a whole-building thermal simulation and to assess performance under hypothetical extreme conditions.

Making changes to different variables in the building thermal models allows the user to assess the impact and effectiveness of various design changes, such as passive solar heat gains and solar shading, passive ventilation, the sizing and provision of services and plant as well as the value of adding renewable systems in a whole-building context.

Building performance can be assessed or predicted with a high degree of confidence through modelling building elements, such as walls and junctions (Figure 6.9). Elemental modelling can be used to calculate the U-values of plane

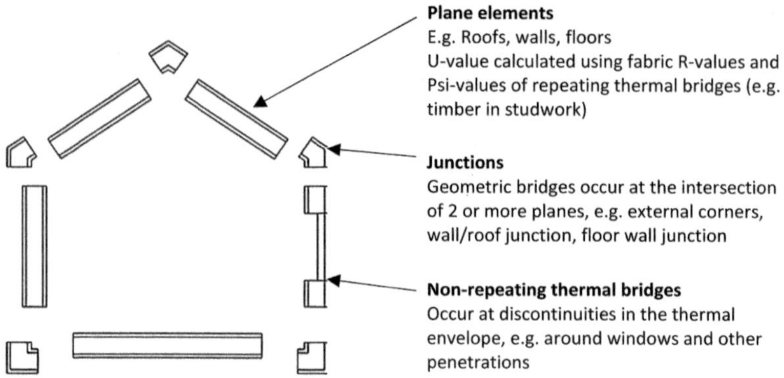

Figure 6.9 Plane elements, junctions (geometrical thermal bridges), and thermal bridges (Thomas in Gorse et al., 2015).

elements (e.g. walls, floor, roofs). Modelling can also be used to determine the magnitude of thermal bridging that occurs at junctions and throughout the building envelope and to determine whether moisture poses a risk to the building fabric.

Detailed and accurate information about the make-up of different building elements and material properties is needed for modelling and calculating building elemental performance. Inaccuracies and assumptions in the model input will lead to inaccurate output values for performance.

Temperature factor: Risk of condensation

Thermal modelling can also indicate the risk of condensation forming at a modelled junction or thermal bridge. See also Figure 6.10. Outputs from a thermal simulation often include a minimum surface temperature of the internal boundary condition. A temperature factor (f_{rsi}) can be derived using the following formula:

$$f_{\text{rsi}} = \frac{T_{Si} - T_e}{T_i - T_e}$$

Where:

f_{rsi} = temperature factor
T_{Si} = internal surface temperature
T_e = external air temperature
T_i = internal air temperature

Where temperature factors are less than 0.75 there is a risk of condensation formation at the junction modelled.

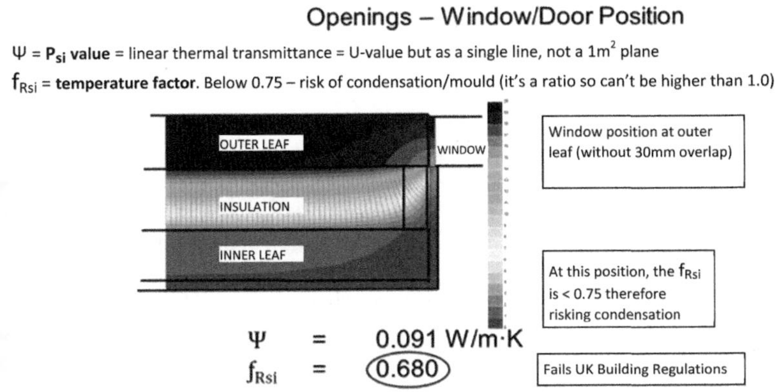

Figure 6.10 Example temperature factor at off-set window position (source: Dr M. Brook-Peat).

Hygrothermal simulation

Hygrothermal simulation software predicts the behaviour of real-world heat and moisture in a modelled building element. It can be used to identify potential moisture problems within a building's fabric over time, such as moisture accumulation, interstitial condensation formation, and mould growth.

Hygrothermal simulation software is often one-dimensional, simulating moisture and heat behaviour in a straight line through a representative model of a wall or other building element. More advanced two- and three-dimensional hygrothermal simulation software is capable of predicting the behaviour of moisture in more complex geometries found within buildings; the more complex the model, the more time-consuming. Software, such as WUFI, can carry out dynamic simulations, simulating changes in environmental conditions over time, potentially identifying problems that may only manifest years after construction has been completed.

Hygrothermal simulation is a complex task that requires a competent user with knowledge of the hygrothermal behaviour of materials and buildings, as well as experience or training in the use of hygrothermal simulation software.

References for "Thermal performance"

Gorse, C., Johnston, D., Glew, D., Fylan, F., Thomas, F., Miles Shenton, D., Fletcher, M., Erkoreka, A. and Stafford A. (2015). Monitoring and measuring building performance. In: Dastbaz, M. Strange, I. and Selkowitz, S. (eds.) *Building sustainable futures, design and the built environment*. Switzerland, Springer International Publishing. ISBN 878-3-319-19347-2.

THERMAL PERFORMANCE OF EXISTING BUILDINGS AND RETROFIT

Chris Gorse and Melanie Smith

Thermal retrofitting of dwellings has been a priority in many countries, including the UK, for over a decade. However, concerns have arisen around their build quality and consequences, so that thermal retrofit presents one of the major energy challenges faced. An understanding of a building's construction and its risk of decay is fundamental for the building's successful design, construction, repair, maintenance, and alteration. Thermal retrofit can change a building's risk for decay. Smith (2018) and others have identified common building pathology-related issues prevalent in thermal retrofit of walls. In addition, a lack of clarity, description, and specification have all been found for many pre-design retrofit surveys.

Improving thermal performance in a building should always start with knowledge of the existing performance. This is not a simple science. Performance depends on construction and conditions. Construction depends on age, design, implementation, defects/degradation/entropy, alteration history, and use. Thus, no two properties have exactly the same construction at any given time. Generalities of course can be used, but if the individual property is to be upgraded, the individual property should be taken into consideration. For thermal upgrades, a pre-retrofit survey is required, and this would take an anticipatory survey approach for the prevention of future problems, as outlined in Chapter 1.

The initial desktop research would consider the expected design of, say, the walls. A few of the thermal properties of different wall constructions are listed in Table 6.9. Some differences noted between measured and assumed U-values may be attributed to the quality of assembly, workmanship, and unique attributes of the wall and surrounding features. Notwithstanding this, the table offers a useful guide and does demonstrate the variance that can be found even in one construction form (Rye, 2014; DECC, 2011; Baker, 2008).

Before the 1940s, UK brick walls were often 9 inches thick (225 mm or greater), bonded by masonry and constructed using lime mortar. Holes and gaps deep inside the walls helped reduce the passage of water from external to internal faces. From the 1920s, the inner leaf and outer leaf of the walls were separated, bonded by galvanised steel ties, to better prevent the passage of water. Such cavities are generally ventilated, even if not by design, and therefore do not help insulate the wall. During the period between the 1930s and the 1960s, blockwork was introduced for the inner leaf of cavity walls (NHBC, 2019), and blockwork in masonry construction is now commonplace. It is the inner leaf which is responsible for both being loadbearing and insulating. The outer leaf is the rain-shield. Currently, new masonry walls are constructed of a range of materials, usually with wide cavities, which are filled or partially filled with insulation.

Table 6.9 Measured and assumed thermal properties of traditional wall construction (Baker, 2011; DECC, 2011; NHBC, 2019; Rye, 2014)

Construction	Measured U-value(from Rye, 2014)	Assumed SAP value (from DECC, 2011)	Measured U-value(from Baker, 2011)
Masonry with granite	1.4 W/m²·K (for 520 mm construction)	2.3 W/m²·K (for 500 mm construction)	1.1 ± 0.2 W/m²·K (for 600 mm) 0.8–0.9 W/m²·K (field measurements)
Masonry with sandstone	1.5 W/m²·K (for 475 mm construction)	2.0 W/m²·K (for 500 mm construction)	1.1 ± 0.2 W/m²·K (for 600 mm) 0.8–1.6 W/m²·K (field measurements) 1.7 W/m²·K (for 600 mm)
Masonry with brick	1.4 W/m²·K (for 220 mm construction)	2.1 W/m²·K (for 220 mm construction)	
Cob	0.7 W/m²·K (for 600 mm construction)	0.8 W/m²·K (for 550 mm construction)	0.4–0.8 W/m²·K
Timber frame	2.0 W/m²·K (for 135 mm construction)	2.5 W/m²·K (for 150 mm construction)	
Concrete blocks, both leaves with 65 mm uninsulated cavity			1.1 W/m²·K
Concrete blocks, both leaves with 65 mm insulated cavity			0.6 W/m²·K
Bricks, both leaves with uninsulated cavity		1.44 W/m²·K, (Anderson from Baker) 50 mm cavity	1.3 W/m²·K, 65 mm cavity, 300 mm construction
Bricks, both leaves with 65 mm insulated cavity		0.3 W/m²·K brick/ block cavity wall	0.3 W/m²·K, 300 mm construction

The images in Table 6.10 show a range of construction types and thermal performance (generalised only) for solid masonry, cavity wall construction, timber-frame, cob, and straw-bale constructions.

Care is required when using given U-values, as the range of values measured in field varies considerably from building to building (Glew et al., 2020; Gorse et al.,

2017; Baker, 2011). Baker's study explores some of the differences experienced in traditional buildings and found that, for the case of 600 mm thick solid stonewall with plaster on laths, a typical performance of 1.1 ± 0.2 W/m^2·K was measured onsite, compared to default U-values provided by CIBSE and EST (Anderson 2006a,b) of 1.38 and 1.7 W/m^2·K, respectively.

U-values for uninsulated cavity walls of a 1930s semi-detached house and a 1970 detached house were measured at 1.3 W/m^2·K and 1.1 W/m^2·K, respectively, compared to default U-values of 1.44 and 1.7 W/m^2·K respectively.

For the case of cavity walls retrofitted with insulation post-2003, the U-value wa 0.3 W/m^2·K which matched the default U-value provided by EST.

Other calculated results discussed by Baker (2011) include those below:

Anderson (2006a):

- 1.38 W/m^2·K for a 600 mm stonewall with a 50 mm airspace and finished with 25 mm dense plaster on laths
- 2.09 W/m^2·K for a 220 mm solid brick wall with 13 mm dense plaster
- 1.41 W/m^2·K for a 220 mm solid brick wall with 50 mm airspace/battens and 12.5 mm plasterboard
- 1.44 W/m^2·K for a brick/brick cavity wall with 105 mm brick, 50 mm airspace 105 mm brick, and 13 mm dense plaster

Energy Saving Trust (2004):

- 1.7 W/m^2·K for a traditional sandstone (or granite) dwelling with solid walls Stone thickness typically 600 mm with internal lath and plaster finish (for the pre-1919 period)
- 1.7 W/m^2·K for cavity walls involving brick and block with external render (for 1919–1975)
- 0.3 W/m^2·K for brick/block cavity walls with insulation (for 2003–present)

These results are considered to be indicative of what can be found, therefore please note that these are not necessarily indicative of a particular wall type's performance. When assessing a wall's performance, the construction and condition of the wall and connecting elements should be considered. Table 6.11 shows observed and typical introduced problems when wall insulation is applied to some existing building types.

Table 6.10 Construction types and modelled or measured thermal performance

Construction types and thermal performance		
Stone walls	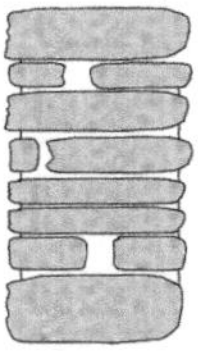	Approximate U-values 1.2–2.0 W/m^2·K (stone) 1.5–2.0 W/m^2·K (sandstone) 1.4–2.3 W/m^2·K (granite)
Stone outer leaf with brick/stone inner leaf and rubble-filled cavity creating a solid wall	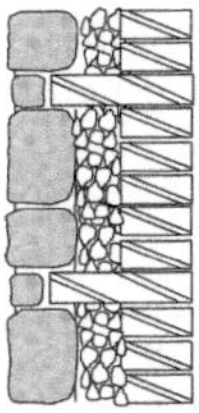	External stone is porous, but water that penetrates drips down the unconnected rubble cavity rather than crossing through it Thicker walls, with sealed finish, tend to have lower U-values Approximate U-values 1.2–2.3 W/m^2·K 0.9–2.4 W/m^2·K (Baker, 2011)
Uninsulated stone/brick with small finger cavity	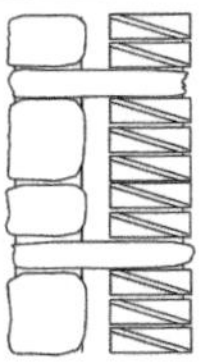	The thin cavities are often not sealed, leading to internal and external air channels through floor joists and voids, service penetrations, etc. Approximate U-value 1.4 to 1.5 W/m^2·K
Solid brick wall construction	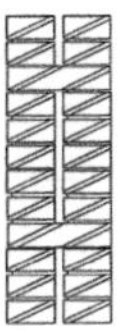	Typical construction in 1800s Two leaves of brick with cross bricks in a variety of bonding patterns. May have 10–25 mm un-mortared internal breaks to prevent moisture passage. Solid lime plaster applied to the internal surface Approximate U-value 1.2–2.4 W/m^2·K
Uninsulated brick cavity wall	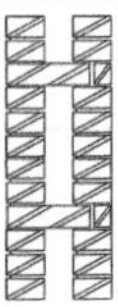	Double brick leaves with 40–70 mm cavity and brick cavity ties. If air flows free through the cavity, heat loss during high winds will be considerable. The break between the leaves further reduces the potential transfer of rain and moisture from the outside environment Approximate U-value 1.4–1.5 W/m^2·K 1.3 W/m^2·K (Baker, 2011)

(continued)

Table 6.10 *(continued)*

Construction types and thermal performance		
Uninsulated brick cavity wall	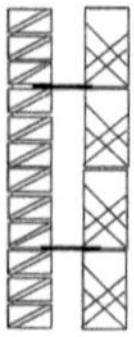	Early cavity walls, 1920s built with steel cavity ties. Ties prone to rusting and often breaking through corrosion, resulting in instability of the two leaves Approximate U-value 1.4–1.5 W/m^2·K 1.3 W/m^2·K (Baker, 2011)
Uninsulated brick block cavity wall	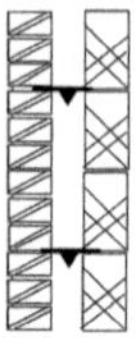	Galvanised-steel, stainless-steel, and plastic wall ties with drip formations were included to avoid corrosion and water passing across the cavity Approximate U-value 1.3 W/m^2·K (Baker, 2011)
Brick block cavity wall with partial fill cavity insulation	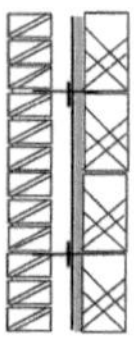	Partial fill insulation was introduced around 1980, glass fibre batts fixed within the cavity and trapped into place using preformed cavity clips Approximate U-value 0.6 W/m^2·K (Baker, 2011)
Brick block cavity wall with partial fill cavity insulation	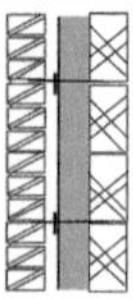	Thickness of partial fill increased and the quality of insulants improved using closed cell PIR and PUR Approximate U-value using 50-mm PIR 0.4 W/m^2·K
Partial fill insulation	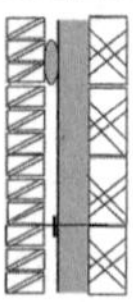	Taped and sealed to provide air barrier, cavity sock used at the head and foot of the cavity to seal the void Approximate U-value 0.3–0.5 W/m^2·K
Full fill	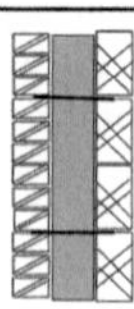	Approximate U-value achieved with mineral wool fill thickness 200 mm 0.15 W/m^2·K 90 mm 0.30 W/m^2·K 75 mm 0.44 W/m^2·K 50 mm 0.52 W/m^2·K Approximate U-value achieved with polybead thickness 50 mm 0.45 W/m^2·K

(continued)

Table 6.10 *(continued)*

Construction types and thermal performance		
Combined high thermal insulants	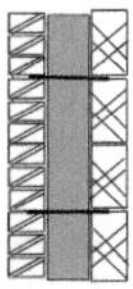	The combination of full fill insulation (140) and light thermally resistant blocks can achieve low U-values: Approximate U-value 0.16 W/m^2·K
Nearly zero	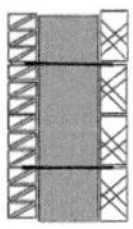	Masonry cavity walls have been built to passive standard using mineral wool, 300 mm cavity, 425 basalt fibre low-conductivity wall ties Approximate U-value 0.113 W/m^2·K
Timber frame with brick cladding	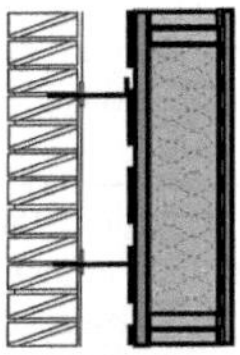	Early form of timber frame Approximate U-value 2.0 to 2.5 W/m^2·K .
Two skins of timber frame	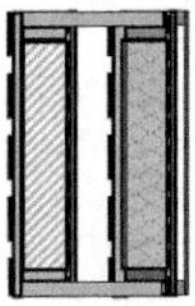	90 mm stud, 50 mm insulation Note that in traditional timber frame the U-values are affected by the timber fraction in the wall Approximate U-value 0.39–0.56 W/m^2·K (Bell and Overend, 2001)
Timber frame with I beams and brick slip cladding	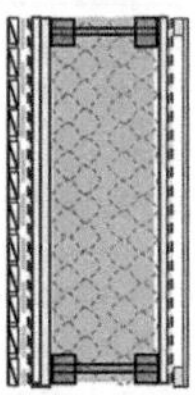	The void between the panels of timber is filled with insulation. Within the panels thermal bridging is limited to the I beams and junctions Approximate U-value 0.113 W/m^2·K
Timber frame built to Passivhaus	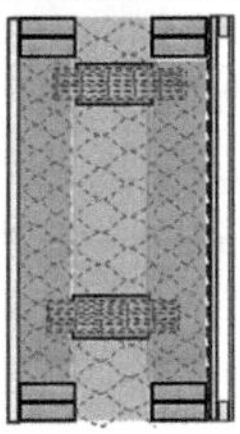	300 mm twin stud, 36 mm × 89 mm studs, 122 mm space, filled with cellulose insulation Internal service void with plaster finish Approximate U-value 0.12 W/m^2·K

(continued)

Table 6.10 *(continued)*

Construction types and thermal performance		
Cob or rammed earth	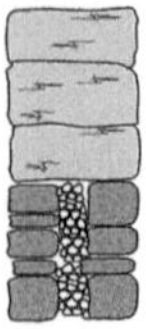	Cob, cleam, clob, or clom: Subsoil with a high clay content, mixed with straw, sand, and water. Typically in-filled between a post and beam structure. Rammed earth: gravel, sand, and clay compacted (between formwork) in layers. Both historically used, still common in some developing countries. Dries a stone-like hardness similar to adobe mud brick. Lower levels can be stonework with rubble fill (Harrison and de Vekey, 1998) Approximate U-value 0.4–0.8 $W/m^2{\cdot}K$ (mud wall) Baker (2011) 0.7–0.8 $W/m^2{\cdot}K$ (cob) (for 500–600 mm wide) 1.2–2.0 $W/m^2{\cdot}K$ (stone)
Straw bale	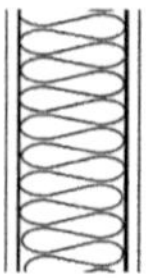	Straw bales are around 450 mm thick, often finished with lime render and lime plaster, or timber (e.g. in ModCell) and seated on a foundation base. Thermal properties are dependent on the straw density. Load-bearing walls will be typically at least 110 kg/m^3, whilst infill can be 80–100 kg/m^3 (Goodhew et al., 2010) Approximate U-values 0.13–0.16 $W/m^2{\cdot}K$ (Rye, 2011) 0.19 $W/m^2{\cdot}K$ (https://strawbalebuildinguk.com)
Retrofit stone wall internally insulated with cellulose or similar	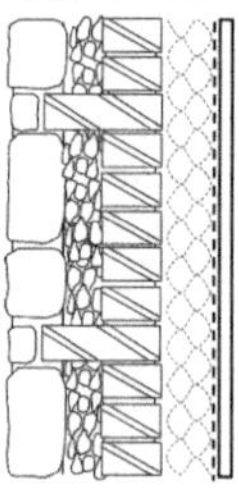	Stone cavity wall, filled with rubble and insulated from the inside with cellulose or similar. Porous stone allows water to penetrate. Rubble fill prevents water passing through the cavity. During winter months, the external skin may retain high levels of moisture. Breathable insulant allows any moisture built up to evaporate through the insulation. Breathable vapour barrier prevents moisture entering the fabric from the internal environment but allows any moisture trapped in the fabric to escape Approximate U-value 0.35 $W/m^2{\cdot}K$ (100 mm cellulose, sheep's wool, or mineral wool)

Table 6.11 Summary of issues and challenges for internal wall insulation (IWI) and external wall insulation (EWI) for solid wall properties (Gorse et al., 2017)

Property type	Intervention	Thermal breaks(TBs)	Services &fittings	Element interfaces	Lack ofinsulation	Ventilation &uninsulated elements	Openings: doors & windows	Party walls
Historic stone built	EWI	Planning – EWI not permitted	Where permitted, potential as below	Where permitted, potential as below	Where permitted, potential as below	Where permitted, potential as below	Where permitted, potential as below	Where permitted, potential as below
	IWI	Increased frost damage to external face	Potential as below	Potential as below	Potential as below	Potential as below	Potential as below	Potential as below
Pre-1919 solid brick terrace and back-to-backs	EWI	TBs introduced to top floor wall/ ceiling junctions	Thermal breaks at insulation cut-outs at external pipes, services, walls, etc., incl air gaps around	Thermal breaks introduced to ground floor/ wall junctions	Lack of insulation to ground floor over unconditioned basement	TBs introduced at uninsulated doors, stair soffits, & spandrels to uninsulated basement	Window & door lintels, jambs, and sills left uninsulated or with reduced insulation	EWI not extended across to neighbouring property: TBs at party wall/ ext. wall jns. EWI extended: fire propagation risk
	IWI	TBs introduced at intermediate floor/ wall junctions	Air gaps around service passing through external walls	Potential as above	Potential as above over basement	Lack of improved ventilation	Potential as above	Potential as above

(continued)

Table 6.11 *(continued)*

Property type	Intervention	Thermal breaks(TBs)	Services &fittings	Element interfaces	Lack ofinsulation	Ventilation &uninsulated elements	Openings: doors & windows	Party walls
1919–1940s solid brick semis and terraces	EWI	TBs introduced to topmost wall/ ceiling junctions, especially where rooms partially in roof space with sloping soffits	TBs at insulation cut-outs at external pipes, services, externally accessed stores, walls, etc.	Potential as above	Lack of improved floor/ roof insulation	Potential as above	Potential as above	Potential as above
	IWI	Potential as above for IWIs	Potential as above for IWIs	Potential as above	Potential as above	Potential as above	Potential as above	Potential as above
1945–1970s system-built, incl. flats, terraces & semis	EWI	TBs introduced to top floor wall/ ceiling junctions	Potential as above for EWI	Potential as above	Potential as above	Potential as above	Potential as above	Potential as above
	IWI	Potential as above for IWIs	Potential as above for IWIs	Potential as above	Potential as above	Potential as above	Potential as above	Potential as above

References for "Thermal performance of existing buildings and retrofit"

Anderson, B. (2006a). *Conventions for U-value calculations* (BR 443:2006). Bracknell, Berkshire, BRE Press.

Anderson, B. (2006b). Thermal properties of building structures. In: *CIBSE guide A: environmental design* (The Chartered Institution of Building Services Engineers, 2006). 7th ed. London, CIBSE. Ch. 3, pp. 3.1–48.

Baker, P. (2008). *In situ U-value measurements in traditional buildings: preliminary results* (Historic Scotland Technical Paper 2). Edinburgh, Historic Scotland. Available at http://www.historic-scotland.gov.uk/technicalpapers

Baker, P. (2011). U-values and traditional buildings In situ measurements and their comparisons to calculated values. Technical Paper 10, Conservation Group, Historic Scotland, Alba Aosmhor. Available at https://www.historicenvironment.scot/.

Bell, M. and Overend, P. (2001). Building regulation and energy efficiency in timber frame housing. COBRA Conference at Glasgow Caledonian University, 3–5 September 2001. RICS Foundation.

DECC (2011). The governments standard assessment rating of dwellings SAP 2009 edition incorporating RdSAP 2009 Table S6: Wall U-values England and Wales RdSAP Appendix S.

Energy Saving Trust (2004). *Scotland: assessing U-values of existing housing. (Energy efficiency best practice in housing: CE84)* London, Energy Saving Trust. Available at http://www.energysavingtrust.org.uk/business/content/download/180012/441763/version/4/file/ce84.pdf

Glew, D., Fylan, F., Farmer, D., Miles Shenton, D., Parker, J., Thomas, F., Fletcher, Hardy, A., Shikder, S., Brooke-Peat, M., M., Sturges. J. and Gorse. C. (2020). Thin internal wall insulation, measuring energy performance improvements in dwellings using thin internal wall insulation. Summary Report. Department for Business, Energy & Industrial Strategy.

Goodhew, S., Carfrae, J., and De Wilde P. (2010). Briefing: challenges related to straw bale construction. *Proceedings of the Institution of Civil Engineers, Engineering Sustainability* 163, December 2010, Issue ES4, pp. 185–189.

Gorse, C. Glew, D. Johnston, D. Fylan, F. Miles-Shenton, D., Smith, M., Brook-Peat, M. Framer, D., Stafford, A., Parker, J., Fletcher, M. and Felix, T. (2017). Core cities green deal monitoring project leeds. BEIS Buildings Energy Efficiency Technical Research, 9 November 2017. Available at https://www.gov.uk/government/collections/buildings-energy-efficiency-technical-research

Harrison, H. W. and de Vekey, R. C. (1998). *BRE building elements: walls, windows and doors*. Watford, UK, BRE.

NHBC (2019). House building: a century of Innovation. Technical Advances in Conventional Construction. NHBC Foundation. Milton Keynes. Available at https://thinkhouse.org.uk/site/assets/files/1367/nhbc1019.pdf

Rye, C. (2011). *The real life of buildings—insitu monitoring and the performance of traditional buildings*. Archimetrics. SPAB, London, UK.

Rye, C. (2014). U Values and retrofitting traditionally built walls, archimetrics. Available at http://www.wtbf.co.uk/newsite/_blog/app/web/upload/Wall%20Insulation%20Conference/3%20Caroline%20Rye%20U-values%20%26%20Retrofitting%20Traditionally%20Built%20Walls.pdf
Smith, M. B. (2018). A study of building surveying praxes to inform design of thermal retrofit. Ph.D. Thesis. Leeds Beckett University

EMERGING THEMES: GAPS AND NEEDS IN BUILDING PERFORMANCE SIMULATION FOR BUILDING RETROFIT

Cormac Flood and Lloyd M. Scott

In an increasingly expanding and developing world, information and data have become more crucial to understanding how to progress and refine processes to achieve better quality of life. Construction has always been a sector driven by results and progression, and while research carried out to establish if the accurate analysis of in-situ performance is possible, a number of findings have been established as a common theme from many case studies:

- Inconsistencies in default resistances for standard U-value calculation method.
- The gap between standard U-values and in-situ measurements is significant – walls are not performing as designed.
- The importance of wind-driven rain inclusion within the simulations – target U-values are not achievable if based on the standard calculation procedure.
- Recognition of the influence of orientation in relation to the U-value performance of walls – each orientation causes the wall to perform differently.
- The issues connected with construction information records – not adequate, with no strict procedure to do so.
- The importance of accurate material data in simulations.
- The U-value fluctuates according to the climate conditions, meaning that the monthly average may be different throughout the year.

Thermal analysis is an important function which assesses the energy performance of the building fabric through calculation, simulation, or in-situ analysis. The accuracy of these figures is fundamental to the projection of design requirements to achieve energy-efficiency targets through the building fabric. Buildings are continually subject to fluctuating internal and external climate conditions such as temperature, moisture, solar radiation, and wind. Wind-driven rain has an effect on the hygrothermal changes of materials depending on the climate and orientation. These variations represent potential key factors that affect and define the actual physical thermal performance and sustainability of the building envelope.

Findings from research challenge traditional understandings of building envelope performance. What is evident is that the external walls perform more thermally efficiently in a dry state. Existing properties can frequently be regarded as "damp". This can be an issue when the property is being refurbished and/or retrofitted; however, the economics of a retrofit project often preclude any drying-out period. Meanwhile the property is expected to perform as it would were it "dry". In contrast to the concept of a "drying-out period", external walls can display signs of a "wetting period" until they reach a moisture equilibrium.

Guidelines on moisture control associated with the preservation and renovation of heritage constructions and the rehabilitation of the building stock include publications by the International Association for Science and Technology of Building Maintenance and Monument Preservation and EN 15026. Many published guidelines do not contain any information on how to deal with small defects in the building envelope. Later publications link rainwater penetration into constructions with rendered facades to damage and consider the effects of small leaks in the exterior finishes of exposed walls. For example, ASHRAE considers that any flaws in the façade should be accounted for in analysis programmes, by simulating 1% of driving rain to penetrate the outer surface of the façade for at least one hour. Obviously, the orientation and geographical position of the façade have an effect on the amount of wind-driven rain the façade is normally exposed to.

In addition, climate projections include more extreme weather conditions, so this should be taken into account when considering current climate records. Construction and building envelope performance is highly weather-dependent. Optimal construction conditions require elements such as precipitation, wind, temperature, and relative humidity to fall within firm limits, while many products and materials are designed to be installed within certain climate conditions, outside of which their performance declines, depending on the severity of the conditions. Building envelopes should be built of materials that can resist the deteriorating effects of the weather, in line with the appropriate building standards. However, building standards do not generally facilitate the inclusion of the aforementioned factors with regard to thermal performance. Additionally, with climate change, the current relationship between building fabric performance and weather is being altered continually, enhancing the agenda for building design and adaptation measures to incorporate current and developing climate patterns.

U-values and the development of correction factors, or of the calculation itself, to account for climate patterns are scarce. To this end, these values (external (Rso), cavity (Rcav), and internal (Rsi) surface resistances) incorporate two factors only – wind speed and emissivity. Although wind and other weather conditions are dissimilar across the entirety of countries, default values for Rso, Rcav, and Rsi have set default values, used nationwide.

There are many private companies and national weather centres offering the latest available climate data to incorporate into project planning. The key point is that the data used currently in assessing the risks to weather are historical. Because the cost and duration of construction activities are heavily influenced by climatic conditions, it is becoming increasingly important that decision-making should be based on future projections, as well as past conditions. Nonetheless, it is important that the relationship between current weather conditions and construction be discussed first to establish a baseline condition against which the impacts of climate change can be compared.

Obviously, any imperfections in an external finish will leave a gap that rain can penetrate. This has been long known, for example in hairline-crazed render, permitting moisture to enter but protecting it from evaporation in sun or wind. The presence of moisture usually increases the U-value, and this needs to be considered for thermal performance. Many conventional assessment methods completely ignore liquid moisture transport; for example, as a result of wind-driven rain. More sophisticated one- and two-dimensional full heat and moisture models are available that allow modelling vapour and liquid flow, that are transient in nature, that consider moisture sources such as wind-driven rain, rising damp, initial moisture, sorption and de-sorption, interstitial condensation, and surface condensation.

The complexity of energy-efficiency improvement of existing dwellings cannot be underestimated. Investments in the homes are particularly labour and cost intensive, thus it is difficult to convince homeowners to implement full or even partial retrofit strategies. Research shows a number of barriers to action regarding domestic energy conservation and the implementation of retrofit strategies, including:

- High upfront costs
- A lack of understanding amongst homeowners about energy use and potential savings from retrofit of their homes
- A lack of disposable income for retrofit works
- Uncertainty surrounding energy prices, particularly with payback periods over five years
- A lack of incentive
- Inconvenience
- A lack of skill in the construction industry to carry out retrofit works
- A lack of a concerted marketing strategy targeting homeowners, in getting its message delivered to consumers

See also Chapter 5, the section “Upgrading thermal performance” and Chapter 5, the section “Discussion surrounding thermal upgrades”. Existing limitations in model simulations include building uncertainty and model uncertainty, and a steep learning curve is required.

7

Fire safety

Melanie Smith

This chapter covers some of the main areas of fire safety. It attempts to give background knowledge into matters related to fire. Further reference should also be made to Chapter 3, "Legal and regulatory frameworks" and the subsections on "Fire safety in occupied buildings" and "Fire safety and accessibility", and Chapter 5, "Retrofitting and refurbishment" and the subsection on "Upgrading fire protection". The sections in this chapter cover:

- Nature and development of fire
- Effects of a fire: physiological and behavioural
- Fire precautions from first principles
- Materials and fire
- Means of escape
- External fire spread
- Fire engineering concepts

NATURE AND DEVELOPMENT OF FIRE

The fire triangle

What is a fire? A fire is a very rapid exothermic chemical reaction between a fuel and oxygen releasing heat and light. Flames are the visible manifestation of the reaction between a gaseous fuel and oxygen.

The "fire triangle" in Figure 7.1 shows the three components necessary for fire to occur.

The significance of this simplistic model is that:
- A fire cannot occur if any one of the three components is missing.
- Removal of any one of the three will terminate the reaction and put out a fire.

Fire prevention is, therefore, about avoiding the introduction of the ever-present ignition source to the ubiquitous fuel.

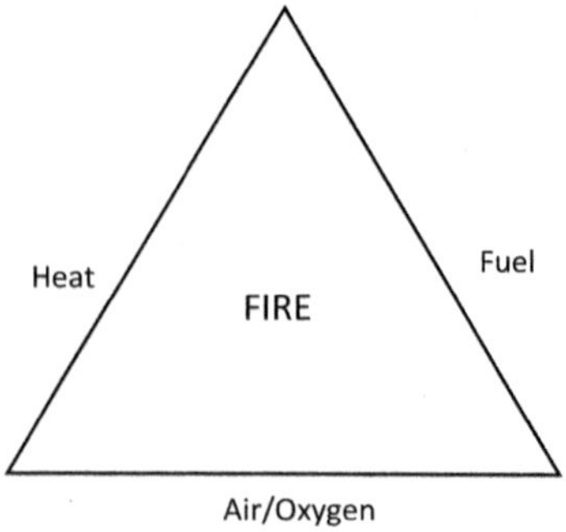

Figure 7.1 Fire triangle.

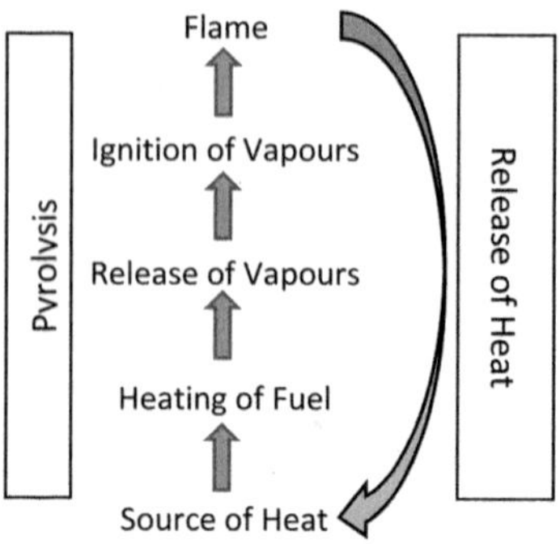

Figure 7.2 Closed feedback loop of fire.

Combustion of solid materials

Solid materials do not "burn". This is not obvious from common experience. Flame is actually a gaseous-phase chemical reaction between a fuel vapour and oxygen. It is not the fuel itself that burns but the vapours given off as the fuel is heated. As an example, when a piece of wood is heated sufficiently, the chemical components of the wood begin to break down, releasing flammable gases. These gases or vapours then ignite to form the visible flame. The combustion of the vapours in the presence of oxygen releases heat. This heat is transferred back to the wood surface, releasing more vapours to be ignited. A closed feedback loop is thus created, see Figure 7.2.

Whilst having regard to the above, **spontaneous ignition** can also occur. Some materials can "self-heat" and do not require an external ignition source, for example a heap of oil-soaked rags kept in a warm cupboard can self-ignite.

Why do solid materials burn more readily if they are split into a finer form? For example, loose sheets of paper ignite more readily and burn more fiercely than a book. The heat sink effect is less, less heat is lost, more heat is more easily utilised in the burning reaction. Secondly, more oxygen is available between the sheets of paper. Extreme examples are **dust explosions**, e.g. of flour – a dust explosion is the rapid combustion of fine particles suspended in the air, often in an enclosed location, e.g. grain, sawdust, coal.

Combustion of flammable liquids

It is not the liquid that burns but the vapour above the liquid surface. Heat is usually required to produce the vapour. Useful measures of the hazard of a flammable liquid are its flashpoint, its flammability limits, and its auto-ignition temperature. The **flashpoint** is the minimum temperature in test conditions at which the liquid gives off sufficient vapour to ignite. For example, the flashpoint of vegetable oils can be 200°C, whilst petrol is typically –43°C.

Some liquids (e.g. castor oil, olive oil) act more like solids and require a substantial amount of heat before they will ignite. These liquids can be called **combustible**. Their flashpoints are over 55°C.

Other liquids (e.g. white spirit, paraffin) need much less heat in order to release enough vapour to ignite. These are termed **flammable**. They pose a much greater threat than combustible liquids. Their flashpoints are between 32°C and 55°C.

More dangerous are **highly flammable** liquids (e.g. petrol, ether) which produce sufficient vapour at normal ambient temperatures that heating as such is not required, but the vapour/air mixture can be ignited by a small source such as a spark. These liquids are particularly hazardous and require special precautions so that sources of ignition do not come into contact with their vapours. Their flashpoints are below 32°C.

Combustion of flammable gases

As stated before, it is the gas that ignites. There is not much difference between the combustion of flammable vapours from a liquid (at normal temperatures and pressure) and the combustion of flammable gases (at normal temperatures and pressure).

Flashpoint as such does not occur because the flammable substance is already a vapour. The flammability limits may give an indication of hazard. Auto-ignition (as in liquids) is less significant and commonly ranges between 200°C and 600°C for most common flammable gases.

Flammable gases are present, for example, in fuels for heating and cooking, and cutting/welding equipment (liquefied petroleum gas (LPG) or acetylene) and tar boilers (LPG). LPG is much denser than air and can collect in areas such as basements, pits, and drains. Acetylene has very wide explosive limits.

Extinguishment

Refer to the fire triangle, Figure 7.1. In order to extinguish a fire, it is necessary to remove one of the components. Or refer to the closed feedback loop, Figure 7.2. The loop must be broken.

Water puts out a fire primarily by cooling the fuel (reducing the heat). The fuel no longer releases enough vapours to allow combustion to be sustained.

Foam applied to the surface of a burning liquid acts as a physical barrier, preventing release of the vapour into the combustion zone above the liquid's surface. The foam also stops the radiation of heat back down onto the liquid surface.

Carbon dioxide, CO_2, displaces oxygen sufficiently to stop the combustion process. So, it directly removes one of the three components of the fire triangle. This is useful where water, foam, or powder might damage the articles to be protected.

Halons interfere with the chemical reactions that occur within the flames (chemical inhibition) and used to be the choice for protecting computer, etc., equipment. Unfortunately, halons also have the highest ozone-depleting capacity of any chemicals in common use, and halon extinguishers were subsequently banned for this use. They have generally been replaced with CO_2.

Powders also use chemical inhibitions, as well as dilution of the fuel vapour, restricted access to oxygen, and the removal of heat. Powders are also very effective with rapid knockdown of flames.

Gaseous fire-suppression systems flood the areas with inert gases acting similarly to CO_2 and halon described above. Occupants must leave before the suppression system is activated.

Classification of fires

Different extinguishing agents are suitable for different types of fire. For instance, it would not be sensible to try to extinguish an electrical fire with water. For this reason, fires are grouped into different classifications according to the nature of the fuel. In the UK, BS4547 gives the following:

Class A involves "normal" (usually carbonaceous) solid materials, e.g. wood, paper, textiles.
Class B involves liquids or liquefiable solids.
Class C involves gases.

Class D involves metals. Requires special extinguishing agents.
Class F involves cooking oils and fat.

There are also electrical fires, a term sometimes used but not classified because electricity is a source of ignition, not a fuel – the main concern here is that the electricity is turned off before extinguishment is attempted. Class C fires most often occur through a gas leak, e.g. from a cylinder. If the flame is extinguished without the leak being checked, there is the possibility of explosion. In most premises, it is the Class A and Class B fires that are the important ones for normal consideration.

The development and spread of fire

Once a fire in a room has ignited, the subsequent development can be rapid. Mathematically, the development is exponential, doubling in size at regular intervals of only a few minutes. This rapid growth is beyond the common experience of most people because it contrasts sharply with the more common, sedate behaviour of a fire in the open such as a garden bonfire.

Fires in enclosed spaces behave differently and with different rates of burning from fires in the open. The presence of a ceiling or roof over a fire has an immediate effect in increasing the radiant heat returned back down onto the fuel (see footage of the Bradford City Stadium fire).

This inability to anticipate the rate of development has proved to be a major factor where there have been multiple deaths in fires, for example, the Stardust Disco fire, the Bradford City Stadium fire, the Manchester Woolworth's fire, and possibly the Grenfell fire. There were failures to realise the need for immediate evacuation.

Fire develops by heat transfer, i.e. conduction, convection, and radiation. The most important means within a building are convection and radiation.

Typical progress of a fire

After the first item in a room is ignited, hot gases rise vertically in a relatively narrow plume into which air is entrained, so increasing the volume of smoke and gases (Figure 7.3).

As the smoke reaches the ceiling, it spreads out in all directions (ceiling jet) and begins to form a rapidly deepening layer below the ceiling. Therefore, with a low-ceiling room or restricted space such as a corridor, loss of visibility may be one of the earliest threats created by the fire. The smoke is not just a particle cloud, it includes hazardous gases which have been released, and burning gases. The main point in life safety is that the occupants get out before this layer descends on them.

As the fire grows, the flames reach the ceiling and are deflected horizontally, radiating downwards over a large area of the room. The downward radiation can

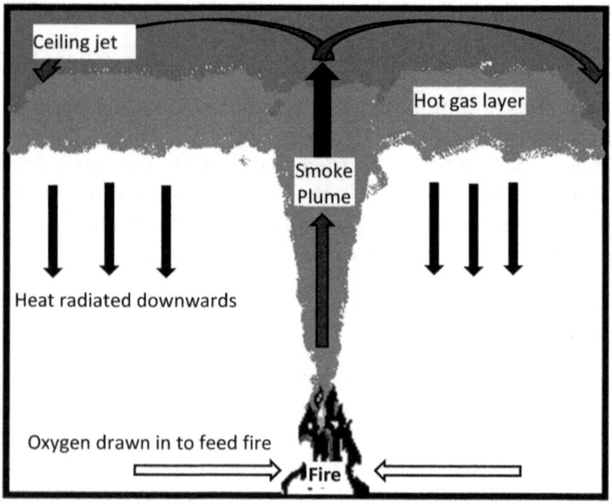

Figure 7.3 Fire smoke plume.

be so strong that a large area of combustible materials in the room will reach a temperature at which these materials spontaneously burst into flames. This is known as **flashover** and may be reached quite quickly. At flashover, virtually all items in the room are alight and survival of occupants within the room is impossible. Flashover typically occurs at around 600°C. See also Figure 7.4.

The progress of a fire can be split into three distinct phases: growth, post-flashover, and decay:

1. **Growth period** – this lasts from the moment of ignition to the time when all combustible materials within the room are alight. The temperature rises relatively slowly until flashover. The duration of the growth period depends on factors such as fuel type and amount, and ventilation. A critical moment is reached when the flames reach the ceiling and the radiant heat transfer back down is dramatically increased. This occurs at around 550–600°C. The remaining combustible materials now rapidly reach their fire points and ignite within three to four seconds. This sudden transition is known as **flashover** and represents the start of the **stable phase** of the fire. If there is insufficient ventilation available to the fire, then the fire may fail to flashover due to oxygen starvation. It may die completely or smoulder quietly. A smouldering

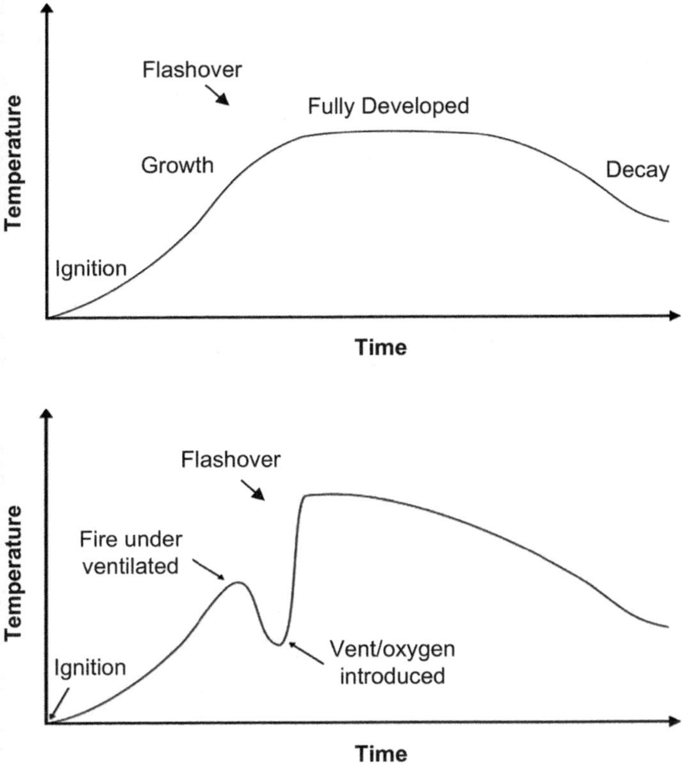

Figure 7.4 Fire development: oxygen constrained and oxygen introduced (www.nist.gov).

fire can be extremely dangerous as the room fills with flammable vapours. If a fresh supply of oxygen is then introduced (for instance by a door opening) the fire may ignite with an eruption of flame. This is known as **backdraught**.

2. **Post-flashover period** – the temperature is very high owing to the involvement of most of the combustible materials in the room. It is at this stage that the highest temperatures will be reached. The flaming is no longer localised but occurs throughout the room. The rate of burning is determined by the level of ventilation and the amount of fuel present.
3. **Decay period** – this arises from the total consumption of most of the combustible materials and continues until there remains no fuel for combustion. The temperature steadily falls.

The plotting of the temperature against time from ignition gives a fire growth curve. Fire growth curves vary depending on the conditions and are useful to fire scientists/engineers considering the consequences of altering these conditions. Work is continuing on design fire growth rate models for fire engineering.

As a fire develops, it produces heat and smoke; both are dangerous and must be designed against.

Heat

The rate of burning is dependent on the fuel and ventilation available, and these determine the amount of heat produced. The quantity of the potential fuel within a building is described as the building's **fire load**. This includes both the fabric of the building and the contents. An estimate of the fire load can give guidance on the likely heat production and fire severity.

It is very difficult to establish accurately the fuel load of a building due to the multiplicity of materials involved in the construction and the usually unknown materials that make up the building's contents. However, NIST and NFPA (for example) have conducted much work on this.

Building fabric

One of our problems is the plethora of terms used to describe the fire safety of materials. It is not always possible to define a material as "safe" or "unsafe" – it all might depend on how and where the material is used. You must be wary of manufacturers who simply assure you that the material has a certificate – you need to know what certificate, the circumstances in which it gained that certificate, what (exactly) was tested, and how this relates to real life and/or the circumstances in which the material is actually used in your project.

The different terms do not make life any easier. The essential characteristics of building materials which can be measured are:

Ignitability – the ease with which a material can be ignited when subjected to a flame.

Combustibility – whether or not a material will burn when subjected to heat from an already existing fire.

Non-combustible materials – (not a universal definition) for example, a material that, under the conditions anticipated, will not ignite or burn when subject to fire or heat; products classified as non-combustible under BS476-4:1970.

Materials of limited combustibility – (England and Wales only) – a material (including "non-combustible materials") which meets the various test criteria set out in the Approved Documents for Building Regulation B.

Fire propagation – the degree to which a material will contribute to the spread of a fire through heat release when it is itself heated; this is concerned with the level of heat emission and the rate of heat release.

Surface spread of flame – whether a material will enable the spread of flame across its surfaces. Fire spread occurs to a great extent along the surfaces. Internal linings (cladding surfaces, walls, ceilings, and floors) form a large area of continuous surfaces for a fire to use. When hot, these linings can heat materials not yet involved in the fire – mainly by radiation.

A point of ignition heats up a small area of the surface sufficiently for flammable gases to be released from the surface. The gases ignite, heating up (by radiation) the adjacent small areas of surface, which release more flammable gases and so on (Figure 7.5). By these incremental movements, fire spread can occur quickly along a surface, independently of the initial point of ignition's capacity to burn into the material.

Lining materials also produce smoke and toxic gases. But these are not measured in the tests nor covered by the regulations. It is something, however, to consider when carrying out fire engineering.

Materials are therefore classified according to their surface's ability to propagate fire spread. In the UK these are classed 1–3 and above, with Class 1 not being a good propagator.

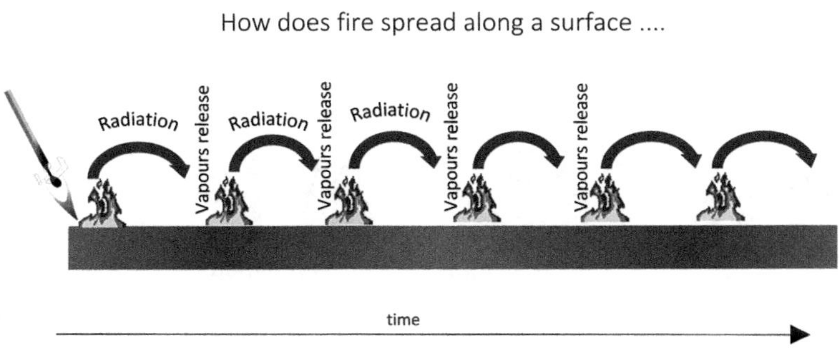

Figure 7.5 Surface spread of flame.

Class 0 – Building Regulations (England and Wales) – this is a classification of surface spread of flame and does not indicate anything else, nor does it over-ride other tests or standards. A Class 0 material must meet Class 1 surface spread of flame and must also limit fire propagation meeting the requirements of BS476-6.

Fire resistance – whether or not a component, or assembly of components (i.e. not just a material), will resist fire by retaining their loadbearing capacity, integrity, and their insulating properties; see Figure 7.6. Construction of known fire resistance is the most fundamental means of protecting escape routes and preventing fire spread. Defined in BS4422 as the "ability of an element of building construction, component or structure to fulfil, for a stated period of time, the required stability, fire integrity and/or thermal insulation and/or other expected duty in a standard fire resistance test".

a. **Stability or load-bearing capacity** is the ability of the element to continue to support its load without excessive deformation.
b. **Integrity** is the ability of the element to continue to contain a fire without collapse or developing splits, holes, or cracks through which flame could easily pass, and without sustained flaming on the side not exposed to the fire.
c. **Insulation** is the ability of the element to resist the passage of heat from the exposed to the unexposed side. If heat can readily be conducted through, this can start a new fire on the other side, and/or burn people trying to escape along the other side.

The legal building regulations or standards in the relevant country will provide the time duration needed for fire resistance of the building element or component.

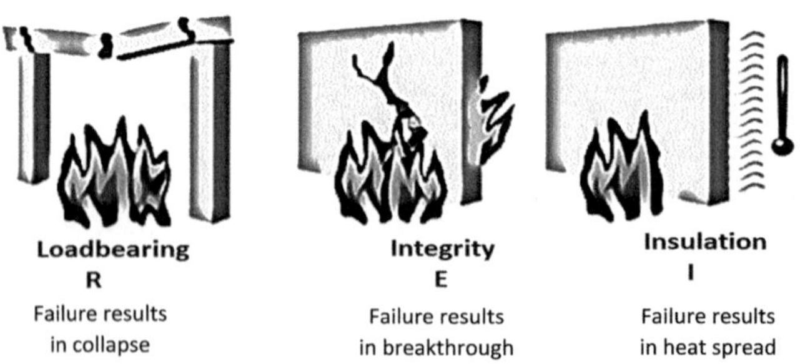

Figure 7.6 The three properties of fire resistance (adapted from Structural Timber Association, www.structuraltimber.co.uk/assets/InformationCentre/eb7.pdf).

Smoke obscuration – the degree to which the material when burning will produce smoke that leads to reduced visibility.

Tests and measurements

There are tests and standards for each of these properties. A material can be good in one area, and hopeless in another. The tests, which permit these various terms to be used, are usually made using small-scale testing on isolated product samples, and not necessarily the full system/component in which the materials are to be used in practice on site. Even where a combined system is tested, minor changes (such as the width of a gap) can have a notable impact on a real building, and these changes might be site purchase or design alterations for practicality or cost.

Additionally, the results obtained in such tests are heavily dependent on the exposure conditions. Even materials classified as non-combustible can affect the development of a fire. Materials with good thermal insulation properties can allow a rapid rise in temperature in the fire compartment. Materials of high heat capacity can soak up heat from the fire and tend to slow its rate of growth.

EFFECTS OF A FIRE: PHYSIOLOGICAL AND BEHAVIOURAL

Physical effects and toxicology

We all know that fire burns. It also does other things to people. Before we get to those awful physical conditions when life is untenable, the atmosphere can get progressively worse as people are trying to escape.

a. The atmosphere may be hot: temperature may exceed 1000°C (see Table 7.1).
b. Toxic and narcotic gases will be present, e.g. carbon monoxide and hydrogen cyanide:
   i. At high concentrations carbon monoxide causes rapid death;
   ii. At lower concentrations it brings about a loss of co-ordination, particularly on exertion.
c. A fire atmosphere will contain a low concentration of oxygen. This in itself can bring about unconsciousness and death, but normally the effects of toxic gases predominate.
d. There may be so many small particles in the atmosphere that vision is severely restricted.
e. The effects of irritants to the upper respiratory tracts and eyes may impede the ability to escape.

Table 7.1 Fire dynamics

Temperature °C	**Response**
37°C	Average normal human oral/body temperature
38°C	Typical body core temperature for a working fire fighter
43°C	Human body core temperature that may cause death
44°C	Human skin temperature when pain is felt
48°C	Human skin temperature causing a first-degree burn injury
54°C	Hot water causing a scald burn injury with 30-second exposure
55°C	Human skin temperature with blistering and second-degree burn injury
62°C	Temperature when burned human tissue becomes numb
72°C	Human skin temperature at which tissue is instantly destroyed
100°C	Temperature when water boils and produces steam
250°C	Temperature when charring of natural cotton begins
>300°C	Modern synthetic protective clothing fabrics begin to char
≥400°C	Temperature of gases at the beginning of room flashover
≈1000°C	Temperature inside a room undergoing flashover

Adapted from NIST (www.nist.gov).

These effects are invariably present all together in fires. Low concentrations of fire gases bring about behavioural changes and incapacitation, so reducing a person's chance of escape. High concentrations bring about rapid unconsciousness and death within minutes.

Escape behaviour

Appreciating the behaviour of people in a fire can help designers and building managers to improve means of escape design, and help people to safely evacuate a building in a fire. Appropriate design for escape needs an understanding of:

1. The normal use of a building (e.g. which routes are most familiar), and
2. The fact that a building is not only a physical structure but an information and communication system.

There is a complex relationship when we consider people in a building, which is made more complex when we add a fire in an arbitrary place. Figure 7.7 shows the different factors involved in fire safety.

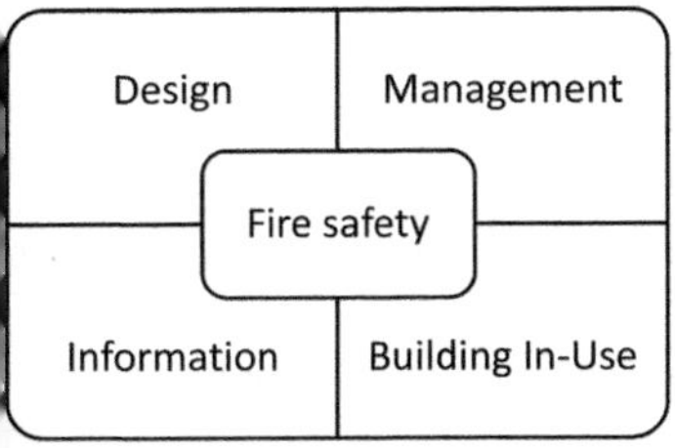

Figure 7.7 Essential factors in fire safety.

Many of the problems for people in fires arise out of an inadequate appraisal of the way in which buildings are managed, used, and comprehended by people in normal circumstances. Certain architectural forms can be seriously disorientating for people in everyday use. The non-correspondence between the outside form of a building and the internal layout, for example, may give difficulties for people trying to escape, e.g. internal corridors with no windows, basements, underground stations.

Environments provide information; information is selectively attended to by people. The anticipations and expectations of people about spatial layouts and dimensions influence the cues that they pick up, before and during movement for escape.

There are two very different models used to predict, describe, and explain people moving around buildings, Model A: Movement, and Model B: Behaviour.

Model A: Movement

The engineering, physical science, or "ball-bearing" model of human movement. This is the predominant model of escape behaviours in fire codes, fire science and engineering, and in media coverage of fires. The assumptions are:

- People behave irrationally, without thought.
- Homogenous crowd of individuals in flight.
- Able to model by non-thinking objects propelled to an exit.
- Panic.

Therefore, for Model A, there is a need to:

1. Keep information about a threat away from people until absolutely necessary – to avoid panic.

2. Determine:
   a. Spread of fire
   b. Speed of smoke spread
   c. Numbers of people
   d. Dimensions of escape routes (positions of exits, travel distances, widths o exits)
3. To assess the time for the "ball-bearings", or people, to escape and the direc- tion of movement.

Model B: Behaviour

The social science or psychological model of human reactions. Assumptions are:

- People are active agents, not ball-bearings.
- People think and act according to the available information, group ties, anc their role within a building.
- Emphasis on patterns of behaviour rather than where behaviour takes place.
- Flight is not a panic reaction, but a rational action of people warned too late tc evacuate in an orderly fashion.

Therefore, for Model B, there is a need to:

1. Provide early information for people to act appropriately.
2. Assess the time for people to escape, determine:
   a. Reaction time
   b. Knowledge of the building
   c. Social constraints
   d. Visual access to exits, and wayfinding

There are a number of "significant fires", and study of the use of these models ir each of these, or consideration of how using a different model could have helpec reduce the casualties, can be a useful exercise. Model A is easier to study, com- prehend, and apply mathematics to. Model B tends to follow "real life". The bes solutions use both models appropriately. See Table 7.2 for a selection of significan fire summaries.

Research into fire progression and escape behaviour continues. One classic study was an experiment on individual and group reactions to smoke seeping intc an unfamiliar office room from under a door. The reactions of individuals alone in the room, in going to and opening the door, were much faster than in a grour of strangers together in the room. This reflects constraint on being the first to act

Table 7.2 Selection of significant fires summaries, UK and Ireland

Selection of significant fires summaries, UK and Ireland (date order)		
Name	**Details**	**Findings**
Ronan Point	1968 London 4 deaths	Partial collapse due to gas explosion blowing out loadbearing walls. Although "few" casualties, the visual effect of the complete collapsed corner of the 22-storey building resulted in loss of public confidence in high rise. • Led to change in Building Regulations for structural adequacy and "disproportional collapse".

Summerland Leisure Complex	1973
	Douglas
	Isle of Man
	41 deaths
		Rapid fire spread in cavities in the external wall construction.
		Not all occupants prioritised escape on the alarm but prioritised travelling to upper floors to reunite with their children.
		Criticisms:
		• No overall duty in respect of fire safety.
		• No staff training in fire safety.
		• Long delay in summoning the fire brigade.
		• No organised methodical evacuation.
		• Locked fire exits and car parked across doors.
		• Misguided actions by staff.
		• Delay in operating fire alarm system.
		• Plastic wall linings used which aided growth.
		  • Led to maximum sizes of cavities and requirements for cavity barriers.
		  • Led to glazed areas being considered as part of wall surfaces.

Woolworths, Manchester	1979
	Manchester
	10 deaths
		Originated in a furniture display area in an open storey that also included a restaurant.
		Criticisms:
		• Delay in summoning the fire brigade. Subsequent tests showed the fire would have reached maximum severity 2 minutes after ignition.
		  • Led to amendments to the 1971 Fire Precautions Act to allow fire authorities to take account of active fire-extinguishing systems in connection with fire certificate.
		  • Public unable to appreciate need for immediate evacuation and died as a result.
		  • Use of keys in glass-fronted boxes for doors on escape routes not to be allowed in future.
		  • Started research on human factors.

(continued)

Table 7.2 *(continued)*

Stardust Disco	1981 Dublin 48 deaths	Interaction between wall linings, radiation from low ceilings, and furnishings which together aided rapid fire growth. Criticisms: • No emergency evacuation plan. • Employees not allocated specific duties in the event of a fire. • Staff as confused as patrons during fire. • Exits locked or gave impression of being locked. • Delay in summoning fire brigade. • Actions of staff uncoordinated and inadequate. • Failure to operate the alarm system. Public were unable to appreciate need for immediate evacuation and died as a result.
Valley Parade Football Stadium	1985 Bradford 56 deaths	"Open-air" conflagration caught on TV. Combustible ceiling, configuration which aided rapid fire growth. Public failed to appreciate need for immediate evacuation and died as a result. • Led to amendments and extended scope of Safety of Sports Ground Act 1975 to include sports stadia with a capacity of 10,000 or more. • Led to Fire Safety and Places of Sport Act 1987. • Introduced certification for "regulated stands".

King's Cross Underground	1987
	London
	31 deaths
		Started under an escalator between the platforms and the concourse.
		Criticisms:
		• Delay in summoning the fire brigade.
		• Poor communication with passengers.
		    • Led to review of safety of underground.
		    • Introduction of Fire Precautions Sub-surface Railway Station (Amendment) Regulations.
		    • Started research on quantified risk assessments.
		    • Trench effect identified, where hot gases will move towards and "adhere" to nearby surfaces and inclined planes due to pressure differentials; the heat releases more ignitable gases from such surfaces, resulting in rapid surface spread of flame; particularly relevant for fires where vertical escalation is possible.

(continued)

Table 7.2 *(continued)*

Sun Valley Poultry	1993 Hereford 2 deaths (firefighters)	Following a number of hazardous fires in buildings with composite "sandwich" panels, this fire focussed attention and research on the panels. Typically, these have cores of rigid foamed plastics to give good thermal insulation, high moisture resistance, and low cost. The facings to these cores are usually of materials exhibiting better fire resistance or lower flammability. The hazard arises because firefighters (particularly) can be travelling below such panels unaware that the cores are on fire. The cores can ignite through fissures in the panels, at joints, and at poorly sealed edge protection. The fire can spread rapidly in the cores and the jointing and support systems are insufficient to prevent the panels delaminating, or collapsing, onto occupants. • Led to changes in building regulations for the manufacture and use of panels in new building works. • Led to changes in fire authority operation procedures. • Further understanding that fires starting in the contents of a building can spread into the structure.

Lakanal House	2009
	London
	6 deaths
		Rapid fire spread across exterior cladding.
		14-storey residential block, 98 flats, single central staircase.
		Criticisms:
		• Criticisms of "stay-put" policy.
		• Criticism of no sprinklers (using cost as justification).
		• Renovations and alterations (previous to fire) had removed fire-stopping material between flats and communal corridors.
		    • Refurbishment to meet fire safety standards prior to fire.
		    • Led to warnings that similar fires could occur, especially due to lack of sprinklers in tower blocks.

(continued)

Table 7.2 *(continued)*

Grenfell Tower	2017 London 72 deaths	At the time of writing, still under public inquiry, police investigations, and coroners' inquests. Rapid fire spread relating to the cladding in the external wall construction. Single central stair, 24-storey, 120 flats. Lower four storeys non-residential. Not sprinklered. Currently being attributed to the cladding type. External wall cladding thermally renovated 2015/16 to PIR type. Currently, criticism of: • Composite panels retrofit to external walls external faces. • Combustible insulation with the composite panels. • Management "stay put" policy. • Alarm system and lack of inter-connectivity. • On-site fire extinguishers. • Fire door condition. • Smoke venting system. • Firefighting lift controls. • Led to fire tests on similarly clad towers.

Research after the King's Cross Fire, based in the Newcastle underground system, shows that some methods of fire alarm produce no reaction in commuters, with no attempt to escape.

Canadian research suggests that there has been a significant under-estimation in international codes of times to evacuate high-rise buildings. Research on exit choice behaviour (when a person has a choice of exists – which does he choose and why) has shown that the route chosen could be predicted by formulae based on the **visibility** of stair entrances, in contrast to the stair **width** or **distance** to exits.

In the Manchester Woolworths fire, the public mainly used one staircase, one that led to the restaurant directly from the street, while the staff preferred the "staff area" staircase. People will travel further to get to a route they know rather than looking around for a better one.

People do not move when an alarm goes off. People wait to see if it is relevant to them. There are different methods of alarm, and these attract different responses. The time to escape should be taken as the time to start to escape plus the time taken to make the escape.

See Figure 7.8, Evacuation time. The pre-movement time will include the **investigation time**. People may be doing something – but that may not be making their way out of the building – they may be collecting their children from the floor above (Summerland fire). Another factor in pre-movement time is whether the occupants are asleep. They first of all have to wake up – then recognise that the alarm is going – then decide to do something about it.

$$T_{evac} = T_{pre} + T_{trav}$$

where T_{evac} = evacuation time

T_{pre} = pre-movement time

T_{trav} = travel time

Figure 7.8 Evacuation time.

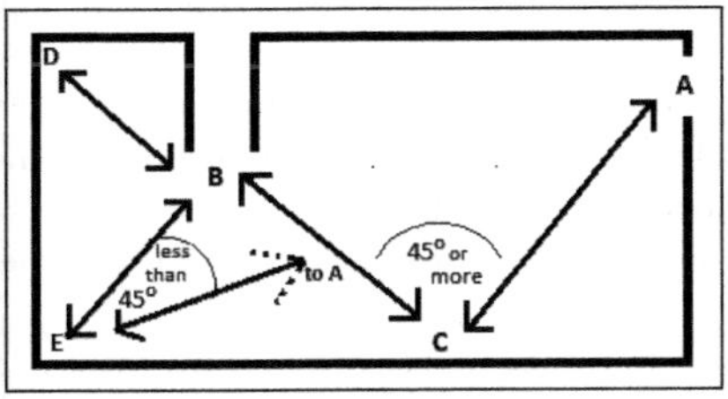

Point A and B are exits
Point C has two alternate routes to exit
Point D has a single route to exit
Point E has a single route to exit

Figure 7.9 Alternate routes of escape – or not. Diagram adapted from Approved Document B (www.gov.uk/government/publications/fire-safety-approved-document-b).

There are different methods of fire **alarm** including a person shouting, an electronic alarm, informative warning (IFW), and visual display alarms. By varying the warning, the pre-movement time can vary dramatically. A person shouting might evoke a quick response in a small building – workshop, small office, etc. However, in a university hall of residence this would be unlikely to result in the required action.

Average response times have been estimated as: bell six minutes, IFW four minutes, and visual display two minutes. However, research following the King's Cross fire showed that evacuation practice using an alarm bell only was completely ignored by passengers. Directive public announcements were much more effective at reducing pre-movement time.

Once people do start to escape, the travel time depends on their distance to travel before reaching a place of safety. Their route of travel may be restricted by the position of the fire. They have more chance to escape, if there are more than one distinct independent routes to safety. For example, two different, not interconnected, stairways at opposite sides of the building.

As implied by the foregoing, both stairways need to be well-known and used by the occupants.

For two routes to be regarded as separate, they should be over 45° from each other, as in Figure 7.9. Point C has two routes; Points D and E have only one route each and are not alternatives.

FIRE PRECAUTIONS FROM FIRST PRINCIPLES

Fire legislation is mainly involved with fire protection. In this, there are also tw(main offshoots – life protection and property protection. Legislation is concerne(primarily with the safety of life and not with the protection of property or protec tion of a company's ability to function. Insurance companies are often interested i property protection. See Table 7.3 for examples of fire protection measures for life property, active and passive.

Paradoxically the measures required to protect life can be simpler, less exten sive, and less costly than those to protect assets, because people can be remove(from the danger by evacuation.

A "Tactics and Objectives" matrix can be set up to help in considerations whe seeking to fulfil the objectives of life safety and property protection, such as th one shown in Figure 7.10 originally created by Paul Stollard.

Prevention – ensuring fires do not start by controlling ignition and fue sources.

Communication – ensuring that if ignition occurs, the occupants are informe(and any active fire systems are triggered.

Escape – ensuring that the occupants of the building are able to move to place of safety before they are threatened by heat and smoke.

Containment – ensuring that the fire is contained to the smallest possible are limiting the amount of property likely to be damaged and the threat to life safety

Extinguishment – ensuring that the fire can be extinguished quickly wit minimum consequential damage to the building.

Table 7.3 Examples of fire protection measures for life, property, active and passive

Life	**Property**
Emergency lighting	Gaseous flooding
Fire resistance of elements (e.g. 30 mins)	Fire resistance of elements (e.g. 2 hrs)
Spread of flame spread/support	Spread of flame spread/support
Sprinklers	Sprinklers
Fire doors	Closed fire doors
Extinguishers	Extinguishers
Enclosed stairways	Fire shutters
Alarm systems	Automated alarm systems
Protected means of escape	
Active	**Passive**
Detection	Structural fire protection
Extinguishers	Compartmentation
Smoke control	Protected means of escape
Sprinklers	
Management	

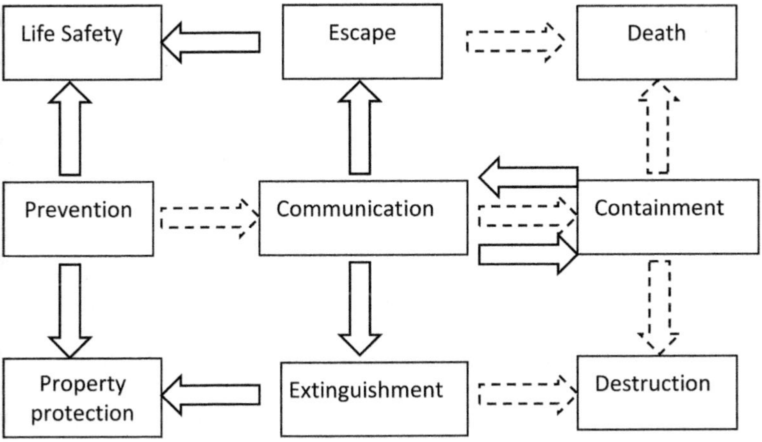

Figure 7.10 Tactics and objectives matrix to fulfil the objectives of life safety and property protection (developed from "Fire from First Principles" by Paul Stollard). The solid line arrows indicate success, dotted line arrows indicate failure.

Considering these in a logical sequence, the first is obviously prevention. Only if this fails are the others needed. Communication by itself, however successful, cannot save lives or protect property but it is the key to the other factors. If communication is successful, the escape and extinguishment can be attempted; if unsuccessful then only containment remains as an available tactic.

The purpose of containment is to prevent the fire spreading beyond defined boundaries and to protect the other areas so separated. The areas to which the fire is confined may be termed as a fire cell, a fire zone, or compartment. Compartment and compartmentation are defined in the Building Regulations Approved Documents B and should only be used in those terms. If you are not making the area as defined there, use a different term.

Protected areas can be a whole building, a complete floor of a building, a vertical enclosure, or a specified part of a building. The object of containment is twofold:

i. Prevent a fire escaping from the compartment of origin, and
ii. Prevent a fire entering into other compartments.

Many countries' building standards and regulations limit the sizes of compartments. In the UK, these maximum compartment sizes can be traced historically. The 1853 London Metropolitan Building Act imposed a limit of 216,000 ft^3 (the cube of 60 ft × 60 ft × 60 ft). This was because the London Fire Brigade, at that time, were confident that they could control a fire within such a volume using hoses with a 60 ft throw and with ladders up to a height of 60 ft, 60 ft being near 20 m (18.2 metres). Current UK limitations are held in HM Government Approved Documents B and Scottish Technical Handbooks. They depend on the building's height, use, the fire resistance duration, and whether or not it is sprinklered.

To form a fire compartment, the walls, floors, ceilings, etc., have to maintain their loadbearing capacity as long as required by the pertinent standards (usually 30, 60, or 120 minutes). The fire might last a lot longer. Compartment walls and floors need not – and generally cannot – be imperforate as they need to be penetrated by doors, service ducts, staircases, etc.

If the building only has to survive until all the occupants have been evacuated, then the necessary structural protection might only have to be a short time, perhaps half an hour. However, if the life safety strategy relies on the provision of places of refuge as in hospitals, or it is necessary for fire fighters to be able to work safely in the building, the protection may be one hour or more. It might be important to the building's insurers that repair, rather than rebuilding, is possible, and this can lead to fire protection times of two hours or four hours.

The construction of buildings is not always so easy in practice than in theory. There are holes, ducts, passages, gaps, etc., introduced, which are often hidden from sight. As a surveyor, you may find that the integrity, of walls, floors, etc., is not complete. For example, a pipe or wire passing through a wall requires a hole bigger than itself, the sides of which create additional passages from one side of the wall to the other. The entrances to these additional passages are covered on each side aesthetically, but not necessarily effectively from a fire-resistant viewpoint. Despite having fire-resistant elements, fire and smoke can by-pass these elements and spread through the building.

A second example, is suspended ceilings. The fire-resistant walls are not always taken right up to the structural floor above and fire can spread quickly along in the void – aided of course by being able to travel along the top surface of the suspended ceiling (Figure 7.5).

Gaps and cavities, and ducts are a major problem – especially in older buildings where they are not easily found. The fire destruction at Windsor Castle in 1992 was greatly helped by floor voids and gaps behind panelling. As building surveyors, what we do with these is to: find, fire-stop, cavity barrier.

A **cavity barrier** is simply construction with a minimum standard of fire resistance, used to seal or sub-divide the cavity, to restrict the movement of smoke or flame within a space. Fire-stops should be provided around such things as pipes or ducts that pass through a fire-separating element and also at junctions where such elements are not firmly bonded together.

Fire doors carry out two functions whilst allowing persons access through the element. A wedged-open fire door, however, provides no function. Fire doors nearly always require a self-closing device. The hindrance caused by these is unfortunately a major complaint of building users. In certain circumstances, an automatic release mechanism that holds the door open in general use, but releases it when activated by automatic fire detection so that the door that closes, is sometimes an answer.

Where possible, a better method is to position lines of fire-resisting separation with the internal sub-divisions of the building, where doors in walls or partitions would normally be closed. If doors are a hindrance in normal use then it is likely that they will be propped open, regardless of whether they have been identified as critical to fire safety. Such practices may well demonstrate the inferiority of the design solution chosen.

Smoke is the main hazard to persons escaping. Although often used at the same positions in fire doors, smoke seals and intumescent strips are very different. Smoke seals restrict the passage of smoke at low temperatures. For hot smoke and gases, intumescent strips swell typically at 150°C and seal the gap around the edge of the door. It is unlikely that a door would achieve a fire resistance of 30 minutes without intumescent strips.

Older doors used to have to be 45 mm thick and have 25 mm rebates (no intumescent strips). Current fire doors do not need to be so thick and use intumescent smoke seals in place of the rebates. If a good fit in its frame, an old door can be upgraded by the provision of strips and smoke seals. If the door is warped or fits poorly in the frame, it may need replacing. It is the door set – i.e. the door and its frame – not just the door, which must have adequate fire resistance. Panelled doors may also need the panels and their joints upgrading.

Materials and fire

Timber

Timber burns – it is combustible – it is a complex hydrocarbon. But the performance of timber in real fires is frequently far superior to unprotected, non-combustible materials such as steel and aluminium. It is also predictable and measurable.
This is because:

a. It has a low conductivity, low thermal diffusivity, and low thermal expansion.
b. To pyrolyse, timber must be heated to around 160–200°C. Due to its low thermal conductivity, getting the heat into timber to result in pyrolysis is not easy.
c. Timber has the inherent ability to protect itself: the build-up of charcoal on the surface of burning timber limits the availability of oxygen thereby insulating the remainder of the section.
d. The rate of char is predictable and varies only slightly with species of timber and not on the severity of the fire. Values of 0.55 mm/min and 0.75 mm/min

are given for softwood and hardwood respectively. So large cross-sectioned timber will survive for long periods in a fire. Thin sections – floorboards, wall panelling, laths, etc. – will quickly burn through. Laminated timber can also act as large sectioned timber.

e. "Sacrificial" timber can be built into the construction to be consumed without affecting the loadbearing capacity of the structural core of the timber.

The use of flame-retardant treatments will not normally slow down the charring rate, but can result in a better spread of flame rating.

Brick

Clay bricks perform well in fire. They have already been fired in the kiln at temperatures over 1000°C so further heat exposure does little damage. Bricks have low thermal conductivity and diffusivity so heat will be slow to soak into the brickwork. However small cracks and fissures will develop or expand, causing some distortion. Brickwork is constructed of mortar as well as bricks, and the mortar joints do not perform so well. The heat has a weakening effect on the joints, and brickwork suffers a certain amount of weakening and distortion over 1000°C.

As well as clay bricks, we use concrete and calcium silicate bricks. Although there are many different types of these three bricks and different types of mortar used for bedding, no distinctions are made in classifying their behaviour.

Stonework

Stone used in construction is generally granite, sandstone, and limestone.

Igneous rocks such as **granites** contain free quartz which has the peculiar property of expanding very rapidly at 575°C and completely shattering the rock. Considerable spalling at the surface may occur in a fire, and thin sections of stone may disintegrate entirely.

Limestones are composed of calcium carbonate which decomposes at about 800°C into free lime and carbon dioxide. This change is gradual with little alteration in volume, and the interior of the stone may be protected by the surface. Water used to fight the fire slakes away the quicklime so formed and will cause the outer skin to fall away.

Sandstone generally comes between granite and limestone in fire behaviour and may shrink and crack in a fire.

Stone is generally a good heat insulator, but is inferior to brick when subjected to continuous heat because of its tendency to split or spall, especially when water is applied. Stonework should always be carefully watched for signs of cracking when working near or beneath it.

Steel

Steel is dense, strong, and has high thermal conductivity and very high thermal diffusivity. It has a high melting point, about 1535°C, so it is unlikely to melt in a normal fire. But it does perform very poorly in a fire. As a rule of thumb, it loses half its strength at 500°C, and two-thirds of its strength at 600°C. This can be reached after 15–20 minutes' exposure to a fully developed fire. It begins to sag and twist in proportion to the amount and direction of load to which it is subjected. These temperatures are by no means abnormal in even a moderate fire. A 10 m steel joist will expand 60 mm for a 500°C rise in temperature, and where built into a loadbearing wall, this expansion could result in collapse.

A loaded steel section therefore has to be protected to ensure that the steel is not heated to these critical temperatures within the period of fire resistance required.

This can be achieved either by insulation or dissipation:

1. Encase in a material with low thermal conductivity and low thermal diffusivity, e.g. brick, concrete, insulating boards, sprayed coatings, intumescent paints; or
2. Dissipate the heat away from the steel by water or by filling a hollow steel section with concrete.

Concrete

Concrete does not melt or soften in a fire. It has a low thermal conductivity and a very low thermal diffusivity.

When concrete is heated, it expands due to thermal expansion of the materials, but the hardened cement paste shrinks as a result of loss of moisture by drying out. As a result, the overall change is not easily predicted and internal stresses can be set up within the concrete. In a severe fire, spalling of the surface occurs and is aggravated by sudden cooling, e.g. by a jet of water. Concrete made of limestones or lightweight aggregates is much less susceptible to spalling than those made with more dense aggregates; hence the fire resistance of structural concrete is classified differently according to the aggregate used.

It is possible to achieve very high levels of fire resistance with reinforced concrete, e.g. up to 4 hours.

The fire resistance of reinforced concrete is determined by the protection of the steel against excessive rise in temperature. This is given by the concrete cover, i.e. the concrete between the surface of the structural element and the nearest surface of the embedded steel. Generally, the greater the amount of cover, the longer the period of fire resistance. However simply increasing the thickness of the concrete cover does not necessarily give a corresponding increase in safety because of the concrete's tendency to spall in a fire. This can reduce the cover, and it may be nec essary to provide some additional reinforcement to counteract this danger if the cover is over 40 mm.

Aluminium

Aluminium melts at 660°C (pure) and less than 600°C (alloys). Critical tempera ture (at which it loses half its strength) is about 200–250°C. There is a very rapid loss of strength in a fire; stability is affected from 100°C. It has a high expansion rate, about twice that of steel.

Aluminium does not perform well in a fire. It is mainly used for glazing and wall cladding. Aluminium door furniture should not be used on a door required to be a fire door.

Lead

Melts at 327°C. Danger from molten metal.

Plaster

Gypsum plaster does not burn but when heated tends to de-hydrate – the combined water is gradually driven off. The plaster can absorb a lot of heat for this, thus providing protection to the substrate.

Glass

Glass is a super-cooled liquid which behaves like a solid at room temperature. It does not burn or melt, but gradually softens at high temperatures. Ordinary glass has very little fire resistance, offering little insulation when intact. It has quite a high expansion rate. In a fire situation, this rate of expansion of the glass, held in its framing, will result in shattering, losing any integrity and stability.

Care must be taken when using glass at a fire-resisting element in a sprinklered building to use a glass which will not be adversely affected by the cooling water. Whatever type of glass is used, the design of the frame is as important as the choice of glazing material. The frame obviously has to last as long as the glass in a fire.

Georgian wired glass can solve the problem of integrity and stability by holding the glass in place. Georgian wired glass, however, still does not offer any insulation. Radiant heat can still pass through the material. Toughened glasses can achieve the same integrity and stability without the unattractive appearance of the wires, yet these also fail to provide any insulation.

Laminated glass, incorporating transparent layers of intumescent material, offers insulating properties. On the application of heat, these layers expand to form an insulating barrier. The disadvantages of this glass lie in its weight, cost, and limitations on external use (sunshine can activate the intumescence).

MEANS OF ESCAPE

Means of escape should be designed with the foregoing in mind. The procedure in Table 7.4 can be used for designing, or for checking a design, for means of escape, when considering passive fire safety.

EXTERNAL FIRE SPREAD

External walls

In a fire, the external walls have two main functions. If the fire is inside the building, the wall must try and stop it getting out. If the fire is outside the building, the wall must try and stop it getting in.

Consider a building in the middle of a 10 hectare field, with no other buildings nearby. Apart from a certain loss of money, the fire getting out of the building is not a major problem – there is nothing else to burn. Similarly, there is little concern about another building causing our building to catch alight – there are no other buildings. Thus, there is little concern about the combustibility or fire resistance of the walls.

Conversely, put the building in a city centre with a paint factory next door one way and an old people's home on the other side. In this scenario, there are concerns about danger from a fire in the paint factory, and once the fire has got a hold on the building – it might endanger the old people. In this situation, there are real concerns with the construction of the building's external walls. It is all a matter of distance between buildings.

It is particularly important when the building is less than a metre from the boundary. The danger of ignition and fire spread is now very great.

There are usually restrictions on how combustible the wall, and its composite materials, are. However, as evidenced from the Grenfell fire, interpretation of these restrictions is not always obvious. This is despite the findings of the 1973 Summerland fire. See also Table 7.2.

The fire resistance of external walls depends on the use, height, and size of the building (and consequently the expected size of any fire and the effects on people). Regulations sometimes state that required standards of fire resistance must be so when tested from either side of a wall, but there can be reduced standards when the external wall is more than 1 m from a boundary.

Table 7.4 A procedure to be used for designing, or for checking a design, for means of escape, when considering passive fire safety

Procedure to design, or check a design, for means of escape, when considering passive fire safety		
1.	Decide the purpose group of the building.	
2.	Decide the fire resistance of the building.	Relate to the purpose group.
3.	Find the maximum travel distance and the direct distance.	Use travel distances when all the obstacles are known.
4.	Identify dead ends to use "one direction only" distance.	
5.	Establish the maximum occupancy.	Find the floor space factor and multiply by the area of the room or storey (excluding stair enclosure, lifts, and sanitary accommodation).
6.	Find minimum number of escape routes or exits permitted from the room or storey.	Relates to floor area and number of occupants.
7.	Locate escape stairs and exits and ensure travel distances from all points in the room or storey are not exceeded.	Include for 45° rule.
8.	Check layout of storey, for example	a. Inner rooms. b. Circulation separate from stairways. c. Lengths of corridors. d. Dead ends.
9.	Using points 6, 7, 11, and 12 will determine the number of stairs. Points 10 and 11 will give the width of the stairs.	
10.	Determine minimum width of stairs.	
11.	Determine required width of stairs:	a. Discount one stair if necessary. b. Over 1100 mm wide, might also use: $P = 200w + 50(w - 0.3)(n - 1)$ where: P = no of people w = width of stair (m) n = no. of storeys c. Find "w" which is the minimum combined width of stairs. d. Share this between the number of stairs (less the discounted one). e. Ensure that this meets or exceeds the minimum width. f. Add back in the discounted one at, at least, the minimum width.

(continued)

Table 7.4 *(continued)*

12.	Determine whether any extra stairs are needed.	E.g. accommodation stairs or firefighting stairs.
13.	Determine whether lobbies are needed to the stairs.	
14.	Ensure all escape stairways discharge to a final exit or a protected route.	
15.	Check on specific provisions for building type.	
16.	Provide unambiguous signage.	
17.	Provide emergency lighting.	
18.	Provide portable extinguishers.	

Most walls have pockets of **unprotected areas** (areas which do not have the required fire resistance), for example, windows that permit heat radiation from a fire to pass through. The way many regulations deal with this is not to gradually reduce the requirements overall as the building is moved further from the boundary, but to retain the required standard and allow pockets of reduced standards. These can be limited by percentage according to how far the building is from any boundaries.

Particular care needs employment when **retrofitting thermal insulation onto external walls**. Many types of insulation are combustible. Most walls are not imperforate. Fire can therefore pass from internal surfaces to external, external surfaces to internal, into interstitial layers, and across party wall boundaries. Regulations may restrict combustible material from being applied to buildings which are tall (e.g. above 18 m) but not to rows of terraced houses. Despite the regulations in place, as a building surveyor you may encounter combustible material taken across party walls in rows of terraced houses. These will be coated in a decorative and protecting render; however, as building surveyors, you are also likely to encounter breaks in the integrity of this render.

Roofs

As well as walls, roofs can cause fire spread from one building to another. Similar to walls, roofs have to have their surface classifications limited dependent on the distance to the boundaries. This is assuming there is actually a space separation between the buildings. When there is only a party wall between buildings, similar restrictions apply. There are currently no provisions to stop fire entering a roof void via the external eaves, although fire stopping is usually required across a party boundary at eaves level.

Party walls are now taken up to roof undersides and gaps between properties filled with mortar or other non-combustible material. As a building surveyor, you may find that not all gaps have been filled, or that the filling has shrunk or fallen out, thus breaking the integrity of the party wall.

For older properties in London, because of the historic party wall London bylaws, party walls were usually taken above the roof level. In the rest of the UK and in other parts of the world, this was not the case. As a building surveyor, you are therefore likely to encounter lengths of terraced properties, with different occupancies, having unseparated roof voids. These can enable fires to travel the length of a number of attached properties, as for example, the Bridge Street fire in Chester, the South Bridge fire in Edinburgh, and the Torbay Road fire in Paignton.

Access for fire services

In Scotland, Mandatory Standards require that every building must be accessible to fire and rescue services. Access requirements increase with building size and height. The higher the building, the greater the extent of the building's footprint needs to be accessible. Vehicle access is needed to the exterior faces of buildings to facilitate high-reaching appliances such as turntable ladders and hydraulic platforms. This also enables pumping appliances to supply water and equipment for firefighting and rescue activities. For domestic properties, vehicle access is required to at least one elevation. Similar provisions are required for England and Wales.

Guidance can be found in the Scottish Technical Handbooks and in the Approved Documents. There are requirements for access routes, turning facilities, operating spaces, water supply, and internal facilities. Internal facilities required include, depending on building height, purpose, layout, floor areas, basement depth, etc.:

- Firefighting stairs
- Firefighting lifts
- Firefighting lobbies
- Heat and smoke control (e.g. natural or mechanical ventilation)
- Fire mains (wet or dry)
- Fire shutters
- Fire suppression

INTRODUCTION TO FIRE ENGINEERING CONCEPTS

Initial questions to consider

Fire predictions

- Will it burn?
- How long will it burn for?
- How much heat will it give out?
- How much smoke will it give out?
- How will the smoke and the fire spread and where to?

- Where are the routes for the fire and products of combustion? Note that in a finished building, these routes may be concealed, e.g. into ducts or spaces in or through floors or walls, formed purposefully or simply left unfilled or fire-stopped.

People predictions

- Who is in the building?
- Where are they?
- Where do they go to be safe?
- How long will it take them?

Fire engineered?

If the summation of the fire predictions is "less" than the summation of the people predictions, i.e. **if everyone is out well before untenable conditions occur**, then we have a successfully engineered building.

In fire engineering, prescriptive second-tier guidance such as the Approved Documents is unlikely to be fully used, because the safety is worked out from first principles. For example, British Standards produce fire engineering standards and these may be used instead. See Figure 7.11 for compliance document status.

Concerns of the fire engineer

a. The physical and chemical processes taking place in the course of fire development.
b. The response of human physiology to the effects of heat and toxic fire products.
c. The response of the occupants to fire alarms, etc., given the capacity of the escape routes and the handicaps imposed by a. and b.
d. The effect of automatic fire detection and suppression systems and of fire fighters.

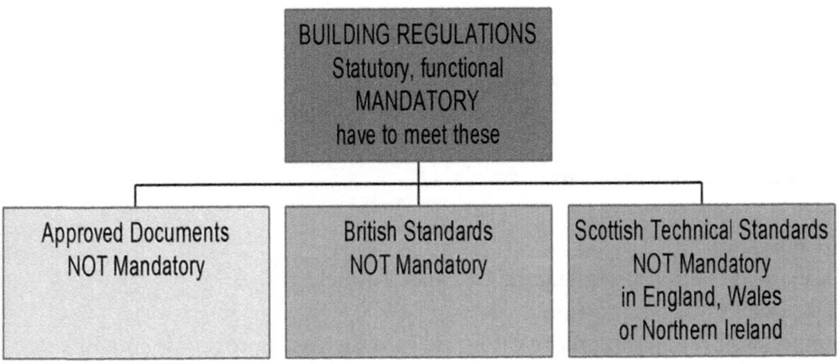

Figure 7.11 Compliance documents.

Fire engineering tools

- Mathematics, e.g. fire size, smoke volumes, evacuation times
- Computer modelling
- Communication systems
- Smoke control systems
- Sprinklers/suppression
- Management systems
- Advice/help from team (building control, fire authority, mechanical and electrical consultants)

Process

1. Qualitative design review: find out everything (QDR).
2. Quantitative analysis: do your sums.
3. Evaluate.
4. Assess against required criteria (e.g. everyone out before smoke reaches 2 m above heads).
5. If (4.) is not satisfactory – go back to (1.).
6. When (4.) is okay, present/report results.

Smoke handling

- Collect smoke and dispense it.
- Very effective for life safety.
- New developments are often fire engineered: evacuation + smoke handling.

Computer modelling

- Very useful in predictions.
- Especially for complex buildings.
- Two basic types – field and zone.
- Zone more simplified but not necessarily less effective than field.

Suppression considerations

- Sprinklers.
- Gaseous flooding. Used, e.g. in Channel Tunnel.
- Evacuate before suppression.
- Suppression is creeping more into guidance such as approved documents.
- There is a lobby for sprinklers in houses.

Mathematics – basic concepts

k – thermal conductivity
c – specific heat capacity
ρ – (rho) density of the material
H – calorific value
Combine to give other properties:
(k·ρ·c) – thermal inertia: measure of time taken for a surface to heat up.
(k/ρ·c) – thermal diffusivity: speed with which heat soaks into material. Helps to determine strength retention.

Stoichiometric ratio

- The ratio of fuel to oxygen, in which the amount of oxygen present is just sufficient to completely oxidate the fuel.
- A stoichiometric mixture is good for explosions, as the ratio of fuel to oxygen is theoretically correct for complete oxidation.

Typical ratio:

- Most hydrocarbons need 15:1 of air:fuel ratio by weight.
- i.e. the weight of air required is be 15 times that of the fuel.
- Less air is need for wood, approximately 6:1, as wood contains air.

Calorific value H

- The amount of heat which can be released by a unit mass of fuel at a particular temperature and pressure.
- Use this and stoichiometric ratio (r) to get: Heat release per unit mass of oxygen H/r.

8

Disasters and the built environment

John Bruen and John P. Spillane

DISASTERS

Disasters are largely divided into two main areas to include natural and man-made disasters, and both can cause a significant loss and disruption of lives, built and social assets, and the economy. There are many different definitions of disaster and its components. This chapter uses those held in the document, "UNISDR Terminology on Disaster Risk Reduction" published by the United Nations International Strategy for Disaster Reduction, Geneva, Switzerland. This defines a disaster as:

> **Disaster:** "A serious disruption of the functioning of a community or a society involving widespread human, material, economic or environmental losses and impacts, which exceeds the ability of the affected community or society to cope using its own resources". (UNISDR, 2009)

Historically, disasters were all considered "acts of god" or "acts of nature" and nothing could be done to prevent their occurrence (Voogd, 2004). More recently, researchers such as Quarantelli and Perry (2005) argue that disasters are a societal construct, and disaster impact has a direct or indirect link to human actions and behaviours and their interplay with the physical and social world and their subsequent vulnerabilities. The social systems that contribute to vulnerabilities are informed by a number of elements that include the political, cultural, historical, socio-economic, and environmental context of a place (Horlick-Jones and Peters, 1991). It follows that, as human actions and decision-making increase the vulnerability to disasters, it is possible for human actions to eliminate or reduce the conditions giving rise to disasters and, consequently, humans have an obligation to try to prevent such conditions.

Natural disasters can take many different forms including floods, tsunamis, earthquakes, drought, storms, heatwaves, landslides, and fires. Climate change is also identified as an attributing factor to natural disasters globally, and it is predicted that its effects will become greater in the future (Paton et al., 2000). The frequency of natural disasters has risen significantly over recent decades, and disasters are becoming more frequent and severe in least developed countries (LDCs) (Smith, 1992; Cannon, 1994; Hidayat and Egbu, 2010). Many LDCs are innately more vulnerable to natural disasters and have experienced disproportionate levels of devastation as a result of disasters, both in human life, cost, and the resulting large numbers of internally displaced populations (IFRC, 2001; Schilderman, 2004). This vulnerability is often caused by common issues experienced in many LDCs, including a lack of preventative action plans and preparedness, development in areas susceptible to natural disasters, unsettled governments, poor construction standards and techniques, insufficient resources and knowledge in post-disaster recovery, and the reconstruction cost relative to GDP in these regions (Toya and Skidmore, 2007; Cannon, 1994).

Human conflicts are man-made planned events that have been the cause of movement of humans and subsequent destruction and displacement throughout history. The number of internally displaced people (IDPs) has been continually rising for over a decade, and forced migration from conflict is at its highest since World War II. Conflicts destroy many aspects of the built environment including housing stock and limit housing production in affected countries.

The built environment and construction sector accounts for a significant part of a country's physical assets and economic development contribution. As such, the built environment is a major part of society, and it is important to develop appropriate built environments that provide the ability to adapt to the threats of disasters. Regardless of preparation, the built environment is vulnerable to the effects of disasters and, as such, disasters have the potential to have a devastating impact on a country and can create significant impacts on social and economic activities. The task of reconstruction after a disaster is an onerous challenge with many stakeholders from a diverse range of backgrounds requiring effective coordination.

Disaster concepts

It is essential to have a full understanding of disasters, and the many aspects and concepts that are utilised, in order to assess, prepare, and mitigate against them in order to inform better decision-making. The main concepts include hazards, risk, vulnerability, and resilience. These are broad-ranging concepts that can relate to many different types of disaster in many different contexts. This chapter focuses on disasters in relation to the built environment and more specifically to housing where appropriate.

Hazards

UNISDR (2009) defines a hazard as:

> a dangerous phenomenon, substance, human activity or condition that may cause loss of life, injury or other health impacts, property damage, loss of livelihoods and services, social and economic disruption, or environmental damage. (UNISDR, 2009)

Hazards are generally categorised into two groups: Man-made/technical or natural.

Natural hazards can generally be divided into five categories as shown in Table 8.1: (1) hydrological, (2) biological, (3) meteorological (atmospheric), (4) climatological, and (5) geophysical

Man-made or technical hazards are also referred to as anthropogenic hazards and can be caused by human action or inaction. The subsequent disasters can be more protracted, gradual, and predictable than natural disasters. Gustin (2008) states that hazards can be viewed as failures of systems caused by organisational, operational, and social faults. These can include technological hazards such as hazardous materials, radiological emergencies, industrial or nuclear power plant blasts, transportation or structural hazards, and computer or cyber virus hazards. Man-made social hazards such as war and terrorism can also contribute significantly to disasters. In relation to the built environment human aspects play a significant role in terms of the devastation caused when hazards interact with humans, e.g. tsunamis, floods, storms, etc.

Vulnerability

UNISDR (2009) defines vulnerability as:

> The characteristics and circumstances of a community, system or asset that make it susceptible to the damaging effects of a hazard. (UNISDR, 2009)

Vulnerability compromises the physical, social-economic, and other political factors that adversely affect the ability of populations to respond to events or hazards. The level of vulnerability of a person or group is made up of their ability to anticipate, cope with, and recover from the impact of a hazard. Global inequalities have resulted in LDCs being inherently more vulnerable to hazards than developed countries, often due to aspects such as high population, lack of land and planning for development, inadequate design, and construction techniques. The lack of preparedness and ability to cope with a disaster can result in vastly different levels of risk, devastation, and loss of human life and infrastructure for similar disasters depending on where they occur.

Table 8.1 Natural hazard categories and some sub-categories and examples

Biological	**Geophysical**	**Hydrological**	**Meteorological**	**Climatological**
Insect infestation	Earthquake	Flood	Drought	Extreme temperatures
Animal infestation	Volcano	• General flood • Flash flood • Storm surge/coastal flood	Storm • Wind storm • Thunder storm • Snow storm • Blizzard • Local storm	Extreme heat/cold
Disease epidemics • Viral • Bacterial • Parasitic • Fungal • Prion	Mass movement (dry) • Rock fall • Landslide • Avalanche • Subsidence	Mass movement (wet) • Rock fall • Landslide • Avalanche • Subsidence	Cyclone • Tropical cyclone • Hurricane • Typhoon • Tropical storm	Extreme weather conditions
Animal stampede	Tsunami		Tornado	Droughts
			Heat wave Cold wave	Wildfire • Forest fire • Land fire

Disaster risk

UNISDR (2009) defines risk as:

> The probability of harmful consequences, or expected losses (deaths, injuries, property, livelihoods, economic activity disrupted or environment damaged) resulting from interactions between natural or human-induced hazards and vulnerable condition. (UNISDR, 2009)

Risk is also conventionally expressed by the equation:

Risk = Hazard × Vulnerability/Capacity

Risk can be perceived differently by different contexts and communities depending on variables including their social circumstances, knowledge of the risk, and ability to cope with the risk or hazard. As such, the context in which risk occurs is an important consideration in calculating risk. Natural hazards cannot be eliminated totally and risk will always exist, and as such disaster can be seen as a function of the risk process that results from a combination of hazards and conditions of vulnerability or lack of capacity to reduce the negative impact of risk. The area of disaster risk is intrinsically linked to the concepts of "disaster risk reduction" and "disaster risk management". Both concepts are essential elements in terms of the housing design and delivery in LDC and post-disaster contexts and are often used in the same context with similar meanings.

UNISDR (2009) defines both concepts, respectively as:

Disaster risk reduction:

> The concept and practice of reducing disaster risks through systematic efforts to analyse and manage the causal factors of disasters, including through reduced exposure to hazards, lessened vulnerability of people and property, wise management of land and the environment, and improved preparedness for adverse events. (UNISDR, 2009)

Disaster risk management:

> The systematic process of using administrative directives, organizations and operational skills and capacities to implement strategies, policies and improved coping capacities in order to lessen the adverse impacts of hazards and the possibility of disaster. (UNISDR, 2009)

Disaster risk reduction is inherently linked to the vulnerability to hazards. In LDC contexts, the capacity of poorer populations to effectively implement disaster risk reduction policies is less for a number of reasons (Table 8.2). This process can result in a downward spiral of increased poverty and vulnerability as livelihoods can be lost as well as homes following a disaster, and resulting in a further reduction of capacities to reduce risk and prepare for disasters or their aftermath.

Resilience

UNISDR (2009) defines resilience as:

> The ability of a system, community or society exposed to hazards to resist, absorb, accommodate to and recover from the effects of a hazard in a timely and efficient manner, including through the preservation and restoration of its essential basic structures and functions. (UNISDR, 2009)

DFID is the Department for International Development, a UK government department responsible for administering overseas aid, promoting sustainable development, and eliminating world poverty. DFID defines resilience as:

> Disaster Resilience is the ability of countries, communities and households to manage change, by maintaining or transforming living standards in the face of shocks or stresses – such as earthquakes, drought or violent conflict – without compromising their long-term prospects. (DFID, 2011)

Table 8.2 Disaster reduction capacities for richer and poorer countries

Comparative analysis of disaster reduction capacities in richer and poorer countries	
Richer countries	**Poorer countries**
Have regulatory frameworks to minimise disaster risk which are enforced	Regulatory frameworks are weak and absent, and/or the capacity to enforce them is lacking
Have effective early warning and information mechanisms in place to minimise loss of life	Lack of comprehensive information systems linked to pre-emptive response
Have highly developed emergency response and medical care systems	Divert funds from development programs to emergency assistance and recovery
Insurance schemes spread the burden of property losses	Those affected bear full burden of property losses and may lose livelihoods

Adapted from White et al., 2004.

Resilience is a far-stretching and complex concept taking into account many other aspects of disaster including risk reduction, risk management, early warning systems, disaster preparedness, post-disaster strategies, capacity building, knowledge sharing, effective policy making, technical solutions, financial/economic infrastructure, and political structures. Resilience building on these and other aspects can be at various levels, e.g. household, local, state, national, or international, depending on the context and what it is that the resilience is being sought for, e.g. natural or man-made hazards. Resilience is the ability to absorb and recover from hazard impacts. It is often seen as the opposite to vulnerability and as such is sometimes used in the same context as capacity.

Most definitions of resilience have four main elements which can be utilised in order to gain a measure of the resilience for a given community or population: (1) context, (2) disturbance, (3) capacity, and (4) reaction (DFID, 2011). The four elements can be utilised to form a framework to enable levels of resilience to be assessed as shown in Figure 8.1.

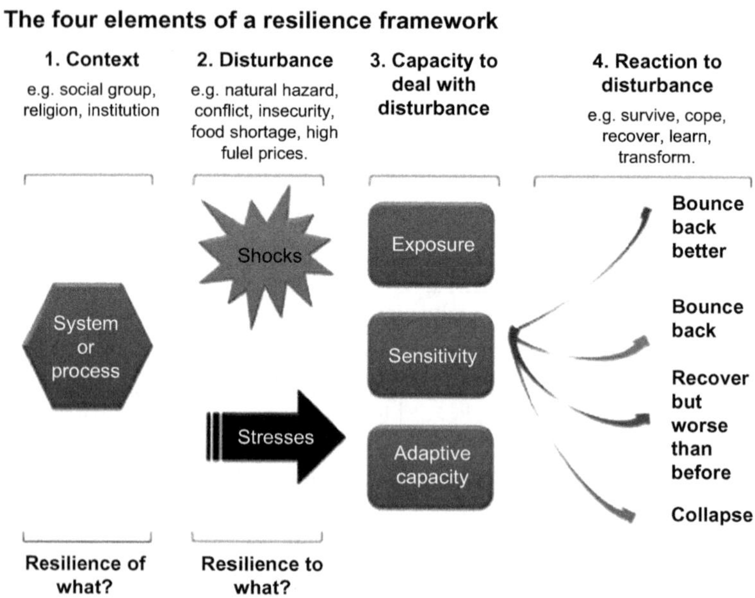

Figure 8.1 Resilience framework (source: DFID, 2011).

DISASTER MANAGEMENT CYCLE

The disaster management cycle, shown at Figure 8.2, is widely accepted and used as a tool for disaster management worldwide. The cycle consists of a set of sequential stages that occur during the unfolding of a disaster at which interventions can be undertaken to help lessen or mitigate against the impact of that disaster. The interventions include both reactive measures during or after the event, and also preventive measures prior to the event (Figure 8.2). The implementation of the cycle is multi-disciplinary in nature and requires collaboration from various disciplines and shareholders. Correct use of the disaster management cycle model can ensure that a holistic approach is taken at the required stages with the required stakeholders to achieve this long-term sustainable recovery goal.

There are three different periods during the cycle: (1) before the event (pre-disaster), (2) impact (disaster), and (3) after the event (post-disaster). There are four stages to the disaster management cycle that occur at these periods. The four main stages are preparation, mitigation, response, and recovery.

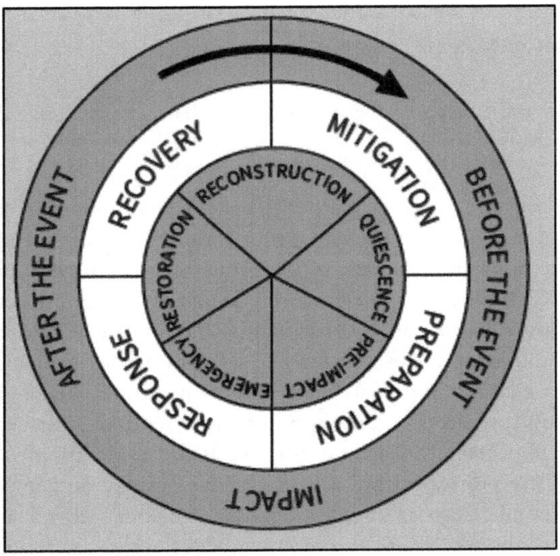

Figure 8.2 Disaster management cycle (source: Alexander, 2002).

1. Preparation

UNISDR (2009) defines **preparedness** as:

> The knowledge and capacities developed by governments, professional response and recovery organizations, communities and individuals to effectively anticipate, respond to, and recover from, the impacts of likely, imminent or current hazard events or conditions. (UNISDR, 2009)

This stage involves taking management actions in advance of a disaster or event occurring. This management stage covers a wide variety of aspects from a macro to a micro level. Appropriate preparedness can significantly reduce the risks of and vulnerability to the hazard or disaster and increase the resilience of a population. Preparedness covers many aspects, e.g. physical/infrastructure and societal preparedness. Practical aspects in relation to preparedness for the built environment and housing include approaches such as early warning systems, adequate infrastructure and structure to withstand the event, contingency planning, hazard mapping, e.g. floods, disaster management plans, and preparing for future events.

2. Mitigation

UNISDR (2009) defines **mitigation** as:

> The lessening or limitation of the adverse impacts of hazards and related disasters. (UNISDR, 2009)

In its simplest sense, mitigation can be seen as approaches to reduce or remove risk to prevent future potential disasters. Practical mitigation actions at a macro and micro level in relation to the built environment and housing can help in relation to unpreventable hazards. Effective mitigation strategies can significantly impact the effect of disasters in terms of human and economic loss and include adequate zoning of land, appropriate building codes and techniques, flood management, and house maintenance. Mitigation and preparedness are intrinsically linked and may overlap in practical terms. It is recognised that poorer countries have a weaker capacity to mitigate disaster impacts but that if pre-emptive action is incorporated into development work in these poorer countries it can have a positive outcome. Pre-emptive action should not be seen as an additional burden, but rather it is justifiable on humanitarian, economic, political, and human development grounds (White et al., 2004).

3. Response

UNISDR (2009) defines **response** as:

> The provision of emergency services and public assistance during or immediately after a disaster in order to save lives, reduce health impacts, ensure public safety and meet the basic subsistence needs of the people affected. (UNISDR, 2009)

The response stage is sometimes referred to as "**disaster relief**" and deals with the immediate threats posed by a disaster. Typical works throughout the response stage include evacuation, short-term relief, medical or transport aid, temporary or transitional shelter, food aid, and search and rescue. The focus of this stage is on the immediate needs of the populations affected by disasters, and humanitarian organisations are often more prevalent at this stage of the disaster management cycle.

This stage concentrates on saving lives and relieving suffering. However, **inappropriate responses** can in fact prolong a crisis, or create a new risk, by:

- Not enabling sufficient local structures to take control.
- Or by the use of inappropriate technical approaches that can affect the long-term recovery and resilience.

4. Recovery

UNISDR (2009) defines **response** as:

> The restoration, and improvement where appropriate, of facilities, livelihoods and living conditions of disaster-affected communities, including efforts to reduce disaster risk factors. (UNISDR, 2009)

The commencement of the recovery stage is an inexact science. The recovery stage commences once the immediate emergency crises have been responded to, and it is usually judged to be when all conditions have returned to a baseline condition. The recovery stage can be seen as: Efforts to return the context back to normality, or better than it was previously. It is generally considered to be in two stages, e.g. short term and long term. The recovery stage provides opportunities to return the affected disaster area to a better or improved state than prior to the disaster by better mitigation against future disasters and better long-term sustainability. The recovery stage should be commenced in sufficient time so as not to miss opportunities to improve on the long-term sustainable development goal that best serve the community affected and reduce the long-term vulnerability o the community. There is an overlap and interconnect between the outlined cycl stages, and the stages do not fit an exact sequential process. The periods of tim spent at these stages will vary depending on the context, and stages will also over lap to reflect the realities of each individual disaster context.

Post-disaster reconstruction

Post-disaster reconstruction forms only one element of the overall recovery pro cess but a very important one. The construction industry has an important rol to play in finding and providing appropriate solutions to the continuous threa of disaster. Ofori (2011) outlines the importance of an improved constructio industry for developing nations in order to equip them for disaster scenarios There is also recognition that the construction industry can play a wider role a it assesses, prepares for, prevents, responds to, and recovers from disasters (Haig et al., 2006). Post-disaster reconstruction covers many different aspects of th built environment depending on the context and disaster that preceded it, e.g transport infrastructure, housing, ports, sanitation, water. The process of recon struction requires significant medium- and long-term strategic considerations i order to capitalise on an opportunity to improve on the built environment tha preceded it. Post-disaster reconstruction represents a large investment per capit of population. As such, in order to obtain maximum benefit of the process, care ful planning and consideration are essential and must be based on an analysis o hazard, risk, and vulnerability.

Housing projects often come as first priority in many post-disaster contexts In LDCs, much of the population will have no home insurance (as they do i developed countries), and it falls on the government and the international commu nity to endeavour to provide housing for the homeless resulting from the disaster Davidson et al. (2007) argue that post-disaster housing has similar challenges t low-cost housing in LDCs, but that a disaster context adds additional challenges t post-disaster houses. Davidson et al. (2007) list these challenges as:

- Donors require quick results for projects.
- Many different international organisations running housing projects at th same time.
- High expectations of reconstruction projects, e.g. sustainability and reductio in vulnerability.
- Many organisations having to compete for the same scarce resources.
- Higher levels of uncertainty.

SUSTAINABILITY

The use of the word "sustainability" is widespread, and although the principle of the concept of sustainability is clear enough, the exact interpretation and definition of it have a large number of meanings to different people. There are numerous definitions of sustainability and sustainable development, and sustainability is one of the most contested ideologies of our time, hence the reason why there is no universally agreed definition. However, the Brundtland Commission (WCED, 1987) definition of **sustainable development** is widely used and cited:

> Sustainable development is development that meets the needs of the present without compromising the ability of future generations to meet their own needs. (WCED, 1987)

Three pillars of sustainability are widely recognised as the guiding principles in terms of achieving and assessing sustainability (Ciegis et al., 2009). The three pillars are interdependent and cannot be viewed in isolation if a holistic picture of sustainability is to be obtained. Often referred to as the triple bottom line of sustainability, the three principles consist of:

- Environmental sustainability
- Economical sustainability
- Social sustainability

Sustainable construction

The construction industry, and the associated materials industry which it relies on, is a major consumer of the earth's natural resources. The construction industry places a massive burden and stress on many aspects of the global environment if not managed correctly due to the innate nature of the industry. Loss of land and food-growing potential results from quarrying for material and reuse for projects, e.g. cities, highways, dams, etc., as well as unsustainable use of forest lands for timber and other raw materials. The construction industry is also a major consumer of non-renewable energy sources and limited materials, e.g. lead, zinc, and copper, in the production of products and materials. Pollutants and waste generation associated with the construction industry, e.g. CFC gases, dust, fibre, and toxic gases, also contribute to the degradation of the global environment and have consequential social and economic implications. Rising global populations have placed further demand on the construction industry, and there are consequent negative effects of many current unsustainable approaches.

The construction industry contributes significantly to the overall socio- and economic performance of a country or region given the far-reaching implications of both its approach and its product. If the construction industry is to play a significant role in the sustainable growth of any country or region, a holistic, considered approach is required in terms of its impact over the three pillars of sustainability. **Sustainable construction** has also been referred to as "green building", ecological building, and sustainable architecture. It is seen as a method for the construction industry to contribute positively to achieving long-term sustainable development and has a number of definitions, for example:

> Sustainable construction is the creation and responsible management of a healthy built environment based on resource efficient and ecological principles. (Du Plessis, 2005, p. 407)
>
> Sustainable construction is an integrative and holistic process aiming to restore harmony between the natural and the built environment, and create settlements that affirm human dignity and encourage economic equity. (Du Plessis, 2002)

Sustainable construction has different approaches and priorities in different countries. Some countries identify the three pillars of sustainability as the framework, but many national definitions focus primarily on the environment or ecological impacts of development. National or region-specific aspects such as population, the economy, natural hazards, land and water availability, energy production and supply, and the building sector and existing building stock can all further influence the national or regional definitions and priorities for sustainable development.

Sustainable construction in LDCs and post-disaster contexts

The majority of the world's population currently live in LDCs, and future population growth projections anticipate that the majority of growth will also be in LDCs. The resulting unprecedented demand for housing and infrastructure to accommodate this mass population has an immense effect on an LDC's construction industry. As such, substantial amounts of capital funds have been allocated to various infrastructure and residential projects in these areas in recent decades. The need for long-term sustainable approaches in the delivery of construction projects to meet these demands from the construction industry is clear (Du Plessis, 2002).

The concept of sustainability has only recently been introduced into LDCs' construction industries and post-disaster contexts, and sustainability and sustainable construction are not yet an essential part of the decision-making process. LDCs and post-disaster contexts often put economic development and recovery above sustainable requirements, with immediate construction demands overshadowing environmental concerns. Othman (2014) notes that a wide range of challenges exist that affect delivery of sustainable construction projects in LDCs and post-disaster contexts. He outlines five main headings under which they fall: (1) human development, (2) technical, (3) managerial, (4) political, and (5) the triple bottom line of sustainability (environmental, social, and economic).

LDCs and post-disaster contexts often have different climates, hazards, and cultural and economic conditions to those of developed countries. However, there are also similarities in issues in terms of approaches to sustainable construction, and it is key that the diversity and commonality are recognised and addressed appropriately when designing or assessing sustainable approaches in each context. The demands of sustainable construction in LDCs and post-disaster contexts can differ greatly from those of developed countries. Perceived higher costs and underlying socio-cultural factors also contribute to the lower levels of social acceptability of sustainable construction in mainstream affordable housing.

Du Plessis (2002) states that the construction approaches from developed countries, which are highly reliant on technical solutions and need adapting by LDCs, are inappropriate and must be expanded to address the social and economic pillars of sustainability. Du Plessis (1999) argues that there is a complete communication gap between the developed and developing worlds in terms of sustainable construction in this regard and that for real sustainable development to exist, developing and developed countries need to communicate and work together to establish first what sustainable development is.

Global inequalities and economic constraints in many LDCs and post-disaster contexts have also resulted in pragmatic governance and decision making in relation to sustainable social and environmental goals being made more difficult. In many instances, due to natural disasters or conflicts, organisations must provide housing to meet the immediate shelter needs of large populations, and this is often given priority over long-term aspects such as sustainable design, participation, and reference to local materials and techniques. Sustainability is thus often perceived as an additional cost to standard practice and is not a necessity but rather a luxury of the rich (Reffat, 2004).

Affordability and sustainability

Affordable housing is an essential concept to LDCs and post-disaster contexts. **Affordability or affordable housing** is a difficult term to define as it can have different meanings to different people and vary widely from country to country or region to region. The term "affordable housing" is often used internationally and represents a number of different subsets of housing that include social housing, low-income housing, financially assisted housing, and government-subsidised housing.

In many LDCs and post-disaster contexts, affordability in the mainstream housing markets is associated with economic sustainability, often with little emphasis on environmental or social sustainability (Randolph et al., 2008). Perceived higher costs and underlying socio-cultural factors also contribute to the lower levels of social acceptability of sustainable construction in the mainstream affordable housing market (Buys et al., 2005). Pragmatic decision making by governments in LDCs or in a post-disaster context to meet the immediate housing needs often fails to fully embrace the principles of sustainability in both the design and delivery of housing, as cost and the maximum number of housing units being delivered as quick as possible are often the priority.

Affordability of housing is usually measured as a relationship between household income and housing cost such as multiples of income or median income for a country or region (Menshawy et al., 2016). However, these mathematical and formula-based approaches, while useful to some extent, result in conclusions which relate to large geographical and socially-diverse areas, rather than endeavouring to adequately address individual contexts and indeed individual households. Consquently they fail to present specific contexts or areas within individual households.

Stone (2006) states that the term "affordable housing" is an unjustified term and that affordability is not a characteristic of housing but rather it is a relationship between housing and people. He argues that for some people, all housing is affordable regardless of price, whilst for others it is not affordable unless it is free. Stone further argues that "affordable" housing can have meaning and use but only if three essential questions are answered:

1. Affordable to whom?
2. On what standard of affordability?
3. For how long?

Incomes in LDCs and some post-disaster contexts are lower and as such, housing can represent a much higher proportion of a family's income. Other aspects relative to LDCs, and some post-disaster contexts (e.g. immense population growth with demand out-stripping supply, higher levels of natural disasters, fewer utilities, and lack of materials, skills, and design/construction knowledge), often serve to raise the cost of housing relative to the incomes in many contexts. As such, many common definitions and measures of affordability, which are used in developed countries with functioning governments and markets and fewer natural hazards, fail to realistically reflect these aspects because they do not take into account the variances on specific or individual contexts.

HOUSING IN LDCS AND POST-DISASTER CONTEXTS: APPROACHES, PARTICIPATION, AND BARRIERS

The provision of housing in LDCs and post-disaster contexts has been an ongoing issue for decades. Many governments in both contexts lack the capacity to adequately design and implement appropriate housing, particularly in a post-disaster context when the additional pressure of the need for immediate shelter is often given priority over long-term sustainable solutions. This, coupled with vast influxes of international aid in the immediate aftermath of natural disasters, has often resulted in governments looking to external international assistance and expertise in relation to post-disaster reconstruction. Common strategies in response to the housing need include subsidy, quality reduction, self-help, and simplification (Johnson, 2006; Katz et al., 2003). Subsidy involves agencies providing payment for part or all of the costs of housing, with some having to be repaid. Quality reduction has been a result of factors including lack of agreement on acceptable standards, political issues, and longevity issues.

Top-down approach strategies for housing provision often prevail, and LDCs and post-disaster contexts are left to what professionals from the formal sector consider the most appropriate solutions. These can often be at odds with the expectations of the future house inhabitants and environmental ideals (Lizarralde and Davidson, 2011; Fallahi, 2007; Bruen et al. 2013).

External assistance in the form of international non-governmental organisations (INGOs) is utilised greatly to assist many local governments in housing provision. However, international assistance does not automatically imply successful outcomes in relation to meeting housing needs. Many INGOs also lack the expertise and strategies in relation to the effective design and delivery of housing in LDCs or post-disaster/conflict contexts.

INGOs often face additional challenges in that they are foreign to the context they are working in, and this has resulted in many introducing inappropriate design solutions that do not cater for the beneficiaries' long-term needs. This is often as a result of time and resource pressure, as well as insufficient knowledge of the local context in terms of its social, cultural, economic, and environmental makeup.

Many current approaches to post-disaster housing view the dwelling as a mere product or output; this often results in inappropriate repetitive constructions and typologies which ironically can have adverse effects on the end users' long-term needs and wellbeing in terms of environmental, social, cultural, and economic development. Current approaches often seek to employ a "one hat fits all" or universal approach to meet housing needs, frequently employing imported industrialised materials, western construction techniques, and typologies more suited to developed countries, as opposed to the surrounding context.

This is due to the perception that these solutions are associated with wealth progress, prosperity, and globalisation. These approaches sometimes result in non-affordable, rigid, and monotonous construction typologies which paradoxically can have negative effects on the end users' long-term needs and wellbeing, e.g. replacement of traditional settlements with modern towns and lack of cohesion between the town's inhabitants leading to negative socio-cultural and economic effects (Adeyemi, 2002). There is need for INGOs and those working for them to have a thorough understanding of and capacity to adhere to the social dynamics at project level of the context they are operating in to provide people-orientated housing.

Participatory design and delivery of housing in LDCs and post-disaster contexts

Many previous approaches to housing shortages and new housing delivery in LDCs and post-disaster contexts often involved a top-down macro-level approach. Many of these failed due to issues including displacement, cost recovery, affordability and replicability, and as a result of failing to engage communities in development projects (Muraya, 2006). Such approaches were employed on the assumption that people did not know their needs and were unwilling or unable to pay for services. This approach does not adequately address the principles of sustainability and affordability in the design and delivery of housing. Change in approaches and policies for housing in LDCs have evolved over the recent past, from being centred on top-down macro-level government-provided housing, to a more bottom-up micro-level self-help approach that focuses on enablement and the involvement of communities and community-based organisations (CBOs). This can result in housing projects where all parties support a people-centred housing process to obtain housing delivery goals.

In order for a participatory approach to be meaningful, it is essential that the appropriate type and intensity of participation are utilised. The term participation is widely expressed under many titles, and its exact application for individual projects is difficult to establish given the diverse nature of projects. Davidson et al (2007) provide a "ladder of participation" (Figure 8.3), noting that approaches at the top of the ladder empower people and communities, offer collaboration, and give communities control over the project. The lower end of the ladder offers beneficiaries a possible consultation (which might be disregarded), or they are merely informed of the shape their project is taking, or they are possibly manipulated into taking part in the project. The lower end of the ladder (rungs 3 to 5) represents approaches where beneficiaries have little or no control over actual decision-making due to the level and type of participation approaches utilised. Davidson et al (2007) state that consulting and informing are often passed off as legitimate forms of participation despite the beneficiaries' decision-making being restricted.

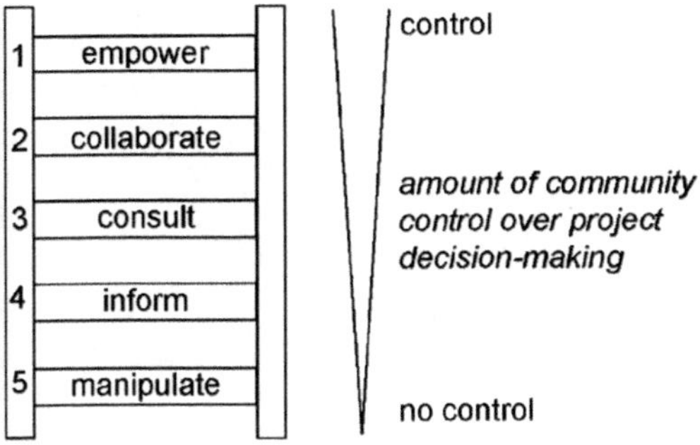

Figure 8.3 Ladder of community participation (Davidson et al., 2007).

Table 8.3 Difficulties in the successful application of effective participatory approaches in LDCs and post-disaster contexts

Challenges jeopardising successful approaches
Insufficient community and municipal interest and difficulties in integrating the community to the design and management of the projects.
Reluctance by the government to give significance to low-income groups or communities.
Lack of definition of what the participation means in a project environment.
Insufficient information.
Insufficient decision-making power.
Difficulties in building mutual trust between agencies and communities.
Reduction of participation to mere seat equity as opposed to active involvement in decision-making.

Various authors (Ishmail, 2005; Davidson et al., 2007; Varol et al., 2011) have identified the difficulties in the successful application of effective participatory approaches in LDCs and post-disaster contexts, as shown in Table 8.3.

In order for participation to be meaningful for a project, beneficiaries:

- Must be involved in the **early stages** of the project, and
- Must have **genuine control** and influence over the decision-making, which
- Must benefit in their **long-term** wellbeing.

Beneficiaries must have **full responsibility** for their own choices and projects, rather than being treated as passive victims receiving aid.

Figure 8.4 Successful post-disaster housing (Sri Lanka) resulting from the participation in the design and delivery process (source: John Bruen and John Spillane).

Figure 8.5 Unfinished dwelling and substandard construction (Sri Lanka) resulting from a top-down approach with little participation by beneficiaries (source: John Bruen and John Spillane).

Figures 8.4 and 8.5 show contrasting results of post-disaster projects in Sri Lanka. Figure 8.4 shows successful post-disaster housing from Sri Lanka because of participation in the design and delivery process. Figure 8.5 gives an example of the negative outcomes of post-disaster housing in Sri Lanka. It shows an unfinished dwelling and substandard construction as a result of a top-down approach with little participation by the proposed beneficiaries.

Barriers to sustainable and affordable construction in LDCs and post-disaster contexts

Common barriers exist in LDCs and post-disaster contexts in the construction industry and in the delivery of affordable and sustainable housing. Some of these barriers are given in Table 8.4.

Table 8.4 Barriers to sustainable and affordable construction in LDCs and post-disaster contexts

Environmental sustainability is often a low priority in LDCs and post-disaster contexts, and the need for immediate shelter is primary for the majority.
Psychological and sociological issues in relation to use of alternative materials and its acceptability by people, i.e. status of certain materials deemed for the poor only.
Effects of globalisation and desire of many to imitate housing approaches of the west leading to inappropriate imported typologies, materials, designs, etc.
Lack of overall holistic approach to sustainable design, i.e. social, cultural, economic, and environmental.
Lack of training and education in sustainable design and construction leading to lack of necessary design and building skills available.
Lack of access to adequate information and knowledge on sustainable development and design.
Inadequate government planning and policy-making.
Perceived higher cost of sustainable building approaches.
Scarcity of professional capabilities, e.g. designers and project managers.
Lack of demonstration examples of best practice sustainable construction approaches.
Disincentive factors over local material production, i.e. government supplying alternative imported materials.
Need to develop cost-effective construction technologies.
Building materials becoming ever more expensive. Need to utilise locally available materials to suit local typologies.
Lack of guidelines available on selection of appropriate building approaches and packages, e.g. materials, methods, designs, equipment.
Unaffordable land and housing prices and lack of land tenure.
Inadequate housing finance systems.
Lack of support of small-scale construction industries.
Inefficient or inadequate implementation strategies.
Lack of research and experimental findings.
Lack of implementation and promotion of research findings on appropriate approaches.
Gradual vanishing of traditional wisdom and knowledge on local material and construction techniques.
Historical legacies of race and class.
Lack of government or political backing to sustainable development.
Inappropriate procurement systems.

(continued)

Table 8.4 *(continued)*

Inability to adopt best practice.
Bureaucratic impediments to the implementation of housing.
Corruption.
Inappropriate building regulations.
Lack of consultation and capacity building with housing inhabitants.
Lack of construction material resources.
Lack of human resources to undertake work.
Health and safety management issues.
Poor housing management and planning policies at both a national and local scale in many countries.

REFERENCES

Adeyemi, A. Y. (2002). Affordable housing production: The influence of traditional construction materials, in: 30th IAHS World Congress on Housing, Housing Construction: An Interdisciplinary Task, Wide Dreams—Projectos Multimédia, 2002, pp. 827–832.

Alexander, D. (2002). '*Principles of Emergency Planning and Management*', Oxford University Press. UK

Bruen, J., Hadjri, K. and von Meding, J. (2013). Design Drivers for Affordable and Sustainable Housing in Developing Countries. Journal of Civil Engineering and Architecture. Oct. 2013, Volume 7, No. 10 (Serial No. 71), pp. 1220–1228.

Buys, L., Barnett, K., Miller, E. and Bailey, C. (2005). 'Smart housing and social responsibility: learning from the residents of Queensland's research house', *Australian Journal of Emerging Technologies and Society* 3(1), pp. 43–47.

Cannon, T. (1994). '*Vulnerability Analysis and the Explanation of 'Natural' Disasters in Disaster, Development and Environment*', ed. A. Varley, John Wiley and Sons, Chichester, pp. 13–30.

Ciegis, R., Ramanauskiene, J. and Martinkus, B. (2009). 'The concept of sustainable development and its use for sustainability scenarios', *Engineering Economics*, 2, pp. 28–37.

Davidson, C.H., Johnson, C., Lizarralde, G., Dikmen, N. and Sliwinski, A. (2007). 'Truths and myths about community participation in post-disaster housing projects', *Habitat International* 31, pp. 100–115.

DFID (2011). 'Defining disaster resilience: a DFID approach paper', Available at https://www.gov.uk/government/uploads/system/uploads/attachment_data/file/186874/defining-disaster-resilience-approach-paper.pdf (Accessed 02-10-17).

Du Plessis, C. (1999). 'Sustainable development demands dialogue between developed and developing worlds', *Building Research and Information* 27(6), pp. 378–389.

Du Plessis, C. (2002). '*Agenda 21 for Sustainable Construction in Developing Countries: A Discussion Document*', CSIR Building and Construction Technology, Pretoria.

Du Plessis, C. (2005). 'Action for sustainability: preparing and African plan for sustainable building and construction', *Building Research and Information* 33(5), pp. 405–415.

Fallahi, A. (2007) Lessons learned from the housing reconstruction following the Bam Earthquake in Iran, *The Australian Journal of Emergency Management* 22 (1) (2007) 26–35.

Gustin, J.F. (2008). '*Safety Management: A Guide for Facility Managers*', Fairmont Press, Distributed by Taylor and Francis Ltd., Lilburn, GA; Boca Raton, FL.

Haigh, R., Amaratunga, D. and Kerimanginaye, K. (2006). 'An exploration of the construction industry's role in disaster preparedness, response and recovery', Proceedings of the Annual Research Conference of the Royal Institution of Chartered Surveyors, Research Institute for the Built and Human Environment, University of United Nations. (2006). 'Exploring key changes and developments in post-disaster settlement, shelter and housing, 1982–2006', United Nations, Scoping study (OCHA/ESB/2006/6). Salford.

Hidayat, B. & Egbu, C. (2010). A Literature Review Of The Role Of Project Management In Post-Disaster Reconstruction. In C. Egbu, ed. Procs. 26th Annual ARCOM Conference, 6-8 September 2010. Association of Researchers in Construction Management, pp. 1269–1278.

Horlick-Jones, T. and Peters, G. (1991). 'Measuring disaster trends, part one: some observations on the Bradford fatality scale', *Disaster Management* 3(3), pp. 144–148.

IFRC (2001). 'World disasters report, focus on recovery', International Federation of the Red Cross and Red Crescent Society. Available at: https://www.ifrc.org/Global/Publications/disasters/WDR/21400_WDR2001.pdf (Accessed 10-15-20).

Ishmail, Z. (2005). 'Evaluating the utility of participation in development projects: how project management techniques contribute to efficiency of upgrading informal human settlements: a case study of the N2 Gateway project', Unpublished thesis, University of Cape Town.

Johnson, M. (2006). Decision Models for Affordable Housing and Sustainable Community Development. Journal of the American Planning Association: The Future(s) of Housing. Available from: https://citeseerx.ist.psu.edu/viewdoc/download?doi=10.1.1.476.900&rep=rep1&type=pdf

Katz, B., Austin, M., Turner, K., Brown, D., Cunningham, M., and Sawyer, N. (2003). Rethinking Local Affordable Housing Strategies: Lessons from 70 Years of Policy and Practice. The Brookings Institution Center on Urban and Metropolitan Policy and The Urban Institute. December 2003. Washington USA.

Lizarralde, G. and Davidson, C. (2011). 'Learning from the poor', IF Research Group, Faculté de l'Aménagement, Université de Montréal, Canada. Available at: www.grif.umontreal.ca/pages/lizarralde_gonzalo.pdf.

Menshawy, A.E., Shafik, S., and Khedr, F. (2016). 'Affordable housing as a method fo informal settlements sustainable upgrading', *Procedia–Social and Behavioural Science* 223, pp. 126–133.

Muraya, P.W.K. (2006). 'Failed top-down policies in housing: the cases of Nairobi an Santo Domingo', *Cities* 23(2), pp. 121–128.

Ofori, G. (2011). '*Contemporary Issues in Construction in Developing Countries*', CIB Routledge, London.

Othman, A. (2014). A conceptual model for overcoming the challenges of meg construction projects in developing countries. *African Journal of Engineering Research* 2(4): 73–84.

Paton, D., Smith, L. and Violanti, J. (2000). 'Disaster response: risk, vulnerability an resilience', *Disaster Prevention and Management*, 9(3), pp. 173–179.

Quarantelli, E.L. and Perry, R.W. (2005). '*What is a Disaster? New Answers to Ol Questions*', Xlibris, Philadelphia, Pennsylvania, pp. 325–396.

Randolph, B., Kam, M. and Graham, P. (2008). 'Who can afford sustainable housing' in: A. Nelson (Ed.), *Steering Sustainability in an Urbanizing World*, Ashgate, Aldershot UK.

Reffat R. (2004). 'Sustainable construction in developing countries', in: Proceedings o First Architectural International Conference, Cairo University, Egypt.

Schilderman, T. (2004). 'Adapting traditional shelter for disaster mitigation an reconstruction: experiences with community-based approaches', *Building Researc and Information* 32(5), pp. 414–426.

Smith, K. (1992). '*Environmental Hazards: Assessing Risk and Reducing Disaster*' Routledge, London.

Stone M.E. (2006). 'What is housing affordability? The case for the residual incom approach', *Housing Policy Debate* 17(1). Available at: www.works.bepress.com/ michael_stone/5/ (Accessed 01-10-17).

Toya, H. and Skidmore, M. (2007). 'Economic development and the impacts of natura disasters', *Economics Letters* 94, pp. 20–25.

UNISDR (2009). '*UNISDR Terminology on Disaster Risk Reduction*', United Nation International Strategy for Disaster Reduction, Geneva, Switzerland.

Varol, C., Ercoskun, O.Y. and Gurer, N. (2011) 'Local participatory mechanisms an collective actions for sustainable urban development in Turkey', *Habitat International* 35, pp. 9–16.

Voogd, H. (2004). 'Disaster prevention in urban environments', *European Journal o Spatial Development* 12, pp. 1–14.

WCED (1987). '*Our Common Future*', World Commission on Environment an Development. Brundtland Commission. Oxford University Press, Oxford.

White P., Pelling, M., Sen, K., Seddon, D., Russel, S. and Few, R. (2004). 'Disaste risk reduction: a development concern. A scoping study on the links betwee disaster risk reduction, poverty and development', Overseas Development Group, University or East Anglia. Available at: www.preventionweb.net/files/1070_drrscopingstudy.pdf (Accessed 02-10-17). Note: The Overseas Development Group has changed its name to International Development UEA. Available at: https://sites.uea.ac.uk/devresearch/about

9

Environmental considerations

Cormac Flood, Chris Gorse, Lloyd M. Scott and Melanie Smith

ENVIRONMENTAL ISSUES

Environmental issues are major concerns in the surveying and construction industry. Action is possible in many areas throughout built environment professions, and for surveying professionals, environmental matters can be a consideration (major or minor) in most work situations.

Environmental issues include, but are not limited to, aspects of sustainability. Law affecting the environment, including climate change, has recently developed strongly, covering local, regional, and international professional concerns. The laws impact on society and commerce, owning, occupying, purchasing, selling, letting, and carrying out works on property. Environmental risks are increasingly important. An obvious area is in construction specification where more sustainable practices and materials can be applied. This, of course, depends on definitions of sustainability. It is the case that one person's sustainable solution can have heavy adverse effects on someone else's.

The issues raised in this chapter cover some of the naturally occurring environmental issues common in the work of a building surveyor. Asbestos and similar deleterious materials are also "natural" but have been processed to use in the built environment and are therefore included in the chapter on building pathology.

Surveyors are affected by:

- The environment
- The law governing the environment and its effects on:
  - The management and use of land and property
  - The development and the re-use of land and property
- The obligations on Chartered Surveyors that arise from these duties

As always, surveyors are obliged to consider their competence to advise clients in assessing environmental considerations and whether they carry sufficient and appropriate professional indemnity insurance (PII) for the works and risks involved. A good overview of the works which can be involved is included in the RICS Guidance Note, "Environmental risks and global real estate" (RICS, 2018). This publication includes a Property Observation Checklist for identifying potential environmental issues for residential, commercial, and rural properties, which can be accessed independently through isurv. The 2018 document supersedes the RICS Guidance Note, "Contamination, the environment and sustainability: Implications for chartered surveyors and their clients" (RICS, 2010) which contained checklists. See also RICS (2020) "Contamination and environmental matters: property observation checklists". This chapter starts by looking at the ground.

GROUND

Foundations and strata, soils, rocks, or made-up ground

There are a number of key points that need to be considered with all foundation systems. Fundamentally:

- The foundation of a building transmits live (active loads, such as people, wind, rain, snow, etc.) and dead loads (structural loads) to the ground.
- The foundation is part of the building's substructure. The substructure is the part of a building that is below the ground level.
- The substructure elements are typically foundations, basements, structural walls, and floors that may be constructed below the ground.
- The ground is the term used to describe the earth's surface. The ground often varies in composition (natural or un-natural make-up).
- Human impact on the composition of the ground is common. The ground is often built up, made up, or laced with services, contaminants, and historic artefacts from prior development.

The substructure provides an interface with the ground, enabling the superstructure to remain stable and its loads (both dead and live) to be safely transferred. The following categories of the natural interface are often used to describe the components and composition of the ground that provides the interface between the building foundation and founding strata. Categories include:

- Rocks
- Non-cohesive soils
- Cohesive soils
- Peat and organic soils
- Made-up ground and fill

Rocks are dense hard, cemented, and rigid geological deposits. Typically, rocks are:

- Granite
- Sandstone
- Limestone

Rocks, generally, will withstand high loads and can be used to found buildings on with relatively little movement or compression.

Soils, in comparison to rocks, are relatively soft and compressible geological deposits; they are loose and not mechanically or naturally cemented geological deposits. Soils include:

- Gravel
- Sand
- Clay

Soils, made-up ground, and fill will under normal conditions compress under the loads of the building and are not suitable strata on which to found buildings.

Foundations are needed to transfer the loads below the compressible soils to substrata where the ground can withstand loads without undue settlement. If a decision is made to locate buildings on made-up ground, the foundation will normally penetrate through the made-up strata, to suitable loadbearing strata. If any load is transferred to the made-up ground and fill, it will be necessary to treat and condition the ground so that it is stable and does not unduly compress.

The building's foundations are designed to transmit loads to the ground without movement that affects the stability of the buildings. When loads are placed on the ground, the ground will resist the force and compress, but where the foundations are sufficient, taking the loads to ground that can sustain the loads, the compression is minimal and does not affect the stability or compromise the integrity of the building. The foundations transfer the loads with minimal and controlled settlement, ensuring the structure of the building, substructure, and services entering and leaving the building remain operational.

The loads of a building can affect neighbouring structures, so when designing foundations careful attention is required to ensure that loads from one building don't unduly impact on the stability of neighbouring structures.

Changes in ground composition and behaviour

Changes to the ground may affect the ability of a foundation to effectively transfer loads. Changes occur in:

- Water levels, water courses, and the water table
- Loads and vibration from traffic
- Movement of plant and industrial operations on neighbouring land and/or buildings
- Additional loads (buildings or otherwise) placed on adjacent ground
- Plant used to construct neighbouring buildings and structures

As the loads placed on the ground do change during the life of a building, the behaviour of the subsoil is affected and will change. When conditions change, the existing foundations of a building may need to be re-designed and strengthened so that they are capable of continuing to transfer the loads of the building. The loads of the building should be transferred without adverse settlement, even when the conditions of the ground change. The inspection and surveying of building foundations and surrounding properties are essential, especially when new buildings are being constructed, trees are removed, or the purpose of the building is altered.

As the loads of a building are placed on the ground, the ground reacts and compresses. A limited amount of settlement is to be expected. However, major, adverse settlement will impact on the integrity of the structure, underground services, and stability of the building.

Settlement and movement

The settlement and movement are characteristically different depending on the nature of the soil and composition.

Typical settlement and movement behaviour on:

- **Non-cohesive soils (gravel and sand)** settlement or other movement generally takes place as the building is erected; this is often described as immediate settlement.
- **Cohesive soils (clay, silts)**: the compression and movement are a more gradual process as water and air are expelled from pores; such settlement is described as consolidation settlement.

Settlement and movement can result from differential loading, different ground conditions, frost heave, trees, and changes in groundwater conditions (e.g. due to water courses or cracked drains). Some of these aspects are also discussed in other chapters, such as under building pathology.

As designers and engineers know that settlement occurs, foundations are designed to limit and control settlement and ensure that any small amount of settlement is consistent across the structure. Throughout the life of the building, settlement will occur; foundations and buildings should be designed so that they do not suffer from differential settlement or settlement that affects the integrity of the building. At all stages, the load need to be effectively transferred so that the loads are placed and transferred to the ground without undue settlement. Combinations of foundation and ground treatment can be used to ensure the building remains stable throughout its life.

Differential settlement (Figure 9.1) can be due to:

1. **Differing building loads**, such as:
   - Dead loads not accounted for – where the building's structural loads are not properly accommodated for within the design and by the foundations
   - Unexpected live loads – loads within the building do change; if services and equipment are installed within the building and the loads or dynamic loads, e.g. vibration, exceed the foundation and soil design strength, then the ground can be overstressed and compress, slip, or fail.
2. **Different ground conditions** under the building, such as:
   - Water courses that destabilise the ground.
   - Weak compressible strata.
   - Adjacent loads, e.g. adjacent buildings, plant, vehicles causing overloading under existing foundations.
   - Changes in vegetation close to the building. Trees and plants will take water out of the ground or if vegetation is removed the volume of water within the soil can increase. The different levels of water in clay soil will cause it to shrink, as water is removed, or swell as the water content increases.
   - Freezing of the ground below or close to foundation can lift or move the foundation and substructure. As water in the ground freezes, it expands and this may cause foundations to lift as the ground below the foundation freezes, or to move as the ground next to it freezes.

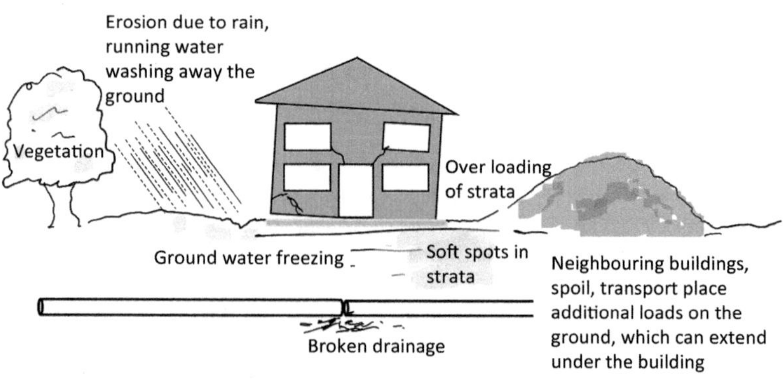

Figure 9.1 Schematic illustrating some of the causes of differential settlement.

Frost heave is caused by the expansion of water in the ground (Figure 9.2), as described above. To avoid frost heave from seasonal changes in groundwater, including the impact of vegetation on water content within the ground, the NHBC Standards (NHBC, 2021, chapter 4.3.3) recommend a minimum depth of 450 mm for strip and trench fill foundation excavations. Obviously deeper trenches are needed for some soil types and ground conditions. The NHBC standard refers here to frost susceptible soils, and construction undertaken during cold weather, however, most foundations are excavated to a minimum distance of 750–1000 mm which does avoid most volume changes due to seasonal movement.

In areas of extreme frost exposure or high levels of rainfall, the depth of foundations may need to be deeper.

Trees and vegetation remove water from the ground. In clay soils, this causes the clay to shrink during periods of high growth. Thus, seasonal movement of the ground occurs.

Where trees present a challenge to foundation stability, action may be required. In extreme cases, physical barriers (concrete or steel barrier walls) may be built between trees and buildings to reduce the impact of the tree roots and extraction of moisture from the soil. Most root action takes place at levels between 1 and 3 m; below 3 m clays are often too dense for roots to easily penetrate the soil. Thus, trees close to the building can affect foundations at this level. However, unless it is deemed that the roots are having an impact on the building's stability, it is not advised to remove trees (Barry, 2017).

See also Chapter 4 and Figures 4.14 and 4.15 and Table 4.11 for reasonable distances of buildings from trees. Removing trees and vegetation can equally present

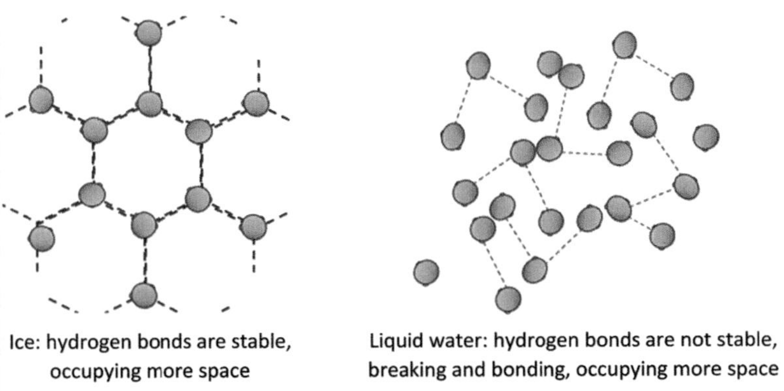

Figure 9.2 Schematic of the different bonding structures of ice and water.

problems if the cohesive soils have become accustomed to high water content. The increase in water as a result of the loss of vegetation can cause the soils to swell, roads and pavements to lift, basement walls to crack, and ground floors to suffer the effects of swelling heaving soils, see Chapter 4, Figure 4.12.

Rocks

Rocks are classified, according to their geological formation, as:

- Sedimentary
- Metamorphic
- Igneous

These are described as in Table 9.1 or by reference to their presumed bearing value (Table 9.2), which is the net loading intensity considered appropriate to the particular type of ground for preliminary design purposes.

Ground bearing capacity and values are based on an assumption that foundations are carried down to the typical strata of unweathered strata and rock (Figure 9.2).

Table 9.1 Rock types

Group	**Rock type**
Sedimentary Formed from an accumulation or deposit of particles, which are laid on top of one another and are composed under pressure of the ground, air, or water above.	Sandstones (including conglomerates) Some hard shales Limestones Dolomite Chalk
Metamorphic Transformation of an existing rock as a result of heat or pressure – metamorphism. These rocks are formed subject to high temperatures and pressures deep beneath the earth's surface; they are not rocks that melt; rocks that do melt form igneous rock.	Some hard shales Slates Marble Quartzite Schists Gneisses
Igneous Formed from molten rock, magma.	Granite Dolerite Basalt Pumice Scoria Tuff

Table 9.2 Bearing capacities of typical strata

Group	**Types of rocks, soils (cohesive and non-cohesive)**	**Bearing capacity (kN/ m^2)**
I: Generally high loadbearing capacity	1. Strong igneous and gneissic rocks in sound condition	10,000
Rocks, schists, slates, and shales	2. Strong limestones and strong sandstones	4,000
	3. Schists and slates	3,000
	4. Strong shales, strong mudstones, and strong siltstones	2,000
	5. Clay shales	1,000
II: Capable of supporting light loads	6. Dense gravel or dense sand and gravel	600
Non-cohesive soils: sand and gravel	7. Medium dense gravel or medium dense gravel and sand	200–600
	8. Loose gravel or loose sand and gravel	200
	9. Compact sand	300
	10. Medium dense sand	100–300
	11. Loose sand	100
III: Cohesive soils: clays	12. Very stiff boulder clays and hard clays	300–600
	13. Stiff clays	150–300
	14. Firm clays	75–150
IV: Weak strata	15. Soft clays and silts	75
	16. Peat and organic soils Will naturally grow, decompose, and move. Not a suitable for founding on.	Foundations carried down through these to a reliable bearing stratum
V: Fill	17. Made-up ground or fill Will compress under self-weight, not suitable for founding on.	Should be investigated and only developed with extreme care
Based on BS 8004:1986.		

Unweathered, hard igneous, and gneissic rocks, when they are in sound condition, have high allowable bearing pressure, and provide excellent material on which to transfer building loads. There is limited potential of foundation failure.

Limestones and sandstones, which are classified as hard and well bedded, can be considered to be stronger than some concretes, and most buildings will not use the full bearing capacity. Limestones, however, can dissolve into solution where water that contains carbon dioxide comes into contact with the limestone, eroding the limestone and reducing the stability of the rock as a founding material.

Schists and slates are rocks with pronounced cleavage of large-grained flakes in sheet-like orientation. Where the rock beds are broken, shattered, or on a steep incline, a reduction in bearing capacity will often result. Where the mudstones and shales are hard and compact, they have a high allowable bearing capacity.

Sandstones and those considered to be soft sandstone are variable, depending on the cementing of the material. The softer shales and mudstones have allowable bearing pressures between the stiffer hard cohesive soils and rock; as with other cohesive soils they are susceptible to swelling and softening with water and shrinkage when water is removed.

Chalk and soft limestone are soft sedimentary carbonate rocks that are porous in nature. They are variable in nature and the allowable bearing pressure is also variable; they can deteriorate when exposed to water and frost, and as such the founding material should be protected with a blinding layer of concrete when the formation level is reached.

Soils: Characteristics

Soils can be classified as non-cohesive or cohesive. The properties of soils that determine characteristic behaviour when used as a founding material include:

- Compressibility
- Cohesion between particles
- Friction between particles
- Permeability

It is convenient to compare the characteristics and behaviour of clean sand, which is a coarse-grained non-cohesive soil, with clay, which is a fine-grained cohesive soil.

Under load, clean sand, which is coarse-grained and non-cohesive, only compresses slightly as the particles rearrange themselves and small amounts of water are expelled. The movement of the particles and settlement are proportional to the load imposed and normally distributed evenly as the loads are applied.

Clay is cohesive, compressible, and will be affected by seasonal changes that change moisture in the ground. However, as clay is impermeable, the soil compresses slowly as the water is expelled through narrow capillaries in the soil. Settlement and movement of clay soils can continue for years. Care should be taken not to overstress clay soils and to ensure foundations are sufficiently deep to not be affected by seasonal change and vegetation that can change moisture levels in the ground.

Where the load of a building is placed on plastic clay, the clay is slowly compressed as air and water are expelled through narrow capillary channels. The movement of the clay under load, as it compresses, also causes heave at the surface of the surrounding ground, as the displaced clay particles are pushed outwards under the load.

Non-cohesive coarse-grained soils such as gravels and sands will compress and consolidate as the particles rapidly rearrange and water is expelled.

Allowable bearing pressures are dependent on:

- Grading of particles.
- Density, degree that particles are packed together.
- Size of particles – larger particles tend to result in larger allowable bearing pressures.
- Groundwater – groundwater and flow can adversely affect bearing pressures by washing out finer particles affecting the grading.
- Lateral support of confinement. Non-cohesive soils need to be confined laterally.

Cohesive fine-grained soils and silts are a result of natural deposit of the fine siliceous and aluminous particles produced from the weathering of rock. Clay has a smooth surface and when wet or moist is greasy and slippery. It has high plasticity, is mouldable, and when it dries, it shrinks.

The main characteristics of cohesive soils:

- Changes in volume: shrink when dry and swell when wet.
- Fine capillary paths allow water to be compressed out gradually.
- Susceptible to seasonal changes in volume.
- When compressed, the expulsion of water and air from clay is gradual, meaning that buildings settle as the construction takes place.
- Settlement continues for some years after construction.
- Trees and shrubs, during growth periods, will remove water in the soil and shrink the clay.

- Seasonal changes in the UK tend to be limited to depths of 1 m, but can be deeper in severe droughts or periods of high and prolonged rainfall.
- Large trees can affect soils and foundations to a depth of 4 m.
- When trees are removed, moisture in the cohesive soils builds up and the soil expands.
- When trees are planted or during periods of high growth, water will be extracted from the soil and the soil will shrink.
- Cohesive soils can take several years to stabilise following changes in the ground conditions and vegetation.

Organic soils and peat, from a civil engineering perspective, are those soils composed mainly of vegetable, or decaying vegetable matter from plants. The soil contains a combination of the decaying vegetable matter (around 75%) with the remaining material being fine sand, silt, or clay. Highly organic soils are compressible, still in the process of decay, and will not be suitable for the foundation of buildings, structures, or founding of roads.

Landfill

Made-up ground, landfill, and fill refer to land that has been raised artificially by the deposit soils, construction materials, backfilling, tipping or waste and refuse, or any other form of fill. It may be ground that has been built up and raised through human and mechanical intervention, being different from the strata naturally formed over millennia. It is the result of depositing materials on top of natural land levels or the filling of previously excavated sites. Generally, such materials are poorly compacted and of an uncertain composition and density. Their variable nature makes them unsuitable for founding buildings, without treating the land. Due to the variable nature of the fill material uneven settlement takes place and the ground will often undulate as a result of this settlement.

Where the loadbearing material is selected, compacted in layers and suitably stabilised, the made-up ground can be loadbearing. Indeed, all flexible roads are constructed by selecting suitable materials, and layering and compacting them so that the ground can withstand the increased load from the traffic. However, many of the made-up sites are a result of fill that has been tipped or deposited loosely into holes and excavations. Landfill sites where waste is deposited are made-up sites which do not provide ground that can be considered suitable for foundations or has meaningful loadbearing properties.

Fill is often used to describe the tipping of material into holes that are a result of mining and quarrying. Even without the loads of buildings, such fill sites will naturally compress and compact, resulting in considerable settlement, which will be variable in nature.

Where information on the landfill is not properly recorded, trail pits and boreholes should be used to gather information on the nature of the material, strata, and depth.

Where buildings or structures are constructed over made-up ground or fill, the foundation should be designed to bridge the fill material or penetrate down to loadbearing strata. Building over fill sites is typically avoided.

Instability

Ground instability can be a result of human activity or natural processes. Landslip on sloping sites can occur where there have been little or no changes to the ground. Rocks and soil are naturally affected by the weather, rain, wind, and erosion, all of which can and do lead to changes in the loadbearing properties of the ground. Mining, excavation, and loading of the ground as a result of human activities can also result in instability where the properties of the soils are changed.

Instability can be considered under the following headings:

- Mining and quarrying
- Landfill
- Landslip
- Surface flooding and soil erosion
- Natural caves
- Surface and underground fissures

Other issues have to be overcome where sites are identified as floodplains, are contaminated, have radioactive material, have seen military action, or have been used as burial sites. The climate is also changing, which is impacting on the suitability of land and soils for development.

A database of useful information can be found at the following websites:

- Environment Agency: https://www.gov.uk/government/organisations/environment-agency
- British Geological Survey: http://www.bgs.ac.uk/
- National Archives: www.nationalarchives.gov.uk
- DEFRA has commissioned studies on landslides: www.defra.gov.uk
- Climate change: https://www.metoffice.gov.uk/weather/learn-about/climate-and-climate-change/climate-change/index

Landslips usually occur on natural slopes where the strata are weak.

- Where strata are weak at the surface, largely superficial landslip.

- Mixed layers of surface strata clay over sand and weak rock.
- Steep slopes – water, changes in climate.
- Steep slopes, where erosion has taken place at the foot of the slope or cliff.
- Sea storms and erosion over time removing supporting strata.
- Overloading.
- Around areas of quarrying and mining.

Surface flooding and soil erosion are becoming more common as the climate changes and a greater proportion of the surface of our land is developed and sealed with materials that do not allow the rain to percolate through the ground. As rain and run-off water has fewer places to penetrate the ground and naturally run into the water table or rivers, the water builds up and remains in built-up areas for longer periods. During periods of heavy or prolonged rainfall, the build-up both at the surface and immediately below the ground is increasingly presenting problems and challenges to our built environment.

Increased volumes of water can change the natural condition of the ground, causing clay soils to swell, and may lead to the erosion of soils and permeable rocks. Areas that flood and those that are designated as floodplains are of increasing concern.

The risk of surface flooding can be reduced by use of passive measures such as:

- Employing sustainable drainage systems (SUDS), allowing water to percolate through the ground and be stored, capturing or slowing down run off.
- Encourage fauna, grass, shrubs, trees, and green roofs; these all slow down run off.
- Provide an alternative approach to surface drainage in built-up areas, drainage, and channels that divert water.
- Planting and maintaining of woodland on hills and in fields is proving important in reducing risk of flooding downstream.

Further information on areas that are at risk of flooding can be found at https://www.gov.uk/check-flood-risk.

Natural caves and fissures occur where there are soluble strata (rocks), such as limestone and chalk. The presence of the voids in the ground can lead to land instability and subsidence.

DEFRA (www.defra.gov.uk) provide information on infiltrations systems and surveys that are useful for assessing ground conditions and the likelihood of land instability.

Extensive mining and quarrying have been carried out in most countries and while the mining of coal has declined, mines continue to work for minerals and precious materials. While recent mining activity is recorded, many of the old mines were undocumented. In areas where minerals and coal were rich, mining was present. In such areas, care should be taken to gather as much information as possible through existing records and surveys to reduce the risk of instability and subsidence.

Information from the British Geological Survey on active and ceased mines and mineral workings is available on https://www.bgs.ac.uk/products/minerals/BRITPITS.html.

Contaminated land

For surveyors conducting surveys for dilapidations, condition, purchase, etc., the demised premises are likely to include the ground on which the property stands. It is therefore important to examine the ground as well as the structure. Where ground contamination has occurred, liability may revert to the freeholder unless it is cleaned up. However, it may not be obvious who is responsible for the clean-up of any contamination. The contamination may be due by a tenant's occupancy of the premises, and they can be held responsible, or a new purchaser or tenant may take on contractual liability for existing contamination.

The UK has a long industrial history so that many areas of land have been used for a wide variety of industrial purposes. These types of activities generally involve the presence, use, or production of a range of chemicals and substances, which can cause contamination of soils and waters if they find their way into the natural environment.

The lack of proper controls, particularly in the 19th and early part of the 20th centuries, means that these substances were not properly handled or contained and many of them have found their way into the ground or water courses at these industrial sites and in their surrounding areas. Thousands of sites have been contaminated by previous industrial use, often associated with traditional processes, which are no longer used.

These sites may present a hazard to the general environment, but there is a growing need to reclaim and redevelop them. The Environmental Protection Act 1990 as amended (UK Government, 2021) gives protection in England, Wales, and Scotland.

"Contaminated land" is land containing substances that could cause:

- Significant harm to people, property, or protected species
- Significant pollution of surface waters (for example lakes and rivers) or groundwater
- Harm to people as a result of radioactivity

Examples of contaminants are:

- Heavy metals, elevated concentrations of metallic elements, e.g. lead, arsenic, cadmium, arsenic, iron, nickel, chromium
- Chemical substances and preparations, e.g. solvents
- Organic substances, e.g. oils, tars, phenols, PAHs, PCBs
- Inorganic compounds, e.g. cyanide, chlorides, ammonia
- Ground gases, e.g. methane, carbon dioxide, hydrogen sulphide, volatiles
- Asbestos
- Radioactive substances
- Faecal, animal, or vegetable matter
- Combustible material, e.g. coal, coke
- Refuse and waste

In the UK, there is a difference between land legally determined to be "contaminated land", and land contaminated but not determined legally to so be. The local council or an environment agency must be consulted when it is suspected that land is contaminated. A surveyor's professional indemnity must specifically include contamination when contaminated land is being considered, otherwise the surveyor is possibly not protected.

The rules for who is responsible for contamination and how to deal with it depend on whether the land is legally considered "contaminated land". The person who caused or allowed the contamination to happen is responsible for dealing with it, unless they cannot be identified or unless the authority investigating the issue decides such a person is exempt and the authority will decide who is responsible for dealing with it instead, e.g. the landowner, or the person who uses the land.

Even if the land is not legally considered "contaminated land", the person responsible for causing it, or the landowner, is likely to be responsible for dealing with the contamination and could face legal action if they do not.

A person developing contaminated land will need to deal with the contamination either before planning permission is granted or else as part of the development.

Severely contaminated land in the UK can be classed as a "special site". The website, Gov.uk, gives examples of these as land which:

- Seriously affects drinking waters, surface waters, or important groundwater sources.
- Is or was used for certain industrial activities, such as oil refining or making explosives.
- Is or was regulated using a pollution permit.
- Has been used to get rid of waste acid tars.
- Is owned or occupied by the Ministry of Defence.

- Is contaminated by radioactivity.
- Is a nuclear site.

The Environment Agency in England, Natural Resources Wales in Wales, and the Scottish Environment Protection Agency in Scotland regulate special sites. The person who caused or allowed the contamination to happen is responsible for dealing with it, unless they cannot be identified or unless the authority investigating the issue decides they are exempt when they will decide who is responsible for dealing with it instead, e.g. the landowner, or the person who uses the land.

Drivers for remediation requirements:

- Growth of environment as a political issue
- Pressure to reduce development of greenfield sites
- Pressure to increase development

Concern for Building Surveying:

- Pre-purchase advice
- Development advice, including change of use
- Asset appraisal
- Approximately 300,000 hectares of contaminated land in the UK
- Skills in co-ordinating specialists
- Costs of remediation – insurance implications

Specialist advice is required for remedial action. See also Tables 9.2 and 9.3. Remedial action for unacceptable risks of contaminated land:

- Treatment (of the source not the receptor, e.g. soil treatment)
- Containment
- Removal and disposal

These broad groups cover many different techniques. Selecting the most appropriate options will often have a significant impact on the viability of a development. Examples:

- Excavation, dredging, and off-site disposal to landfill
- Pump and treat (using, e.g. activated carbon, flocculants, sand filters, air filters)
- Isolate/encapsulation
- Bioremediation
- Thermal decomposition

- Thermal desorption
- Surfactant enhanced aquifer remediation
- Solidification and stabilisation
- In-situ oxidation
- Soil vapour extraction
- Nano-remediation
- Self-collapsing air microbubbles

Table 9.3 Ways of identifying contaminated land before testing

Identifying contaminated land	
Initial suspicions	**Confirming suspicions – site investigation**
Street name, e.g. Gas Works Lane	Desk study
Local knowledge	Site reconnaissance
Obvious signs	Walkover survey
Old maps, old aerial photos, historical reviews, public records	Main investigations and reporting
Environment Agency	Extent determined by use/proposed use of the land
Industrial past – sites likely to contain contaminants	Guidance, e.g. British Standards, BRE Guides, CIRIA guides
Problems with natural/very old/very localised contaminants	Risk assessment
Adjacent sites	Determination of: source–pathway–receptor
Council searches and the Contamination Register	

Table 9.4 Risk assessment stages for contaminated land

Contaminated land – risk assessment stages		
1	Hazard identification	Establish possible source–pathway–receptor links
2	Hazard assessment	Identify pollutant linkages and carry out exploratory site investigation
3	Risk estimation	Establish scale and possible consequences
4	Risk evaluation	Are the risks acceptable or unacceptable?

Ground gas: Radon, methane, and carbon dioxide

Ground gases are not restricted to contaminated land. They can be present in greenfield as well as brownfield sites. They can originate from waste deposits in landfill sites, or occur naturally. When building below ground, as well as protecting the structures against water ingress, it is also important to consider any risk of penetration of ground gases, which may be hazardous to health. These include radon, methane, carbon dioxide, mercury, and volatile organic compounds (VOCs) (Table 9.5).

Extensive guidance can be found on Public Health England's webpage UKRadon for radon (PHE, 2020; UKRadon, 2020). See also the Health Protection Agency (2010) website and Table 9.6. For gases other than radon, consult for example: NHBC, e.g. "Guidance on evaluation of development proposals on sites where methane and carbon dioxide are present", CIRIA, e.g. "Assessing risks posed by hazardous ground gases in buildings", British Geological Survey (BGS), Institute of Petroleum, and Chartered Institute of Waste Management. Approved Document C (HM Government, 2013) is the relevant building regulation guidance in the UK.

Table 9.5 Properties of gases

Gas	Flammable properties	Physiological properties	Smell
Carbon dioxide	Non-flammable	Toxic with O_2 depletion	Odourless
Carbon monoxide	Non-flammable	Toxic	Odourless
Methane	Explosive	Asphyxiating	Odour
Nitrogen	Non-flammable (but used in explosives)	Toxic with O_2 depletion	Odourless
VOCs	Flammable	Toxic	Odour

Table 9.6 Radon exposure limitations for humans

Percentage of properties above Action Level	Probability of radon area	"Radon-affected area"
Less than 1%	Lower	No
Above 1% but less than 10%	Intermediate	Yes
10% and above	Higher	Yes

Source: the Health Protection Agency and UK Government publication "Radon limitation of human exposure" (Health Protection Agency (2010).

Landfill gases are usually methane and carbon dioxide, with VOCs giving the characteristic smell. They are generated by micro-organism action on biodegradable waste. Organic-rich soils, peat, and river silt sediments can produce methane and carbon dioxide, sometimes known as marsh gas, swamp gas, or bog gas. Coal-mining areas can have higher levels of carbon dioxide and nitrogen, and reduced oxygen (hence canaries in coal mines). VOCs can result from spillages of oil, petrol, and solvents.

Radon is a natural contaminant, a naturally occurring, radioactive gas that seeps up from the ground. Radon is the largest source of public radiation exposure in the UK and the second largest cause of lung cancer in the UK. Every building holds radon, but mostly in low concentrations. The chances of a higher level depend on the type of ground it stands on. It is in greater concentrations where there are harder underlying rocks (e.g. some granites, ironstones, phosphatic rocks, shales rich in organic materials, limestone) but clay layers tend to deter it. Thus, the prevalence is greater in say West Yorkshire and the Pennines, than say the Vale of York. The Health Protection Agency and the British Geological Survey publish UK maps showing where high levels are more likely. Radon's ultimate source is uranium. All rocks contain some uranium, and the uranium content of a soil will be about the same as its derivative rock.

Many properties with high levels of radon have yet to be identified and of those that are confirmed, large numbers have yet to be remediated. The safe levels are not absolute and are under review. Recommendations include a maximum level for action of 200 Bq/m^3. If the concentration is above this Action Level, the Target Level for remediation is to reduce the concentration to below 100 Bq/m^3 for existing homes, and for the design and construction of new homes.

The amount of radon in a house is affected by factors in addition to the presence of uranium in the underlying soil. Properties in areas with drier, highly permeable soils and bedrock may have high levels of indoor radon; these can be present in sloping sites, coarse glacial deposits, and bedrock with fractures and cavities. Additional affecting factors include the building's construction, ventilation provision, provided barriers, etc.

Radon can also enter a home through the building's water supply. Private water supplies are more vulnerable because district supplies are usually aerated or otherwise treated, which disperses the radon. Areas with high levels of uranium in the underlying rocks are those most likely to have high levels of radon in groundwater. For example, granites in various parts of the United States, such as New Hampshire, are sources of high levels of radon in groundwater supplied to private water systems.

Generally, expert advice should be sought where ground gases and/or radon are suspected or known. Site investigation surveys may be required including: desk survey, on-site testing, boreholes, and monitoring and analysing flow rates of a specific hazardous gas. These result in a gas screening value site survey (GSV) to arrive at a theoretical "characteristic gas situation" (CS). The CS value is a threat level, numbered 1–6, 1 being very low hazard potential, and 6 being the highest. Once the CS has been defined for a site, the minimum protection score requirement points to suitable protection systems for the site by considering a product's protection score contribution. At the time of writing, these are defined in BS8485.

The gas protection system should consist of at least two different elements:

a. A structural barrier element (e.g. floor slab and substructure), *and*
b. A membrane, or
c. A ventilation/dilution element, or both.

The elements work independently and collaboratively. A single element should not be used because there would be no backstop to protect against ubiquitous defects in an element.

Gas barriers across external wall cavities are common for sites that contain naturally occurring gases such as methane or radon. Since such membranes necessarily bridge the cavity, a low-level cavity tray can be required. The cavity tray has to be placed very close to ground level, leading to problems of insulating a very small strip of wall. Lack of insulation here can lead to extensive thermal bridging around the property's perimeter. See Figure 9.3.

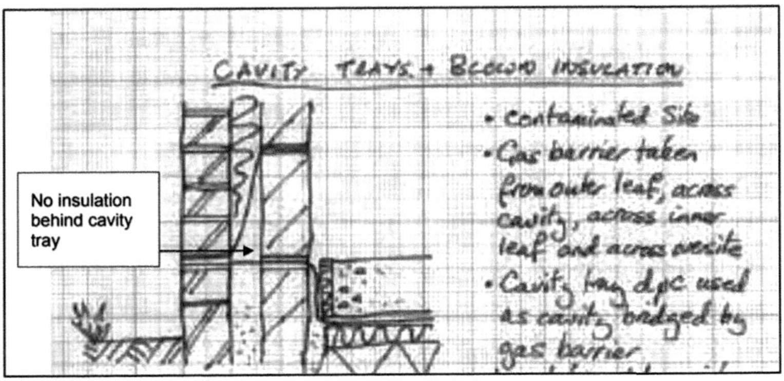

Figure 9.3 Typical site sketch of on-site observations of thermal bridging due to difficulty in insulating behind cavity tray over gas barrier.

Table 9.7 Climate change issues with built environment impacts

Climate change issue	**Built environment impact**
Higher sea levels	More coastal flooding, more coastal erosion, loss of cliff-top properties
More heavy rainfall, more intense precipitation	More river/watercourse flooding, higher groundwater levels, flash floods, landslides, damage to property, insurance costs
Increased summer drying, drought	Property cracks due to ground shrinkage, subsidence, water supply issues
Increased summer temperatures	Overheating in properties, heat stress

Table 9.8 Possible flood damage in individual buildings

Flood damage in an individual building
Health and safety
Sewage
Hydrostatic action – lateral and uplift pressures
Hydrodynamic action – force of flowing water (will include erosion)
Low-level plaster deterioration, contamination
Saturated materials
Electrics damage
Salts in walls
Timber floors deterioration, contamination
Timber fixtures and fittings deterioration
Lifting floor, wall tiles, etc.
Insulations damaged, contaminated

FLOODING

Climate change is expected to have impacts due to, for example, those issues shown in Table 9.7. Possible flood damage is shown in Table 9.8.

Updates on flooding risk are available on various websites, including, at time of writing:

Climate Change Committee, UK, https://www.theccc.org.uk
Gov.UK, https://flood-warning-information.service.gov.uk/long-term-flood-risk/map

Climate Central, https://sealevel.climatecentral.org
Scottish Environment Protection Agency, https://www.sepa.org.uk/environment/water/flooding/flood-maps

Surveyors can have various roles and stances in flooding matters, due to, for example:

- The different professions involved: planners, designers, insurers, surveyors, contractors, etc.
- Lack of definitive agreement in the guidance.
- Decisions may be required quickly, possibly without time for full assessment and consideration.
- Quick drying versus slow drying argument following flooding.
- It is a specialist area – contact needed with the insurance companies involved with the works.
- Advice for flood-proofing for landlords and landowners.

Desktop surveys can identify flood risk. Flood maps produced by UK authorities are intended only as a guide – not designed to be accurate at the individual property level. For Great Britain, flood probability information maps are produced by Gov.uk (2020) for England, Northern Ireland Planning Portal, Scottish Environment Protection Agency, and Resources Wales. There are different maps for flooding from rivers or sea, and for risk of flooding from surface water. Surface water flooding happens when rainwater does not drain away through the normal drainage systems or soak into the ground, but instead lies on, or flows over, the ground.

A flood risk assessment may be required for a particular site; there is advice for this on the authority sites stated above. The UK Environment Agency has defined four flood zones referring to the probability of river and sea flooding; however, these do not take account of the possible impacts of climate change and consequent changes in the future probability of flooding. Therefore, the most up-to-date information and advice should be found and used, for example consulting websites of the Environment Agency, Gov.UK, Planning Authorities, CIRIA, etc.

Reinstatement of property following a flood:

- Draining
- Initial damage assessment
- Drying
- Effect of water on materials, tracking of moisture
- Structural damage
- Renovation of interiors
- Electrical installations
- Consideration of flood-resilient walls, floors, windows/doors, services, drainage, etc.

Follow the pathways forwards and backwards to find adverse effects beyond the flood-ing levels as moisture and contaminates wick or flow: source – pathway – receptor.

Protecting property from floods and their effects involves two approaches o "flood resistance" and/or "flood resilience".

Flood resistance: the ability of a building to resist the entry of flood water fron the outside to the inside.

Flood resilience: the ability of a building to resist damage (exterior and interior as a result of flooding.

See also BS 85500. This is due for an update at the time of writing (BSI, n.d.)

Protecting property from floods and their effects:

- Planning issues – not building on floodplains, directing development away from high-risk areas, etc.
- Emergency protection measures – sandbags, flood boards, airbrick covers, etc.
- Routes of flood water entry
- Property insurance issues
- Storing data, etc., off-site or above flood risk height
- "Replacing timber floors with solid" debate
- "Replacing plasterboard wall finishes and internal walls with solid" debate

FLORA AND FAUNA

Invasive weeds

In the UK, the Wildlife and Countryside Act 1981 and the Environmenta Protection Act 1990 list invasive plants that must not be encouraged in the UK For buildings, Japanese knotweed is the most invasive.

Main reasons why Japanese knotweed is a problem:

- It spreads easily via rhizomes and cut stems or crowns.
- It out-competes native flora.
- It is difficult and expensive to control or eradicate.
  - Can involve a repeated spraying regime over three years and re-landscaping
- It can cause structural damage to structures.
  - Damage to buildings and foundations.
  - Damage to paving and tarmac areas.
  - Damage to flood-defence structures.
  - Damage to archaeological sites.

- Obstructs visibility and access on roads.
- Reduction in land values.
- Accumulation of litter in well-established stands.
- Aesthetically unsightly.

It is listed under the Wildlife and Countryside Act 1981 as a plant that is "not to be planted or otherwise introduced into the wild". The Environmental Protection Act 1990 also lists it as "controlled waste" to be disposed of properly. There is a legal obligation not to cause it to spread if it occurs on your (your client's) land.

Advice is offered by an RICS Professional Information paper and an Environment Agency Code of Practice.

When treating with chemicals, the following is required:

- Anyone spraying must hold a certificate of competence for herbicide use or work under direct supervision of a certificate holder.
- Carry out a Control of Substances Hazardous to Health (COSHH) assessment.
- Get permission from Natural England if the area is protected, for example Sites of Special Scientific Interest.
- Get agreement from the Environment Agency.

Protected species

Rare or threatened species are protected in Great Britain and many other countries, regardless of where the species are found. A range of species are protected, including:

- Bats
- Badgers
- Great crested newts
- Otters
- Water voles
- Wild birds

It is an offence to disturb or to harm protected species, or to damage any structure that they use as shelter. The law protects all wild birds, their nests, and their eggs; birds' nests must not be destroyed or disturbed while they are in use.

Identifying whether these species are present on a particular site can be difficult, and often only comes out at planning stage when suggested by objectors.

If found or suspected, obtain advice from government conservation bodies, e.g. Natural England, the Environment and Heritage Service (Northern Ireland), Scottish Natural Heritage, and the Countryside Council for Wales.

For example, all species of bats and their roosts (whether the bats are present or not) are protected. Bats can be useful, as woodworm beetles and other insects are a favourite food.

If the presence of bats is suspected (e.g. by fresh bat droppings in the roof void), professional advice from government conservation bodies should be sought:

www.eurobats.org/about_eurobats/introduction_to_agreement
www.netregs.org.uk/environmental-topics/land/nature-conservation/protected-and-priority-species

The law does not prevent pest control in a property where a bat roost is present, however, controlling wasps, insects, and rodents may affect bats. Concerns can arise if a chemical product is proposed in or near a bat roost, for example, as an insecticide to control pests or a timber treatment to protect against insects, fungal growth, or weathering. Chemical products cannot legally be used in or near a known bat roost if there are bats present.

Guidance is also given at:

www.gov.uk/guidance/bat-roosts-use-of-chemical-pest-control-products-and-timber-treatments-in-or-near-them

The best time to apply treatments is usually between autumn and spring. Only licenced handlers are permitted to handle bats; there is a risk of rabies from bites.

Sites or Areas of Special Scientific Interest

Sites of Special Scientific Interest (SSSIs) or Areas of Special Scientific Interest (ASSI) in the Isle of Man and Northern Ireland are those areas of land that are considered to best represent natural heritage in terms of flora (plants), fauna (animals including mammals, birds, insects, fish, etc.), geology (rocks, soils), geomorphology (landforms), and a mixture of these natural features. They have formal conservation designation.

Most SSSIs are in private ownership and it is an offence for anyone to intentionally or recklessly damage their protected natural features. The aim is to effect appropriate management of the site's natural features and that decision-makers are aware of the SSSI designation when considering changes in land use or other relevant activities.

There are certain things you cannot do on SSSI land without consulting and getting consent first from Natural England, Scottish Natural Heritage, Natural Resources Wales, or NI's Department of Agriculture, Environment and Rural Affairs. Each SSSI/ASSI has an individual list of activities, known as "operations", which need consent – this can be downloaded from the site's page on Natural England, etc.

Advice is given by:

www.nature.scot
www.gov.uk/guidance/protected-areas-sites-of-special-scientific-interest#check-if-you-need-consent
https://naturalresources.wales
https://www.daera-ni.gov.uk/topics/land-and-landscapes/areas-special-scientific-interest

CLIMATE AND CONSTRUCTION

Cormac Flood and Lloyd M. Scott

Climate change

Mitigating climate change through operational energy reduction in existing buildings is a high priority for policy makers in Europe and elsewhere (Brown et al., 2014). Brown et al. (2014) reasoned that the necessity to limit climate change demands considerable reduction in global warming potential due to operational energy in countries where fossil fuels currently dominate, e.g. UK, Germany, USA. Observations show that global average temperatures have increased by 0.85°C (in the range 0.65–1.06°C) since 1850 (IPCC, 2013). The atmosphere and oceans have warmed, the amount of snow and ice has diminished, the sea level has risen, the sea chemistry is changing, and the concentrations of greenhouse gases (GHGs) in the atmosphere have increased (IPCC, 2013). According to the IPCC (2013), it is extremely likely that more than half of the observed increase in global average temperature from 1951 to 2010 was caused by an anthropogenic increase in GHG concentrations and other anthropogenic forcings.

Changes in Ireland's climate are in line with and similar to relevant global trends. Temperatures have increased by 0.8°C since 1900: an average of 0.07°C per decade over that period (Dwyer, 2013). As a result of the slow response time (inertia) of the climate system, changes are projected to continue and increase over the coming decades. Even if GHG emissions could be stopped immediately, some impacts, such as sea level rise, are projected to continue up to and beyond the end of this century (Desmond et al., 2016).

Buildings contribute almost half of the world's CO_2 emissions (EPA, 2015; Koroneos et al., 2014), the predominant greenhouse gas responsible for bringing about these changes. As a result of these greenhouse gases, the climate projections for Ireland, for example, include increasing average temperature, more extreme weather conditions including rainfall events, an increased likelihood of river and coastal flooding, water shortages, particularly in the East of the country, changes in the types and distribution of wildlife species, and the possible extinction of some vulnerable wildlife species. The main anthropogenic cause of increased levels of atmospheric carbon dioxide is burning fossil fuels to generate energy (IPCC, 2016). In response to this compelling scientific evidence, the Irish Government have set legally binding targets to reduce carbon dioxide emissions (carbon emissions hereafter). The aims of the Climate Action and Low Carbon Development Act 2015 are to decarbonise energy, transport, and housing by 2050. However, it has been argued that these targets are not enough to prevent catastrophic changes to the climate.

Ireland is a party to both the Convention and the Kyoto Protocol, ratifying the Kyoto Protocol, enforcing it as legally binding in 2005. Although Kyoto commitment period 2 (20-20-20) is now in effect, with even more stringent CO_2 reduction targets, predictions have indicated that Ireland will have reached its 2020 target (62.8 Mt CO_2), provided that performance upgrade schemes in accordance with SEAI (2019, 2020) are followed and delivered. Examining CO_2 emissions per dwelling, the average Irish dwelling in 2010 emitted 30% more CO_2 than the average dwelling in the UK, and emissions were 94% more than the average for the EU-27 (Dennehy and Howley, 2013). Additionally, CO_2 emissions in Ireland are predicted to have increased by 0.9% above 2010 levels (61.3 Mt CO_2), translating to a 39% increase above 1990 levels (55.6 Mt CO_2) by 2020.

With dwellings accounting for approximately 25.3% of Irish carbon emissions, it has been recognised that the domestic sector can make a significant contribution to these mandatory reductions. Whilst newly built dwellings are designed and constructed with consideration to climate change, it has been acknowledged that there is an urgent requirement to adapt existing dwellings for both the future climate and to reduce carbon emissions resulting from their use (Hamdy et al., 2017). It has been documented that over two-thirds of existing UK dwellings and 60% of Irish dwellings are expected to still be in use in 2050, while 80% of European housing stock for 2030 is built today (Di Giuseppe et al., 2017). Therefore, it could be argued that there is no choice about addressing the energy efficiency of existing dwellings. However, reducing carbon emissions is not the only issue related to existing dwellings and their elevated energy consumption. There is also the issue of energy security, which thus provides an additional driver for reducing energy consumption from existing dwellings.

Energy security

Energy security is described by the International Energy Agency (IEA) as "the uninterrupted physical availability of energy sources at an affordable price". The EU is the world's largest importer of energy, importing 53% of its energy supplies at an annual cost of €400 billion. It has a particular reliance on imported natural gas (66%) and crude oil (>90%). For example in Ireland, indigenous production accounted for 32% of the country's energy requirements in 1990. However, since the mid-1990s import dependency has grown significantly, due to the increase in energy use together with the decline in indigenous natural gas production at Kinsale since 1995 and decreasing peat production. Imported oil and gas accounted for 77% of total primary energy requirement in 2015, compared with 50% in the early 1990s. Ireland's overall import dependency reached 90% in 2006. It has varied between 85% and 90% since then; it was at 88% in 2015.

Energy efficiency of buildings has therefore become an important issue for all stakeholders, including homeowners, with strong links between improvements in the performance of buildings, energy savings, and a reduction in GHG emissions (Hens, 2015; Verbeeck and Hens, 2005; Tommerup and Svendsen, 2006).

In Ireland, the domestic sector accounted for 24.9% of all energy consumed in 2016, which was second only to transport. Accounting for around 60% of this domestic energy use, space heating has a significant impact on overall consumption and thus demand within Ireland (Kema, 2008; Koroneos et al., 2014). It can therefore be concluded that if energy demand for space heating is reduced, this could have a significant impact on both meeting climate change targets and increasing energy security in Ireland. Furthermore, these energy efficiency improvements should make a significant contribution to alleviating fuel poverty within the most vulnerable dwellings.

Fuel/energy poverty

While climate change and energy resources are a primary focus, fuel poverty appears to be a peripheral issue for groups such as the Climate Change Committee (CCC, 2020), whose suggested focus on price-based policies geared towards combating climate change will put a further estimated 1.7 million households into fuel poverty. This highlights the importance of a collective approach to addressing all three pillars of sustainability. According to CARDI (2011), fuel poverty is defined as a situation where someone is unable to afford to heat their home to a level that is healthy and safe. In Ireland and the UK, fuel poverty is classified as a situation where more than 10% of disposable income is spent on energy use to provide adequate heat in the dwelling (Hirsch et al., 2011; DCENR, 2016). The main factors attributed to fuel poverty are the increasing cost of energy, low household income, energy-inefficient homes, and lack of access to the required energy sources (Palmer et al., 2005; CARDI, 2011). Additionally, there can be a direct correlation between the age and type of a dwelling and its energy performance, although as stated elsewhere correlations must be made with caution and with understanding of each particular building. To reinforce the concept of a correlation, it has been established that retired and unemployed people account for the majority of the population affected by fuel poverty in Ireland (DCENR, 2015). Furthermore, DCENR (2015) confirmed that 28% of Irish dwellings are in fuel poverty at the 10% income threshold, of which poor lower BER-rated, oil-heated detached and semi-detached dwellings were worst affected. The European Commission estimates that energy poverty may affect up to 54 million people in the EU alone, or around 11% of the population. Ireland has set a target for alleviating fuel poverty through a new strategy to combat energy poverty and policy.

Climate and construction

Climate conditions affect many aspects of construction projects during construction. However, upon completion, buildings are subject to continued exposure to elements such as precipitation, relative humidity, wind, ultra-violet radiation and temperature, which may have significant impacts over time. The connection between these conditions, the U-value, and construction performance is dependent on accurate understanding of current and developing climatic trends.

Construction and building envelope performance is highly weather-dependent. Optimal construction conditions require elements such as precipitation, wind, temperature, and relative humidity to fall within firm limits. Many products and materials are designed to be installed within certain climate conditions. Outside of these ideals, performance declines, depending on the severity of the conditions. Building envelopes should be built of materials that can resist the deteriorating effects of the weather, in line with building regulations. However, notwithstanding this, building design in line with many building regulations does not currently facilitate the inclusion of the aforementioned factors with regard to thermal performance. Additionally, with the advent of climate change, the current relationship between building fabric performance and weather is being altered, continually enhancing the agenda for building design and adaptation measures to incorporate current and developing climate patterns.

According to ISO 6946 (2007), "If a problem in an existing building is being investigated, any data measured at the site of the building shall be used; otherwise data from a similar location to that of the building shall be used". This statement conflicts with many current practice guidelines that currently mandate the use of "*default*" figures accounting for weather conditions taken from building regulation guidance. Current research of the U-value and the development of correction factors or the calculation itself to account for climate patterns is scarce. External (Rso), cavity (Rcav), and internal (Rsi) surface resistances incorporate two factors only – wind speed and emissivity. Taking Ireland as an example, wind and other weather conditions are dissimilar across the entirety of Ireland; however, values for Rso, Rcav, and Rsi of 0.04, 0.18, and 0.13 $m^2/K{\cdot}W$ respectively are the default, nationwide. While BSi (2015) elucidates that the standard values are "appropriate in most cases" and that actual (rather than default) values could be calculated using "detailed procedures for low emissivity surfaces, specific external wind speeds and non-planar surfaces", ISO 6946 is engrained within the Irish building regulations, which enforce resistance values currently and historically throughout the years in Technical Guidance Document Part L.

There are many private companies and national weather centres offering the latest available climate data to incorporate into project planning. The key point is that the data used currently in assessing the risks to weather are historical. Because the cost and duration of construction activities are heavily influenced by climatic conditions, it is becoming increasingly important that decision-making should be based on future projections, as well as past conditions. Nonetheless, it is important that the relationship between current weather conditions and construction be discussed first to establish a baseline condition against which the impacts of climate change can be compared.

To assess the current physical impacts of weather on construction, the Irish climate will be discussed briefly below. Key factors that have the potential to affect construction are rainfall (precipitation), wind speed, and temperature.

Precipitation

Precipitation is one of the most disruptive factors during construction. Weather conditions and, more importantly, rain are often reported as two of the main causes of project delays and unscheduled changes. Over the lifecycle of a building, permanent waterproofing of affected areas may be required, as moisture is an element that will penetrate the building shell of the best-designed building fabric. However, once moisture breaches the building shell, drainage provisions can be used to direct it to the exterior. For example, moisture can penetrate brick through capillary action and the action of wind-driven rain. Unless these actions are checked by detail designs such as rain screens, cavity walls, and wall flashings, moisture can reach the interior and condense on interior walls. In addition, any imperfections in finish will leave a gap that rain can penetrate.

Wind

Wind-driven rain is one of the most important climatological factors to affect the hygrothermal performance and durability of building envelopes. Wind can blow rain through gaps as small as 0.15 mm (Hens, 2015), although if moisture breaches the building shell, the aforementioned detail designs including DPCs and weep holes in cavity walls can re-direct it out from the building fabric. In most instances, wind can dramatically multiply the effects of moisture, reducing cooling times by accelerating the removal of heat and increasing the drying capacity of building materials/fabrics by accelerating the removal of moisture (Crissinger, 2005). Additionally, wind can drive moisture into the structure by increasing the pressure on the film of moisture on the surface of the structure and make cold temperatures feel colder by accelerating the evaporation of perspiration. See also BSI (2007).

It is anticipated that projected increases in extreme winter wind speeds and rainfall will contribute to an intensification of wind-driven rain in Ireland by 2050 (Nolan et al., 2012). It is considered the most critical exterior environmental load factor in over 90% of building envelopes and can increase the amount of moisture present in a structure by more than 100 times due to vapour diffusion (Karagiozis et al., 1997, 2003). In heat-air moisture numerical simulations, wind-driven rain is categorised as one of the most important boundary conditions. In conjunction with the temperatures internally and externally (which dictate the rate of heat flow through the building fabric), this will influence material temperatures and conductivity values.

Temperature

Temperature is a critical parameter in construction. If bricks are dry due to high temperatures when laid, they may absorb water from the cement at a rate that also results in a loss of bond strength, the consequence of which may be leaks at weakened joints. This may lead to more significant risk of moisture penetration through the methods listed previously, thus reducing thermal performance. Similarly, temperature can affect how construction materials dry out when subjected to construction moisture, moisture which is absorbed through wind-driven rain, or moisture absorbed through capillary action.

Temperature can cause a change in material performance. All materials have a varying set of properties, defined by the characteristics of the material itself. Work by Kodur and others (2003, 2011) shows that thermal conductivity decreases in various concretes due to thermal expansion resulting from temperature increases. Thus, materials closer to the warmer interior surface may suffer from decreased thermal efficiency due to thermal expansion, while materials closer to the external surface may suffer from wind-driven rain and increased moisture content, again increasing conductivity values, which in turn might combine to result in an increased U-value. The climate relative to the building thus has a considerable role to play in establishing how materials might perform in a building envelope.

REFERENCES

Barry, P. (2017). *Clay soils, subsidence, heave, trees and roots—part 2*. Peter Barry Chartered Surveyors. Accessed from https://www.peterbarry.co.uk/blog/clay-soils-subsidence-heave-trees-roots-part-2

Brown, N. W. O., Olsson, S. & Malmqvist, T. (2014). *Embodied greenhouse gas emissions from refurbishment of residential building stock to achieve a 50% operational energy reduction.* Building and Environment, 79, 46–56.

BSI (n.d.) British Standards Institution. London, UK. Accessed from https://shop.bsigroup.com/

BSI (2007). *BS EN 15026:2007—Hygrothermal Performance of Building Components and Building Elements—Assessment of Moisture Transfer by Numerical Simulation.* British Standards Institution, London, UK.

BSI (2015). *Building Components and Building Elements—Thermal Resistance and Thermal Transmittance—Calculation Method (ISO 6946:2007).* British Standards Institution, London, UK.

CARDI (2011). *Understanding Fuel Poverty in the Older Population.* Dublin, Ireland: Centre for Ageing Research and Development in Ireland (CARDI).

CCC (2020). Climate change committee. Accessed from https://www.theccc.org.uk/

Crissinger, J. L. (2005). Design and construction vs. weather. Interface.

DCENR (2015). *Bottom-Up Analysis of Fuel Poverty in Ireland.* Dublin: DCENR (Department of Communications, Energy & Natural Resources).

DCENR (2016). *A Strategy to Combat Energy Poverty 2016–2019.* Dublin: DCENR (Department of Communications, Energy & Natural Resources).

Dennehy, E. & Howley, M. (2013). Report 2013. *Energy in the Residential Sector.* Dublin, Ireland: SEAI.

Desmond, M., O'Brien, P. & McGovern, F. (2016). *A summary of the state of knowledge on climate change impacts for Ireland.* MaREI Centre, Environmental Research Institute, University College Cork (ed.) EPA Research Programme 2014–2020. EPA, County Wexford, Ireland: Environmental Protection Agency.

Di Giuseppe, E., Iannaccone, M., Telloni, M., D'Orazio, M. & Di Perna, C. (2017). *Probabilistic life cycle costing of existing buildings retrofit interventions towards nZE target: methodology and application example.* Energy and Buildings, 144, 416–432.

Dwyer, N. (2013). *The status of Ireland's climate, 2012.* Wexford, Ireland: Environmental Protection Agency. EPA, County Wexford, Ireland. Available from: https://cora.ucc.ie/handle/10468/2969

EPA (2015). *Greenhouse Gas Projections to 2020.* Environmental Protection Agency. EPA, County Wexford, Ireland.

Gov.uk (2020). Accessed from https://www.gov.uk/

Hamdy, M., Sirén, K. & Attia, S. (2017). *Impact of financial assumptions on the cost optimality towards nearly zero energy buildings—a case study. Energy and Buildings*, 153, 421–438.

Health Protection Agency (2010). *Radon limitation of human exposure.* UK Government publication. Accessed from https://www.gov.uk/government/publications/radon-limitation-of-human-exposure

Hens, H. (2015). Wind-driven rain: from theory to reality. Proceedings of Thermal Performance of the Exterior Envelopes of Whole Buildings XI, 2010a Florida. Oak Ridge National Laboratory, 1–10.

Hirsch, D., Preston, I. & White, V. (2011). *Understanding Fuel Expenditure: Fuel Poverty and Spending on Fuel*. Bristol: Centre for Sustainable Energy.

HM Government (2013). *The Building Regulations 2010*. Approved Document C - Site preparation and resistance to contaminants and moisture (2004 Edition incorporating 2010 and 2013 amendments). NBS. UK.

IPCC, 2013: *Climate Change 2013: The Physical Science Basis*. Contribution of Working Group I to the Fifth Assessment Report of the Intergovernmental Panel on Climate Change [Stocker, T.F., D. Qin, G.-K. Plattner, M. Tignor, S.K. Allen, J. Boschung, A. Nauels, Y. Xia, V. Bex and P.M. Midgley (eds.)]. Cambridge University Press, Cambridge, United Kingdom and New York, NY, USA, 1535 pp.

IPCC, 2007: Climate Change 2007: *Impacts, Adaptation and Vulnerability*. Contribution of Working Group II to the Fourth Assessment Report of the Intergovernmental Panel on Climate Change, M.L. Parry, O.F. Canziani, J.P. Palutikof, P.J. van der Linden and C.E. Hanson, Eds., Cambridge University Press, Cambridge, UK, 976pp.).

Karagiozis, A., Hadjisophocleous, G. & Cao, S. (1997). *Wind-driven rain distributions on two buildings*. Journal of Wind Engineering and Industrial Aerodynamics, 67–68, 559–572.

Karagiozis, A. N., Salonvaara, M., Holm, A. & Kuenzel, H. (2003). *Influence of wind-driven rain data on hygrothermal performance*. Eighth International IBPSA Conference. Eindhoven, Netherlands.

Kema (2008). *Demand side management in Ireland: Evaluating the energy efficiency opportunities*. 1st ed. A Study by Kema for Sustainable Energy Ireland, January 2008. Sustainable Energy Authority Ireland.

Kodur, V. & Khaliq, W. (2011). *Effect of temperature on thermal properties of different types of high-strength concrete*. Journal of Materials in Civil Engineering, 23, 793–801.

Kodur, V. K. R. & Sultan, M. A. (2003). *Effect of temperature on thermal properties of high-strength concrete*. Journal of Materials in Civil Engineering, 15, 101–107.

Koroneos, C., Karatsiori, V. & Grgič, T. (2014). *Energy efficiency analysis of a refurbished dwelling: a case study in Ljubljana, Slovenia*. International Journal of Sustainable Engineering, 7, 214–221.

NHBC (2021). *NHBC Standards 2021*. National House Building Council. NHBC, Milton Keynes, UK.

Nolan, P., Lynch, P., Mcgrath, R., Semmler, T. & Wang, S. (2012). *Simulating climate change and its effects on the wind energy resource of Ireland*. Wind Energy, 15, 593–608.

Palmer, J., Campbell, R., Boardman, B. & Saunders, J. (2005). *Fuel Poverty Research Centre Scoping Study*. Oxford, UK: Environmental Change Institute.

PHE (2020). UKRadon, Public Health England. Accessed from https://www.ukradon.org/

RICS (2010). *Contamination, the environment and sustainability: Implications for chartered surveyors and their clients*. RICS Guidance Note. 3rd. edition, 2010. Royal Institution of Chartered Surveyors, London. Superseded by RICS (2018).

RICS (2018). *Environmental risks and global real estate. RICS Guidance Note. RICS professional standards and guidance, global*. Royal Institution of Chartered Surveyors, London.

RICS (2020). *Contamination and Environmental Matters: Property Observation Checklists*. Royal Institution of Chartered Surveyors, London.

SEAI (2019). *A Homeowner's Guide to Wall Insulation*. The Sustainable Energy Authority of Ireland (SEAI), Dublin, Ireland.

SEAI (2020). The Sustainable Energy Authority of Ireland. Dublin, Ireland. Available at https://www.seai.ie/

Tommerup, H. & Svendsen, S. (2006). *Energy savings in Danish residential building stock*. Energy and Buildings, 38, 618–626

Verbeeck, G. & Hens, H. (2005). *Energy savings in retrofitted dwellings: economically viable?* Energy and Buildings, 37, 747–754.

UKRadon see PHE (2020).

UK Government (2021). *Environmental Protection Act 1990*. UK Government. Available at: https://www.legislation.gov.uk/ukpga/1990/43/contents

10

Sustainability

John L. Sturges

This chapter will deal first with what sustainability is, and what it is not. It will then proceed to a section dealing with the management of sustainability, and conclude with a section discussing how sustainable the main materials of construction really are.

SUSTAINABILITY: AN INTRODUCTION

The term sustainable implies that something is capable of being continued for an extended or prolonged time period. This term is sometimes confused and used interchangeably with environmental impact mitigation, but the terms refer to two different things; impact mitigation is only one part of achieving sustainability. In the context of industry, sustainability has three elements or dimensions:

- Environment
- Society
- Economy

Many writers on sustainability talk in terms of capital – environmental or natural capital, social capital, and economic or financial capital. To be sustainable means conserving all three forms of capital, what Elkington (1997) refers to as "the triple bottom line", to differentiate it from the usual financial or monetary "bottom line". By natural capital is meant the world's natural resources (Helm, 2015), and some of these, in the form of materials and energy, we use and exploit in our construction and operation of buildings. Besides the materials and energy that we consume, we are also sustained by various natural processes, for example, the photosynthesis that occurs in green plants and trees providing the essential oxygen in the air we breathe, the food we eat, and the CO_2 and water vapour that we exhale. The hydrological cycle continually purifies our water supplies so that precipitation is the pure H_2O that is essential for life. The systems that provide photosynthesis and that purify our water are part of this natural capital, and remind us that we do not control nature, but that we are part of it and subject to its laws. Despite our reliance upon it, this natural capital is all too often taken for granted, and no monetary value is put upon it. It is vitally important that we adopt the mindset that properly values this natural capital.

This notion of natural capital will be very important later in this chapter when we consider the sustainability of construction materials. Helm (2015) explains very clearly that conventional economics gives monetary values to materials, and these values reflect the scarcity or abundance of the material, and the cost of extracting it, transport, and processing. This "bottom line" economics has been responsible for the environmental mess that we have created over the past hundred years or so. Helm proposes that we ascribe a realistic value to the material that reflects its true worth in maintaining life on Earth. These values are inevitably higher, and often much higher than those derived from conventional economics. However, such values can help to avoid building "white elephant" and vanity projects that provide no long-term benefit to society. If we are proposing to build some infrastructure that will provide such long-term benefit then we can cost it in the usual way using natural capital values. We can then calculate the cost of the materials as natural capital and compare with the monetary value of what the infrastructure will provide. If the benefit is less than the lost value of the natural capital, we do not go ahead with the building. Such an approach would avoid vanity projects that confer no benefits to society but consume excessive amounts of valuable resources. Viewed from the opposite standpoint, what is the point of building something that benefits very few people at the cost of degrading the natural systems that keep us alive?

Economics that deals in these terms can be viewed as sustainable economics. Realistic values are put on assets whose value greatly exceeds their purely utility price. Social capital should similarly be conserved. This means that it should be properly valued and not exploited or abused, as happens all too often. Workers should be engaged in worthwhile work that they can believe in, properly rewarded for their efforts, and not required to act illegally. These are tenets of social sustainability.

The manufacture and use of materials have economic consequences, and as industrial managers know, you will not be truly productive unless your workers feel valued. Social sustainability implies that everyone subscribes to the policies involved in achieving environmental and economic sustainability. Economic sustainability implies that materials and resources are valued by their intrinsic worth as natural capital and not by their current monetary value or market price. Adherence to just the financial "bottom line" has caused many of the problems that we currently face. Decisions involving the commitment of material and energy resources are usually made according to purely financial considerations. For industry to be sustainable, it should have minimal environmental impact, consume minimum raw materials, and produce minimum waste. The gaining of raw materials, the disposal of mining and manufacturing waste arisings, and the disposal of products and materials at the end of their service lives should not blight the landscape or deplete resources to the extent that future generations are left with nothing to work with. True sustainability should strictly have zero impact; we clearly do not achieve that at present, but that is the ideal towards which we must aim.

Our industrial and economic activities do impact the earth's environment and ecosystems in numerous ways, and if we are to operate sustainably, we must significantly reduce this impact. To do this we must gain some understanding of how the earth and its ecosystems work, and the following section summarises the main impacts that we are making at the present time.

Areas of concern

Climate change and its linkage to global emissions of carbon dioxide are the topics of current debate in the news media, and rightly so. However, there are other impacts that are similarly threatening but which are overshadowed by the climate change debate. A team of leading environmental scientists at the Stockholm Resilience Centre, chaired by Professor Johan Rockström, has produced a list of nine major areas of current concern (Rockström et al., 2009). This team was attempting to define the permissible limits to our impact in these areas with estimates of how closely we are now approaching them, and indeed whether we may have already exceeded any of them. The nine areas they proposed were:

- Biodiversity loss
- Carbon dioxide and climate change
- Nitrogen use and misuse
- Land use and misuse
- Fresh water supplies
- Toxic substance release
- Aerosol particle release
- Ocean acidification
- Ozone layer damage

Biodiversity was put at the head of the list because it implies serious damage to the earth's ecosystems, including both plant and animal life. It represents damage to the life-support system, and the damage is caused by the eight factors listed below biodiversity loss. Of the remaining eight, four are wholly caused by humans and the other four can be caused naturally as well as anthropogenically. Those due to humans are nitrogen pollution and ozone layer damage. Rockström's report concluded that we have already exceeded the safe limits in three of the eight areas, namely, the rate of biodiversity loss, emissions of CO_2, and release of excessive amounts of nitrogen. These eight areas will now be considered.

With its reference to ocean acidification, the list reminds us that the oceans are a major part of the earth's ecosystem. They cover 71% of the earth's surface, but for too long the seas have been regarded as an inexhaustible supply of seafood, and the most convenient repository for all the waste produced by our modern society. The shipping industry is one of the least regulated industries in the world. Illegal fishing, dumping of toxic waste, piracy, people trafficking, etc., all occur on a regular basis (Urbina, 2019), but they take place out of sight, and the perpetrators, even when identified, are often shielded from prosecution by corrupt agencies. The "flag of convenience" system still operates. However, since the hydrological cycle carries a good deal of waste into the seas, it behoves all those running construction projects to ensure that all waste generated is disposed of ethically via reputable and reliable companies.

Climate change

Everyone is now aware of the current debate about climate change, whether they subscribe to the notion that humankind is responsible or not. Scientific opinion is now firmly behind the belief that climate change is anthropogenic in origin. From the evidence of paleoclimate evidence, including air bubbles trapped in ice cores taken from the Greenland and Antarctic icecaps, carbon dioxide in the atmosphere has not exceeded 300 ppm (parts per million) during the ice age cycles of the past 800,000 years or so. Before the Industrial Revolution, the global average amount of CO_2 was about 280 ppm (NOAA, 2020; Jones et al., 2020). In 2019, the atmospheric content of CO_2 had increased to around 410 ppm. Carbon dioxide is known to be a greenhouse gas, and this increase is held to be responsible for global warming, and is often the first thing called to mind when people consider our impact on life on Earth.

Consideration of Earth's history and its ecosystems has led to some remarkable developments in our understanding of life on our planet, including a notable contribution from James Lovelock who put forward his Gaia hypothesis (Lovelock, 1979). This idea postulates that all the life-forms on Earth effectively co-operate to maintain the conditions for life to continue; in effect, the whole world is one living organism. While the idea was initially ridiculed (it no longer is) it stimulated a lot of research into our planet's ecosystems, and as a result, there is now serious concern about several other impacts (Rockström et al., 2009). These include a reduction in biodiversity (and possibly a sixth mass extinction), fresh water supplies, use and misuse of land, nitrogen pollution, release of toxic substances, and aerosol particles into the environment, acidification of the oceans, and destruction of the ozone layer high in the earth's atmosphere.

Some of these impacts are solely anthropogenic, and some can be due to natural causes as well as anthropogenic in origin. If we contaminate water supplies, or misuse or contaminate land, then these are solely due to human agency. Nitrogen pollution and damage to the ozone layer due to the release of chlorofluorocarbons (CFCs) into the atmosphere again are solely caused by human agency. The release of CO_2 can be caused by the combustion of fossil fuel or by gas emissions from hot springs or volcanoes. There are around 1,500 active volcanoes, and one will be erupting somewhere in the world at any given time. Volcanoes will also emit aerosol-sized particles (sizes from 1 μm or 10^{-6} m down to 1 nm or 10^{-9} m), and such particles are also emitted in large amounts during the combustion of fossil fuels. If the atmospheric content of CO_2 is increased then some of it will dissolve in sea water, resulting in increased acidification of the seas. Similarly, an increase in the ambient temperature of the atmosphere will result in warming of the oceans. The specific heat capacity of water is around a thousand times greater than that of air. Finally, the release of toxic chemical species and aerosol particles into the atmosphere can be caused by industrial processes, but they also occur when volcanoes erupt.

The net result of these adverse effects is to cause pollution of natural habitats and the death or extinction of life-forms, plant and animal, that live in the affected areas. Since we are dependent on so many other life-forms, extinctions affect us, and this is the reason for the increase in concern.

The construction industry is by far the largest consumer of materials in the world, using more than all other sectors of industry put together. The manufacture and processing of materials consume vast amounts of energy (Ashby, 2009). Each material has its own value of embodied energy. This is the energy consumed in producing unit quantity of the material (e.g. GJ/tonne). There is also an embodied carbon value (the weight of CO_2 emitted in producing unit weight of material). These values are not intrinsic properties of the material, but they reflect the energy efficiency of their manufacture. If the manufacturing process becomes more efficient over time, these values may decrease. If the materials are transported to site, then their carbon footprint goes up, and the bigger the distance the greater the increase. Specific materials are selected because of their properties, and if one material is substituted by another with "superior" properties, it is important to also assess its environmental impact besides the specific desired property.

It is not difficult to visualise the links between large-scale production of materials by industry and the release of the greenhouse gas CO_2, the acidification of the seas, the emissions of toxic species in the form of dust, gases, and liquids, and the emission of aerosol-sized particles with their adverse impact upon our environment.

Nitrogen pollution

Nitrogen pollution and ozone layer damage require a bit more explanation. The element nitrogen is essential for life, being an important constituent of proteins. The earth's atmosphere consists of 79% nitrogen as a diatomic molecule N_2. These two atoms are bonded together by very strong bonds, so that nitrogen is effectively chemically inert. There are just two natural ways that this atmospheric nitrogen can be chemically "fixed" and made available for nutritional purposes, and these are by lightning strikes during thunderstorms and by the action of enzymes stored in the root nodules of leguminous plants such as peas and beans. Over 200 years ago, Thomas Malthus realised that the human population would be limited in size because of starvation caused by food shortage. He was quite correct; this is how nature works. Fortunately, the answer to the nitrogen shortage problem was found in huge cave systems in South America in countries such as Chile and Peru. Huge populations of bats and birds had lived in these caves since time immemorial, leaving immense deposits of guano, scores of metres thick. This guano is very nitrogen-rich, and during the 19th century, a whole fleet of ships plied between Europe and South America transporting guano for use as fertiliser by the agricultural industry. The guano was turned into ammonium nitrate for use as fertiliser.

To cause nitrogen chemical bonds to be broken to make nitrogen react is an extremely endothermic process (i.e. it takes the input of huge amounts of energy). In the early 20th century in Germany, a process to synthesise ammonia (NH_3) from nitrogen and hydrogen on a laboratory-scale was invented by Fritz Haber. In 1913, in collaboration with an engineer, Robert Bosch, they developed the Haber–Bosch process to synthesise ammonia on an industrial scale. This can be used to make ammonium nitrate which is an excellent fertiliser, but ammonium nitrate is also the basis of explosives. During World War I, the Haber–Bosch process was used to manufacture explosives because Germany was cut off from supplies of guano by the Royal Navy blockade. Artificial fertilisers were only used in large quantities after World War II. They were very effective to the point where they have been over-used by farmers. Excess fertilisers are washed off fields by rainfall and find their way into streams and rivers and end up in the sea. The nitrogen-rich sea water gives rise to algal blooms, and the algae deplete the water of oxygen, leading to the development of "dead zones" where no life can survive.

Aerosol particle releases

Aerosols are small particles that can be liquid or solid, falling in the size range between an upper limit of one micrometre (10^{-6} m or 1 μm), and a lower limit of one nanometre (10^{-9} m or 1 nm). Aerosols can be either natural or anthropogenic in origin, and because of their tiny size, once aloft they can remain in the atmosphere for very long periods as they do not settle quickly under the action of gravity. Natural sources include forest fires, volcanic activity, tornadoes, dust storms, etc. Anthropogenic causes include emissions from transport systems (aircraft, passenger and goods vehicles on land, ships), quarrying operations, crushing and grinding operations, emissions from industrial operations, etc. Table 10.1 shows that these

Table 10.1 Atmospheric aerosol fluxes

Natural source	Average flux ($Mt \cdot a^{-1}$) Ref. A.	Flux range D < 25 μm Ref. B.
Primary Windblown dust		1,000–10,000
Mineral dust	917	
Mineral dust	573	
Forest fires		3–150
Sea salt	10,000	1,000–10,000
Volcanoes	30	4–10,000
Biological	50	26–50
Secondary Sulphates from DMS		60–110
Sulphates from volcanic SO_2	20	10–30
From biogenic VOC	11.2	4–200
From biogenic NO_x		10–40

(Source: Henderson & Henderson, 2009).

Note: DMS = dimethyl sulphide
VOC = volatile organic compounds

Ref. A: Seinfeld, A. and Pandis, S.N., (2006), *Atmospheric Chemistry and Physics*, 2nd Edition, John Wiley, New York.
Ref. B: Brasseur, G.P. et al. (1999), *Atmospheric Chemistry and Global Change*, Oxford University Press, New York

fluxes can contain considerable amounts of fine particulates. For example, there can be over 900,000,000 tonnes of fine mineral dust in the atmosphere each year The finer their size, the longer they remain aloft.

The figures in Table 10.1 are average values, but they can show huge increases after episodes of volcanism, large-scale forest fires, industrial accidents, etc.

Toxic substance releases

Toxic substances can be released both naturally and by anthropogenic agency. For example, volcanoes can release oxides of sulphur and nitrogen as well as halogens such as chlorine. All these can be released by human industrial activity; people have for many centuries extracted metals from the Earth including mercury and cadmium, both of which can be lethal if inhaled or ingested.

Human releases of toxic substances fall under three headings:

- Pesticides, drugs, and other human-made toxins
- Plastic materials
- Radioactive materials

There are now over 1 billion cars on the roads of the world. They emit diesel and petrol fumes and damaging particulates, and they also produce dust in the form of tiny rubber particles, and dust from braking systems. Electric vehicles containing heavy batteries emit more particles from tyres and braking systems than conventional vehicles, to some extent cancelling out their engine emissions advantage. This argues in favour of minimising transport operations as far as possible. "Industrial agriculture" is also responsible for much pollution; nitrates in artificial fertilisers, insecticides, and weed-killers are used in huge amounts, and these get washed off the land and into streams.

Plastic materials also become discarded and finish up in the seas as well. They cause great harm to marine life, birds, and amphibians of various species. The nuclear weapons and power industries were created after World War II, and tests of weapons were carried out in the atmosphere until the 1960s ban. Physicists did not understand the effects of radiation on populations at large in the early years of the nuclear age. Atmospheric dust probably still contains some nuclides with long half-lives from these early tests.

Ozone layer damage

Ozone layer damage is caused by the release of CFCs developed for use as refrigerants for food preservation. When a refrigerator is scrapped, the CFCs it contained are released into the atmosphere. At first, this was not considered to be a problem, until it was discovered that CFCs survived in the upper atmosphere (altitude c. 25 km) for many decades. They take part in various photo-chemical reactions that result in the conversion of ozone (O_3) to normal diatomic oxygen O_2. The triatomic ozone was very effective in filtering out the very short-wavelength, very damaging ultraviolet radiation in sunlight. The use of CFCs was banned by the Montreal Protocol in the late 1980s by most countries in the developed western world and Japan. Other countries finally signed the protocol by 2015.

Ocean acidification

As has already been mentioned, *Homo sapiens* are land-based, and we have tended to view the Earth's ecosystem as limited to what is on land, ignoring life in the seas. There are numerous ecological links between marine and land-based life. Earth's rivers discharge their waters into the seas, and it is this that makes the oceans saline. According to the Natural History Museum (2020), ocean salinity is mainly caused by rain washing mineral ions from the land into water. We know that the seas contain approximately 3.5% salinity, and that more than 5 or 6% would make sea water unable to support life. Among the plant and animal life in the seas are organisms that remove excess salinity and so maintain conditions for life. This realisation was one of the factors that drove James Lovelock (1979) to put forward his Gaia hypothesis in the 1970s as introduced in the subsection "Climate change" above. Billions of tonnes of raw materials and manufactured goods are transported by sea to destinations all over the world. These sea lanes only occupy a small fraction of the whole ocean area. Unfortunately, the shipping industry has hitherto been subject to minimal or no regulation, and those operating fishing boats and transport vessels have caused a large amount of environmental damage. Besides the salts, organic matter, dust and sand, and unauthorised chemical discharges that reach the seas come billions of items of plastic waste. Most of this waste is discharged from ten of the largest rivers on Earth. Fine pieces (of plastic) are mistaken for food and are picked up by birds and fed to their chicks, often with fatal results.

The oceans contain coral reefs, and while these occupy only 0.1% of the oceans' surface area, they contain 25% of the biodiversity in the seas; they truly are biodiversity "hot-spots". Most reefs are located in the seas' photic zones, and they operate by photosynthesis, but a few are located deeper, where they cannot access sunlight, and these operate by chemosynthesis. However, their health is threatened by increases in ocean acidity and by warming of the seas. Recent reports indicate that the upper 700 metres of the world's oceans have warmed by 0.3°C, and that the pH of the seas has decreased from 8.2 to 8.1. These effects sound minimal and not causes for concern. However, a simple calculation reveals that the thermal energy required to warm the upper 700 metres of the seas is sufficient to manufacture all the world's steel at the current (2019) rate for the next 15 years or so. Furthermore, while a pH reduction of 0.1 sounds insignificant, because the pH scale is logarithmic, this really represents an increase in acidity of 30%. The increase in temperature and acidity is sufficient to threaten the survival of crustaceans and any species that form shells for protection.

Land water resources, their use and misuse

The area of land in the world is fixed, as is the volume of water at 1.4 billion cubic kilometres. We cannot significantly increase these figures. The human population of the Earth now stands at over 7.5 billion and is still increasing. Unfortunately, the locations of desirable land and plentiful water supplies are not at all evenly distributed, nor do they coincide with the great centres of population. Decisions about how we use the land available to us are taken by governments as well as individuals.

Similarly, as the population increases, decisions are taken regarding the provision of water supplies for drinking, for agriculture, for the generation of electricity, and for cooling the power plants. These water-supply decisions require careful analysis before action is taken. Too often in the past, decisions have been taken which have had disastrous consequences. The decision to divert the waters of the Abu Syr and Abu Darya rivers in Kazakhstan to support the growing of a cash crop – cotton – has devastated the Aral Sea, creating a toxic wasteland. Around the location of what is left of the Aral Sea, the health of the human population is dire. According to the World Wildlife Fund (WWF, 2020) some 70% of the earth's fresh water goes toward agriculture production, and cotton uses much more water to produce 1 kilogram of cotton compared with all grain, fruit, and vegetables. Further down the line, the fashion and clothing industries now have a serious and adverse effect on the Earth's environment. For these reasons when making decisions about the provision of water supplies or the construction of large-scale infrastructure, it is important to think in natural capital terms. Irreplaceable resources are being used. Resources should not be used for vanity or "white elephant" schemes.

Just as coral reefs are biodiversity hot-spots in the seas, tropical rainforests are biodiversity hot-spots on land. Unfortunately, the rainforests are threatened with deforestation due to land clearing for agriculture, illegal logging, creation of monocultures of palm oil trees, rubber trees, etc. These forests provide services to us in terms of CO_2 sequestration and oxygen production and are vital natural capital.

Traditionally, the mining and mineral extraction industries have had a bad reputation from an environmental point of view, leaving abandoned mine workings in an unsightly and dangerous condition. Fortunately, recent legislation requires them to undertake site clean-up and remediation, and many have adopted exemplary practices where the initial survey of the proposed site includes an evaluation of the local wildlife and a complete record of the plant species present. Some companies have taken samples of plants to be cultivated and preserved for restoration when mining has finished. Some companies have undertaken the creation of infrastructure, roads, and buildings that can be taken over and used by the local population at the end of mining operations. These actions represent a positive contribution to promoting sustainability. The natural capital is not wasted but invested.

These paragraphs illustrate the hundreds of linkages that exist between humankind's actions and the health of our environment both terrestrial and marine. They remind us that in ecosystems everything is connected to everything else.

MANAGING SUSTAINABILITY: INTRODUCTION

An important aspect of management in all branches of industry and commerce is **the management of change**. This is true in the service sector, manufacturing and civil engineering sectors, and in the construction and operation of buildings.

The branch of management that has evolved to handle the management of change is known as **project management**. This is now a mature discipline with its own professional body that had its origins in World War II in what was known as operations research (OR). In wartime, people and materiel were expensive resources in limited supply. The achievement of military objectives required the deployment and use of expensively trained people and expensive weapons systems. The achievement of an objective would place these resources at hazard, and therefore it was important to achieve the objective with minimum losses. A team of highly talented scientists was put together, and they devised various numerical methods of planning and efficiently allocating resources to military operations.

After the war, it was recognised that this approach could be adapted to peacetime situations, to enable projects, especially large ones, to be accomplished most efficiently, and so project management was born. It forms a very important part of the management of sustainability, as we need to bring about changes in a sustainable way.

Project management

Project management is a systematic way of managing projects to bring about change. Before commencement, the work to be done is analysed in detail so that the work can be broken down into work packages. The project is given a timeline, so that the work packages can be accomplished in a timely manner, the objective being that the whole project is not unduly delayed because an essential task has not been completed. Numerous project management software packages have been developed to enable the manager to view the progress of the project in real time. If one task is unavoidably delayed, the ramifications and knock-on effects can be seen immediately. Decisions to allocate more resources to the task can then be taken to ensure overall delivery on time.

How can sustainability be managed? Building construction and management have major environmental, social, and economic ramifications. Building construction impacts the earth and its various ecosystems in numerous ways; it is therefore important that we have some understanding of how our world and its ecosystems function. We must accept the fact that we do not control nature, but we are just a part of it. The natural world is a complex system of inter-connected ecosystems that all work together to maintain life on Earth. This is an insight of Lovelock (1979) who sees the world as one living entity, as previously discussed. As humans we are part of this entity, and if we alter or damage one part of life, this will have wider impacts than those that are immediately apparent.

All projects will have a timeline running from start to finish, divided into several stages. The first stage is the concept stage, where the idea of the project is conceived. This is followed by the definition stage, where the work to be done is defined. Next is the implementation stage where the project tasks are carried out. The work is completed in the close-down stage when the project or system is handed over to those who will operate it. This is the operation stage, and finally when the system has completed its operation, there is final termination.

The most useful concept is that of the project timeline. Any project has a life beginning at the concept stage; it is here that the ideas for achieving the objective are first examined and analysed. Following such analysis, the project will be defined. Here the chosen concept will be turned into concrete, quantified proposals and then the project will be implemented, and finally, the facility brought into being by the project will be handed over and the project will be closed down.

Figure 10.1 shows a schematic project timeline, and its relationship to upstream and downstream activities. It shows that the project is central, with the project manager having total control over what happens. Upstream and downstream activities are shown, but the project manager does not control these; he/she can exert some influence but has no control. For example, upstream are the sources of energy, materials, and other resources that will be required. If the project involves the construction of a building, the project manager can ensure that the supplies of timber come from a sustainable source. Within the project the manager has control but is subject to inputs/influences from government and from his/her own corporate management. The people in the project team will also have some input.

The project may have been the creation of an industrial production line, which is operated until a new product is designed or devised. Figure 10.2 shows the timeline for a project to set up a manufacturing facility to make a product. The product could be the manufacture of a vehicle for a transport system or modules

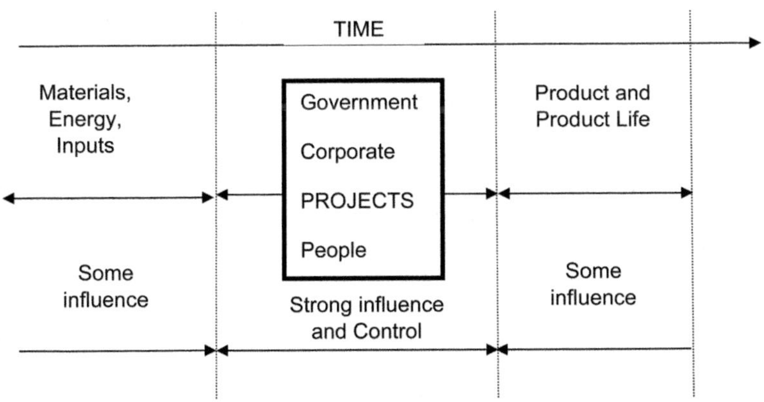

Figure 10.1 Showing the timeline for a project, and the relationship of the project to upstream and downstream influences.

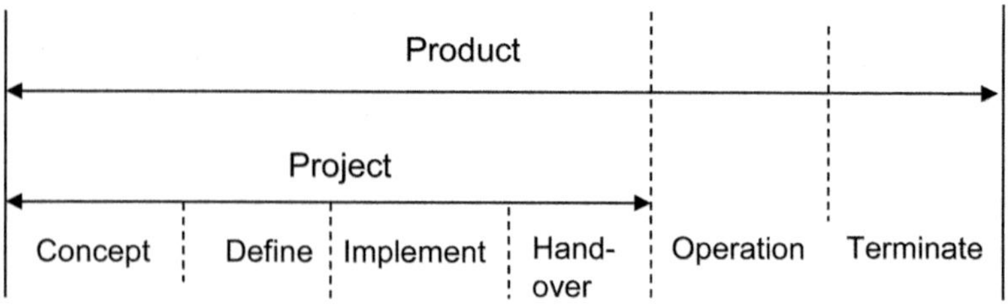

Figure 10.2 Timeline for a project where a product is created.

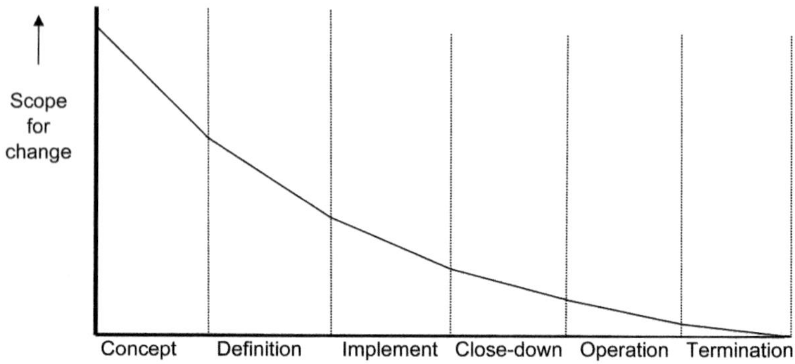

Figure 10.3 Showing the reducing scope for influence and major change as a project proceeds

for pre-fabricated buildings. It shows the times from initiation to completion of the project and to the manufacture of the product.

All the big decisions relating to and governing the project will be determined in the concept and definition stages. Two important points emerge from this. Firstly, it is very easy to make changes to the project during the concept and definition stages, and it becomes increasingly difficult to make changes as the project moves toward completion. This is shown in Figure 10.3.

Secondly, changes can be made very easily at minimal cost in the early stages, but changes become increasingly expensive as the project moves towards completion. See Figure 10.4.

The foregoing shows that ideally all decisions relating to the sustainability of the project should be made in the concept and definition stages, where the scope for making easy, low-cost changes is greatest.

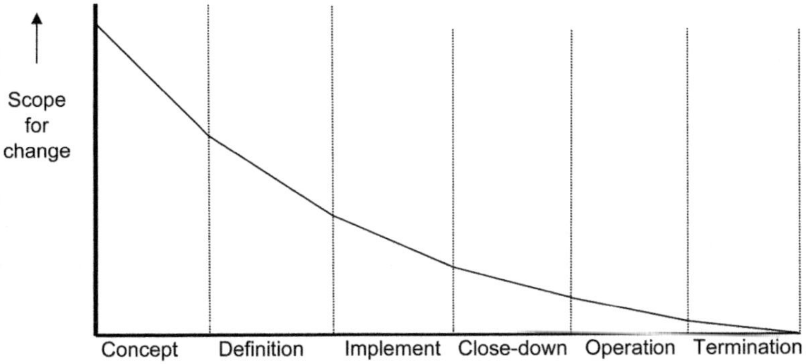

Figure 10.4 Showing how the cost of changes with time after project initiation.

Inter-dependence of all life

The term ecology was only coined in the 19th century, but there are now more life-scientists working in this area than ever before. It covers not only symbiotic relationships between different species of plants but multiple inter-dependencies between plants, animals, insects, fish, and bird life (Wohlleben, 2017). Therefore, it is coming to be recognised that engineers need to have some understanding of the world's ecological systems.

In this connection, it will be helpful to recall the areas of concern set out by the Stockholm Resilience Centre (Rockström et al., 2009):

- Biodiversity loss
- Carbon dioxide and climate change
- Nitrogen use and misuse
- Land use and misuse
- Fresh water supplies
- Toxic substance release
- Aerosol particle release
- Ocean acidification
- Ozone layer damage

Biodiversity loss was placed first in the list because it implies serious damage to the earth's ecosystems, including both plant and animal life, i.e. to our life-support system. This damage can be caused by the next eight factors in the list. Of these, two are wholly caused by humans, two can be caused by humans, and the other four can be caused both naturally and by humans. Table 10.2 shows the causes of environmental problems, both natural and anthropogenic.

Table 10.2 Origin, natural or anthropogenic, of environmental problems

Factor	Cause – anthropogenic, natural, or both
Carbon dioxide climate change	Anthropogenic and natural
Nitrogen pollution	Wholly anthropogenic
Land use and misuse	Anthropogenic
Fresh water supplies	Anthropogenic
Toxic substance release	Anthropogenic and natural
Aerosol particle release	Anthropogenic and natural
Ocean acidification	Anthropogenic and natural
Ozone layer damage	Wholly anthropogenic

Emissions of CO_2, and releases of toxic species and aerosol-sized particles can occur from volcanoes and hot springs and also from industrial gas and particle emissions. Extra CO_2 in the atmosphere will lead to some being dissolved in sea water, turning it more acidic. Nitrogen pollution and ozone layer damage are solely due to man. Large quantities of artificial fertilisers are manufactured and used and ozone layer damage is caused by emissions of CFCs used as refrigerants and as propellants for aerosol sprays. These are now generally banned worldwide as stated in the section "Ozone layer damage". With release rates now decreasing, the ozone layer should repair itself in the coming decades.

Principles of sustainable operation

If we are to operate in a sustainable way, there are some general principles that should be borne in mind, and the preceding section has pointed to what these might be. Study of natural ecosystems teaches us that nature consumes huge quantities of resource, but it produces no waste. This is what environmentalists are referring to when they speak of the need to "close the loop". All too often we obtain resources from one location, manufacture our product, and are left with manufacturing waste, and the problems of disposal of manufactured items at the end of their service lives in a different location. By contrast in nature at end-of-life, dead plant and animal materials are cleared up and effectively recycled by other life-forms, plant and animal. Victorian industries produced spoil and slag heaps from mining and metal extraction activities which sometimes had disastrous consequences, such as the events at Aberfan in 1966. Serious air and river pollution also resulted as industrialists sought to maximise their profits by "externalising" the costs of clean up. Mine workings were often left in a very dangerous state when the mine was worked out. Externalising of clean-up costs has historically been one of the major causes of environmental degradation (Scruton, 2012).

The Victorians' legacy of industrial waste reminds us of the dangers of anthropocentric thinking and its dangers. We have since learned to our cost that we do not rule the natural world, but we have yet to learn how to minimise our impact and live more sustainably.

In today's world, raw materials and manufactured goods are transported around the world by air, sea, and land in huge quantities. Since 1960, air travel has grown exponentially, especially for tourists. The impounding of water for drinking, irrigation, and hydroelectric power generation has proceeded apace since the early 20th century, but some of our dam-building/water-management decisions have had disastrous consequences. Since the volume of water in the world is fixed and limited, and the world's population is increasing, the provision of fresh water supplies is a growing problem.

Sustainability checklist

The following is a list of general guidance notes to more sustainable working:

- Minimise transport costs by sourcing labour and materials locally as far as possible.
- Use renewable and recycled materials as far as possible.
- Bear in mind that while timber is a renewable material, it is also an important part of Earth's (and our) life-support system.
- If timber is used, use timber from sustainably managed sources, i.e. Forestry Stewardship Council (FSC) timber.
- Use renewable energy supplies and avoid fossil fuels as far as possible.
- Minimise energy use.
- Minimise water consumption and avoid the release of contaminated water into drains, streams, and water courses.
- Minimise/eliminate waste production.
- Recycle and re-use materials as much as possible.
- Reduce waste and avoid waste incineration and land-filling of waste. Aim to segregate waste at source on site.
- Reduce/eliminate emissions of carbon dioxide, oxides of sulphur, and nitrogen.
- Reduce/avoid emissions of aerosol-sized particulates.
- Design items to be capable of dismantling and ease of servicing and maintenance.
- Avoid planned obsolescence.
- Avoid the "externalising" of costs, especially "clean-up" costs.
- Banish from everyone the mindset that it's "someone else's problem". We must ALL take individual responsibility.
- When waste is to be disposed of, it is vitally important to ensure that it is dealt with by reliable and responsible people.

These are general principles to adhere to in the interests of minimising our impact. This is very important, but it must be realised that sustainability means leaving the world in a similar state to how we found it. Sustainability is more than just impact mitigation; it is an ideal, but we must work towards achieving it if we are to bequeath a viable world to our descendants. Only when we do this can we claim to be truly sustainable. As a global society, we are currently nowhere near the achievement of sustainability.

Checklist for project managers in front-line situations

The following suggestions were made by Tom Taylor, the joint founder and retained consultant with Buro Four (Taylor, 2011). Tom Taylor has also been Vice-President of the Association for Project Management (APM). They are suggested as possible interventions to be made by project managers in the course of their work in managing projects. We have seen that there may be opportunities for intervention by the project manager both upstream and downstream of the project. A few such interventions are suggested in .

Finally, the foregoing discussion indicates very strongly that project managers should not think in purely terrestrial terms. The oceans represent 71% of the earth's ecosystem, and possible effects on marine life must be taken into account in any decision making.

CONSTRUCTION MATERIALS AND SUSTAINABILITY: AN INTRODUCTION

The construction industry sector is by far the largest consumer of materials, consuming more materials than all other sectors added together by a factor of three or four. For this reason, those in the industry need to be aware of the impact that this has on the earth and its numerous ecosystems, i.e. they need to have a view of the "bigger picture".

Steel and concrete production

Currently (2019), two materials predominate in terms of the volumes used in construction and these are concrete and steel, and they both have a large environmental impact. Concrete is made using cement, and the production of cement is carried out in a kiln operated at temperatures around 1,450°C. Steel is made from iron, and iron blast furnaces are also operated at very high temperatures in excess of 1,500°C. In both of these processes limestone is a major feedstock, and in both processes fuel is burned to generate the high temperatures required to drive the chemical reactions. The result is that there are two sources of carbon dioxide production in each

Table 10.3 Possible intervention by the project manager both upstream and downstream of the project

Responsibilities – from the beginning to the end
By identifying the client's current corporate, social, and other responsibility policies, and how they are applied, or may change in the project definition and delivery periods and thereafter in use. If the client does not have such policies, do they need help in devising them? Or if they are operational, or dated, do they need to be revised and brought up-to-date to suit a forward-looking, capital project or programme – especially if it is a "change" project.
Briefing for your project
Encourage and support clients/customers and sponsors/champions to incorporate sustainability into their statements of requirements – and apply them throughout the project. This can be a specific section of the briefing document of process, and/or can be picked up in individual aspects. It can also be specific in measurable terms or have stimulation of opportunities.
Modelling for your situation and culture
By recommending, understanding, and promoting appropriate models at key stages, emphasising the triple bottom line (economic, social, and environmental) and whole-life costings, with an awareness and knowledge of financial benefits.
Selection of your team participants
By incorporating sustainability criteria into selecting team members, contractors, suppliers, and specialists for their general outlook, credentials, and project-specific approach. And then writing obligations into appointments and contracts.
Strategies for your situation
By considering sustainability in the fundamental options of scoping, phasing, sequencing, sourcing, procurement, contracts, etc. Also by considering projects with programmes and portfolios for relationships, consistency, and effectiveness to optimise sustainability results.
Benchmarking for your sector/industry/location
What is good practice? What is best practice? What is cutting edge? How is performance to be measured?
Legislation
By identifying the current legislation and standards and how to comply with them, and whether to exceed them. Also the future trends during project, at handover, in the longer-term.
Financing
By defining the business benefits of environmental decisions; by securing monetary incentives and grants; by avoiding taxes, penalties, and charges.
Design proposals on your projects
By stimulating designs that creatively respond to statements of requirements, statutory standards, and good practice for sustainability. And then checking or auditing achievements.

(continued)

Table 10.3 *(continued)*

Specifications for your project
By conveying through the specifications that components and assemblies are sourced and delivered in sustainable ways, with proof.
Change decisions
By addressing sustainability aspects when change is under review, for example, when making tactical decisions that the change review includes considered value engineering and not just short-term, cost-driven capital expenditure effects, financial value, urgency, or expediency.
Delivery stage on your project
By reviewing, updating, confirming, promoting, and implementing the predetermined project sustainability arrangements, corporate standards, and good practice – including waste avoidance, packaging, sequencing efficiencies. Deliver the goods. Search for additional benefits.
Management arrangements you can adopt
By adopting effective project managerial and operational arrangements covering logistics, meetings, communications, consumables, re-use, and recycling – at collective, corporate, and individual citizen levels.
Sustainability communications as a theme
By the team conveying the sustainability aspects of the project throughout its duration to stakeholders including to end-users/occupiers at handovers, through documentation, briefings, and follow-up support.

$CaCO_3$	$\rightarrow$	CaO	+	CO_2
LIMESTONE	HEAT	QUICKLIME		CARBON DIOXIDE

Figure 10.5 Calcination reaction of limestone.

process; CO_2 is produced due to the combustion of fuel, and CO_2 is produced due to the calcination of the limestone in the burden. We can quantify this by a quick stoichiometric calculation. The atomic weights of calcium, carbon, and oxygen are 40, 12, and 16 respectively. Therefore, the molecular weight of limestone ($CaCO_3$) is 100. The calcination reaction is shown in Figure 10.5.

If we calcine 100 tonnes of limestone, we produce 56 tonnes of quicklime, and 44 tonnes of carbon dioxide. The calcination of 1 billion tonnes of limestone will produce 440 million tonnes of CO_2. The manufacture and use of cement currently run at around 4.5 billion tonnes per annum. The production of this quantity will require the calcination of at least 5 billion tonnes of limestone, and this will produce 2.2 billion tonnes of CO_2. To this figure, we must add the CO_2 generated by firing the kiln. From these figures, we can see that the production of 4.5 billion tonnes of cement makes a massive contribution to global CO_2 emissions.

The story is a similar one in the case of steel production. Globally, the manufacture and consumption of steel run at 1.6 billion tonnes per annum. Limestone is added to the blast furnace burden to provide CO_2 to act as a reducing agent, and to aid the production of slag, which acts as a flux floating on top of the molten iron and helping the removal of impurities. In addition to this there will be the CO_2 generated in heating the furnace. As with cement production, the generation of CO_2 will be in the billion-tonne order of magnitude.

The huge increase in global cement and steel production seen over the past two decades is due to Chinese infrastructure development. China now produces half of the world's steel and well over half of the world's cement. In the past four or five years, they have poured as much concrete as was poured in the USA in the entire 20th century. It is well-known that CO_2 is a greenhouse gas, and CO_2 emissions are held responsible for climate change. See the section "Climate change". As stated, the CO_2 content of the Earth's atmosphere has increased from around 300 parts per million (ppm) in 1900 to 410 ppm in 2019, and the construction industry has played a significant part in generating this increase.

Sand (fine aggregate) is an important ingredient for concrete-making. It is surprising given the quantity of sand that there is in the world that supplies are becoming a problem for makers of concrete in some areas. Sea sand and desert sand have more rounded grains than river sand because of tide and wind erosion. River sand has more angular grains which confer higher strength on concrete made with it. In addition, there are no problems of salt contamination with river sand.

Alternative construction materials

Two other important materials used in construction are polymers and timber. The use of timber predates the use of steel, concrete, and polymers, all three of which have been invented in their present forms only during the past 200 years. Along with natural stone, timber is one of the very oldest construction materials. Synthetic polymers were invented during the 20th century, and their use in construction really dates from immediately after World War II.

Polymers in construction

Currently, around 400 million tonnes of polymers are manufactured each year. As a class, polymers are hydrocarbon-based, being manufactured from oil; they are low density, with low stiffness and strength properties compared with steel and concrete. Polymers are chemically and thermodynamically extremely stable, and they resist corrosion and chemical attack. The polymerisation reactions by which they are made are highly endothermic, so polymers as a class have high embodied energy values. This is clearly shown in Table 10.4.

Table 10.4 Embodied energies of polymers compared with steel and concrete

Material	**Embodied energy (MJ/kg)**
High-density polyethylene	82
Polypropylene	82
Polyethylene terephthalate	85
Polystyrene	101
Poly vinyl chloride	66
Steel	12
Softwoods	8
Concrete	1

Source: Ashby et al., 2014.

Their high energy nature means that they are expensive, and their intrinsically low strength and stiffness of properties means that they are not used as structural materials in construction. The development of fibre-reinforced polymers has been driven by the need for materials combining high stiffness with low density. Firstly, glass fibre-reinforced plastics were developed and then more recently, carbon fibre-reinforced polymers have been developed combining light weight with stiffness properties half the value of steel (materials such as Kevlar). Kevlar particularly is very expensive (possibly £20,000 per tonne or more) and is used for rather exotic applications including military aircraft, Grand Prix racing cars, and sports equipment. A few pedestrian footbridges have been built using glass-reinforced plastics. The main applications for polymers in construction have been uPVC window frames for double-glazing, for plugs and fastenings, for damp-proof membranes and sheeting, for rainwater goods, and for electrical purposes for socket outlets, switch-plates, ducting, cable sheathing, etc. Around 400 million tonnes of polymers are produced each year, which does not sound much compared with the weights of steel and concrete, but because polymers are low-density, their volumes in cubic metres are comparable with those of steel.

One of the main problems that has arisen with polymers concerns their use for single-use items such as supermarket shopping bags and in packaging. In these applications their durability properties are not used, and indeed their durability becomes a problem because when discarded they do not biodegrade. Their low density means that they easily float in water, and they find their ways into rivers and streams and end up in the oceans, where their effect on marine wildlife is devastating. They can also enter the food-chain, and eventually present us with health problems. Floating plastic items eventually coalesce into huge floating gyres, the largest of which is in the Pacific Ocean and which contains tens of millions of tonnes of plastic waste. This problem has been known about for decades, but it took Sir David Attenborough's "Blue Planet II" TV series to finally bring to the attention of the general public on a scale where action is now being taken.

The mis-use of plastics is a major environmental disaster, because most plastics are inherently capable of being recycled. While plastic waste is now a major environmental problem, some solutions are to hand. Single-use plastic bags and magazine covers can be chopped up and heated and extruded and moulded to make useful products for the construction industry. Items which are highly durable such as kerbs, fence posts, chemical filtration units, etc., can be made. The British Board of Agrément (BBA) rates these products with a service life of over 100 years. These applications take full advantage of the chemical stability and durability of polymeric materials. It is ironic that transient, single-use products provide the feedstock for a recycling process that finally takes full advantage of their chemical stability and inherent durability.

Products such as window frame units for double-glazing are made incorporating U-V filter materials such as brilliant white titanium dioxide and carbon black, giving them service lives of up to three decades. Furthermore, such uPVC can be returned and recycled at the end of its service life, and uPVC windows and doors are therefore worthwhile products. The recycled uPVC can be greyish in colour, but the window and door sections can be extruded with a fresh coating layer of brilliant white plastic, or any other desired colour. Their use for doors and windows is an environmentally sound application of these materials.

Regarding their use in construction, an important point to mention is that because polymers are hydrocarbon-based, they burn easily in most cases, and can represent a major fire hazard if used inappropriately in buildings They present a two-fold hazard; firstly, their ease of combustion and secondly the toxic species that can be produced in smoke when some polymers burn. Substances such as hydrogen cyanide and hydrogen chloride gases can be emitted; these gases can be lethal and damaging to internal organs.

Timber as a construction material

The damaging effects of steel and concrete production have led engineers to seek alternative materials for construction. Land values in city centres have driven the construction of ever-taller buildings, culminating in the Burj Khalifa in Dubai which is 828 m tall. Some of these buildings are vanity projects, but they exacerbate the adverse environmental impact by their necessary use of steel and concrete. Engineers are seeking alternative materials for tall buildings. The world's population is growing, leading to a pressing need to build more housing units, and this has also driven the search for alternatives.

Timber is a renewable material and therefore, in principle, a sustainable one. Natural timber has highly anisotropic mechanical properties because of its grain structure. Cellulose vessel cells run predominantly in the vertical direction along the trunk of the tree. A much smaller number run in the radial direction, and none run circumferentially.

When timber is cut the vessel cells are full of moisture, and the timber must be seasoned to dry out most of the moisture. This drying out causes the timber to shrink, but again, the shrinkage is not isotropic. Zero shrinkage occurs longitudinally, maximum shrinkage occurs circumferentially, and an intermediate amount occurs radially because of the alignment of the vessel cells. The anisotropy has two principal effects; the strength and stiffness of timber vary widely depending on the fibre direction (longitudinal, radial, or circumferential), and when the timber is cut and seasoned, its moisture movement will vary from zero to a maximum depending on the fibre direction. The humidity of the air can vary when timber is put into service, and the timber will always seek to be in equilibrium with its surroundings. It will absorb moisture in wet weather and lose moisture in dry weather. This is the reason that round sections can become oval, square sections can become diamond shaped, floorboards can "cup" and so on. Another source of problems is that of knots. The presence of a large knot in the centre of a wooden beam can seriously weaken it.

This anisotropy problem led in the 19th century to the invention of plywood. Veneers are cut circumferentially and then laminated with the grain direction alternately aligned at right angles. An odd number of lamellae are used, and they restrict each other's ability to swell or shrink when local humidity changes. Glulam sections were also developed to obviate the adverse effects of knots in timber sections. Their use also removes the need for stress-grading of the timber. Finally, over the past two or three decades, the development of "engineered timber" has widened the potential use of timber for many applications. With engineered timber, thicker sections are laminated and glued together with grain directions alternately aligned, so the components have more isotropic structural properties. These items can be manufactured under factory conditions, with high dimensional accuracy, and with less wastage of timber. More of the timber is used than when solid timber sections are produced.

Over the past decade, there has been increasing interest in constructing tall buildings using engineered timber. The 9-storey, 29 m tall Stadthaus in London was an early example, but taller structures have since been built. The advantages claimed for timber construction include:

- Using prefabricated timber units made in a factory from engineered timber enabled faster erection.
- The structure was considerably lighter in weight.
- Light weight meant much less massive foundations.

- Timber is an excellent insulant; no extra insulation was required.
- Overall cost savings.

This is made clear by the details of density, stiffness (elastic modulus), thermal conductivity, and thermal diffusivity for steel, concrete, and timber set out in Table 10.5.

However, timber differs from steel and concrete because timber is a vital part of the Earth's ecosystems that we rely on to keep us alive. This is discussed further below.

The embodied energy of concrete is lower than that of cement because of the large content of aggregates and sand. Table 10.4 shows quite clearly that polymeric materials are high-energy materials compared with the other materials of construction. The use of steel and concrete raises concerns over the CO_2 emissions during their manufacture. Engineered timber is seen by many as a material of the future, with much better environmental credentials. It is interesting that around 25 years ago, the Chinese placed restrictions on the cutting of timber because they could foresee China becoming deforested in a few decades. They now import huge amounts of timber from Russian Siberia. The fully mechanised tree-felling machines can grasp a large 2 metre diameter tree trunk and then a large circular saw cuts through the trunk and the tree is carried to where its branches can be cut off and the log added to the pile to be transported out to the lumber mill and for further onward transport. This machinery enables the rapid harvesting of trees but is so heavy it compacts the forest soil to an excessive degree.

Properly designed timber buildings can survive in service for long periods of time. There are wooden churches in Russia that are over 500 years old. This is partly due to the climate, as the weather there is too cold for termite populations to survive. The city of Venice exists because the buildings all stand on foundations of wooden piles driven into the mud. St. Maria de la Salute at the end of the Grand Canal built around AD 1700 stands on 1.1 million wooden piles.

Table 10.5 Showing values of density, stiffness, thermal conductivity, and thermal diffusivity for steel, concrete, and timber

Material	**Density ($tonne{\cdot}m^{-3}$)**	**Stiffness Young's modulus ($GN{\cdot}m^{-2}$)**	**Thermal conductivity ($W{\cdot}m^{-2}{\cdot}K^{-1}$)**	**Thermal diffusivity ($m^2{\cdot}s^{-1}$)**
Steel	7.8	200–207	6.0	$15{,}959 \times 10^{-6}$
Concrete	2.6	45–50	1.4	0.663×10^{-6}
Timber	0.5–1.0	9–16	0.11	0.074×10^{-6}

Before opting to increase the use of timber for tall buildings, we must remember that it comes from trees, and trees are a vital part of our planet's life-support system. About 12,000 years ago, when humankind began to move from being hunter-gatherers to being farmers, there were around 6 trillion trees on the Earth (Lewis & Maslin, 2018). Today that figure stands at 3 trillion, with the Amazon rainforest alone containing 400 billion trees. We have already consumed half of the world's trees, as discussed more in the subsection "Trees as part of the life-support system" below. Forestry has become a major global industry and is the subject of many misconceptions. The notion is widespread that if we cut trees down all that is required is to plant more trees, but it is not so simple. We should now look at the part that trees play in the Earth's life support system.

Trees as part of the life-support system

Deforestation due to cutting of timber for fuel and the construction of buildings and boats became apparent in England and Japan around 500 years ago. This marked the beginning of our awareness of the problem of sustainability, although it was not given that name for another four or five centuries. In England and Japan measures were put in place to prohibit the unauthorised felling of trees. In Japan the Shogun also implemented a re-forestation programme, but in England, large-scale tree planting did not happen until after World War I, when the Forestry Commission was founded. Large tracts of land were planted with coniferous trees as these were viewed as faster-growing than deciduous species such as oak and beech. We now have large plantations of mature conifers in England, Scotland, and Wales.

A century ago, the decision to plant conifers was almost certainly taken in ignorance of all the differences between conifers and deciduous trees; however, we now know that tree species are not exactly interchangeable. Their respective effects on local climate and plant and animal species diversity were not understood at the time. While we now know much more about the lives of trees, we are still learning.

Unfortunately, in some parts of the world, forests (especially rainforests) are being cleared in the interests of gaining more land for agriculture besides illegal logging. This is true of the Amazon basin. In Indonesia, forests are being cleared and replaced with palm oil plantations, and to a lesser extent, rubber tree plantations. This is also a very mistaken policy, because it drastically reduces biodiversity in these areas. Nature avoids monocultures. Today, 95% of Columbia's lowland rainforest has already been cleared. The rate of forest clearance in the Amazon has recently risen again after some years of reduction. It is believed that around 2 million species of plants, animals, insects, etc., live in the Amazon rainforest. Forest clearance has a very damaging effect on the planet's overall biodiversity.

Over the past 12,000 years, we have halved the number of trees on the planet. Most of the most durable species are deciduous hardwoods, which take a long time to grow to maturity. Forests do much, much more than furnish growing timber, they provide habitats for millions of insects, birds, and small and large animals, as well as countless species of smaller plants and fungi. In other words, their part in maintaining our planet's necessary biodiversity is huge and of inestimable value. Old-growth forests are not monocultures, but diverse. Our modern propensity for creating monocultures of palm oil and rubber plantations is destructive. Clearing rainforest to create agricultural land is similarly destructive. The soils that support the Amazon rainforest are not rich, and when exposed will produce crops for less than a decade before rainfall erosion washes them away, creating the "need" for more forest clearance.

There has been a tendency to think of deciduous and coniferous trees as if they are interchangeable. Historically, coniferous trees evolved and populated the great boreal forests spreading across North America, Northern Europe, and Northern Russia. The climate in these regions is colder with long winters, and consequently, shorter growing seasons. Conifers do not shed their dark green, needle-like leaves. The dark green of their needles means that they absorb more solar radiation and this has a warming effect. Conifers also retain moisture more, so the air in coniferous forests is drier. The way that conifers manage their moisture intensifies the warming effect of their dark needles. With a short growing season, they can begin photosynthesis as soon as winter is over.

Deciduous trees grow in large forests further south, where the climate is warmer with longer growing seasons. Deciduous trees are lighter in colour so they absorb less radiation, but they can transpire a great deal of moisture. Ancient beech forests for example can transpire up to 2,000 m^3 of water per square kilometre, which cools the forest air for a considerable distance beneath their crowns. Deciduous trees shed their leaves in the autumn. Because they present a skeletal structure to winter storms, single deciduous trees suffer less damage than single conifers.

Conifers tend to be faster growing than deciduous trees. Certainly, the densest and most durable types of timber are deciduous, such as teak and mahogany. These species have fewer vessel cells in their microstructures, so they suffer from moisture movement effects to a very minimal degree compared with most other types. Their slow growth and excellent durability are the reasons for their high market prices.

Despite the great inroads made into the world's stock of trees, we are only now beginning to understand how they live together in forest conditions. We have known for a long time that numerous species of fungi thrive in forests, and for a long time it was thought that their function was to recycle dead wood. However, we now know that they form huge mycelium bodies which spread out a huge network of hyphae that connect the root systems of all the trees in the vicinity (Wohlleben, 2018). These hyphae enable trees to communicate with each other, and they can detect if a tree is struggling and transfer nutrients to that tree. It is as if the trees "know" that they are better together than standing alone. They have no neurological system in the animal sense, but they behave collaboratively rather than competitively in many ways. They are also able to warn each other if water supplies become difficult.

Other plants behave similarly, including sunflowers. If their roots detect very fertile soil, they produce fewer roots so as to leave plenty of soil nutrients for others growing nearby. If they detect soil of low fertility, then they produce more, wider-spreading root systems. The danger is that we over-harvest these natural resources before we have really discovered how they live. For example, we have only recently learned how long it can take to re-generate an old-growth forest if one is clear cut. Wohlleben (2018) cites an example from Germany. More than a century ago, oak forests were planted on land on Lüneburg Heath in North Germany. The land had been previously used for agriculture and it was assumed that the trees would grow and that the network of fungi and microbial life would re-establish themselves in a few decades. However, they discovered that even after the lapse of more than a century, there were large gaps in the list of species that inhabit ancient woodlands (Fichter et al., 2014). The lack of the missing species poses threats to the long-term health of the "new" forest, because the necessary nutrient cycles of birth and decay are not yet working correctly. In addition to this, the soil was found to still contain excess nitrogen from the agricultural fertilisers previously used there. This finding adds weight to the argument that it is not merely a matter of planting new saplings once mature trees have been felled. Just as the minerals and ore deposits in a mine can be worked out, so an ancient forest can be "worked out". The loss of mineral resources means that they are not available for future generations, but the loss of a forest means that its life-support services are not available to succeeding generations, a much more serious loss.

Clay bricks

In the UK in many parts of the country, brick-built semi-detached houses are ubiquitous. Brick has hitherto been a widely accepted material of construction, especially for house-building. The last two decades have seen a sharp rise in immigration to the UK, with a consequent steep increase in the need for affordable housing. In recent years, the rate of construction is less than half the rate achieved in the early to mid-1950s when there was a drive to replace war-damaged and slum housing. In 2018 around 250,000 housing units were constructed, an increase on the rate achieved for some years. But even if house-building doubled its current rate, it could not keep up with the rapidly growing need for affordable housing. The brick-making industry could not keep up with the demand for bricks if the construction rate were to double.

As building materials, bricks possess many excellent qualities. They can be made with a wide range of properties, both strength-wise and aesthetically. Facing bricks are the most common type and are used for vertical walling where an aesthetically pleasing finish is required. They are made from wet clay and are extruded and wire-cut. As a result, they are the most porous type, and should only be used for vertical walling, and never in situations where buildings will be exposed to driving rain. Because they are porous, they are water-absorbent, and if frost occurs, the damp surfaces will explode. Their surface appearance will be lost, and the brick will then continue to erode. For similar reasons, they should never be laid stretcher face upwards on top of walls in situations where frost can damage them. Such bricks are no use as paviours because of their susceptibility to frost damage.

Paviour bricks are made for this express purpose. They are much less porous. The most durable brick type is the Engineering Types A and B. These bricks typically absorb only 1 or 2% moisture in testing. They are made by compression-moulding and contain much less porosity when they go to the brick-kiln where they are subject to a two-stage firing process. In this process, the bricks emerge from the first firing, and then they enter a second, short, high-temperature stage where the bricks are heated until the surface layers just begin to melt. They are then removed and cooled. The incipient melting has the effect of sealing porosity and prevents such bricks from absorbing more than 1 or 1.5% of water by weight. Such bricks will not suffer from frost damage and may even be laid as a damp-proof course in a wall. In former times, bricks were moulded, dried, stacked, and covered in fuel that was then burned. Those bricks closest to the hottest part of the fire became "well burned" and acquired a blue colour. They were found to be the most durable because of incipient surface melting, and our modern engineering bricks make use of this knowledge. Like modern engineering bricks, these "well-burnt" bricks are very durable. The North Bar entrance to Beverley in East Yorkshire was built using "well-burnt" bricks in 1406, and this brickwork is still in excellent condition (Pevsner, 2002).

Traditional house-building requires a skilled workforce. The Code for Sustainable Homes (Dept. for Communities & Local Government, 2008), issued and since withdrawn, represented a move towards high-energy-efficiency house-building. Adherence to this code demanded skill levels higher than many in the industry possess. Training for personnel was not provided and no mechanisms were put in place to police the system, i.e. to ensure that the houses built actually met the required standards. The present level of demand for affordable housing compared with the present levels of construction point towards the need for a paradigm shift. The future may lie with pre-fabricated building systems. Building with brick may well continue, but it will have to be supplemented by pre-fabricated construction of some kind.

Selection of materials

Prof. Martin Ashby and his colleagues in the Department of Engineering at Cambridge University have devised and developed a method for the rational selection of materials. This was published nearly 30 years ago (Ashby, 1992), mainly directed at engineering rather than construction materials, and emerged into an expert system, using computers. Sturges (1999a,b) realised in 1992 that environmental impact indices could possibly be incorporated into the method which could then work for the selection of materials for construction. The growing importance of sustainability and environmental impact also drove Ashby and his colleagues simultaneously to develop their methodology for construction materials as well as general engineering materials (Ashby et al., 2014). This method is now software-based, and highly developed. It can be used by designers working in all fields of engineering, including civil engineering and construction.

The properties of thousands of materials are set up in a database that can be interrogated by the user who prioritises the most important design requirements. Among the materials, which are mostly human-made, is timber of various species and in various forms. Ashby's method is the best available method for rational materials selection, but it does not take account of the fact that timber is different from other materials by virtue of its important position in the planet's life-support system. We have learned a great deal about the life of trees and their importance in the evolution and sustaining of life on Earth over the last 20 years that was not known about when Ashby inaugurated his methodology.

Conclusions

We have reviewed the main materials of construction, discussing their respective advantages and problems. Steel and concrete, timber, and clay brick can all be used to build load-bearing structures. Concrete is the most widely used material of construction (20–25 billion tonnes per annum) because it is the cheapest way to get compressive strength. It is poor in tension, and so much of structural concrete is reinforced with steel. Steel frame construction is also widely used as it offers speedy erection, with frame members being cut to shape and drilled off-site. However, both these materials carry a huge carbon emission burden.

Timber is a natural, renewable material, and trees sequester carbon dioxide from the atmosphere. However, they are an important component of the earth's life-support systems, and they sustain huge biodiversity both plant and animal. Over the past 12,000 years or so, we have cut down half of the world's trees (Lewis & Maslin, 2018), and to cut down the rest would be human-race suicide. We have also replaced biodiverse forests with monocultures of palm oil trees and rubber trees. This policy also destroys biodiversity.

Clay brick is another old-established building material, but whether it can be the material to deliver housing units on the scale currently required is an open question. Recycled plastic products can be used by the industry, but not for structural, load-bearing purposes.

The conclusion from this is that no material comes without some environmental penalty. Timber is an excellent renewable, and potentially sustainable material, but if it is harvested at a rate that would meet current requirements, the rate of deforestation would increase with serious consequences.

REFERENCES AND FURTHER READING

Ashby, M. F. (1992). "*Materials Selection in Mechanical Design*". Pergamon Press, Oxford.

Ashby, M. F. (2009). "*Materials and the Environment*". Butterworth-Heinemann, Elsevier, Amsterdam.

Ashby, M. F., Shercliff, H. & Cebon, D. (2014). "*Materials*". 3rd Edition. Butterworth-Heinemann, Oxford.

Association for Project Management (2018). "*APM Body of Knowledge*". 6th Edition. Association for Project Management, Princes Risborough.

Berners-Lee, M. (2019). "*There is No Planet B*". Cambridge University Press, Cambridge.

Dept. for Communities and Local Government (2008). "*Code for Sustainable Homes*". Drafted by the BREEAM Centre, BRE, Watford.

Elkington, J. (1997). "*Cannibals with Forks. The Triple Bottom Line of 21st Century Business*". Capstone Publishing, Oxford.

Epstein, M. (2008). "*Making Sustainability Work: Best Practice in Managing and Measuring Corporate Social, Environmental and Economic Impacts*". Greenleaf Publishing Ltd, UK.

Fichter, K., Weiß, R., Bergset, L., Clausen, J., Hain, A. and Tiemann, I. (2014). *Analysis of the support system for green start-ups in Germany*. Oldenburg: University of Oldenburg.

Gibson, L. J. & Ashby, M. F. (1988). "*Cellular Solids, Structure & Properties*". Pergamon Press, Oxford.

Helm, D. (2015). "*Natural Capital. Valuing the Planet*". Yale University Press, New Haven and London.

Henderson, P. & Henderson, G. M. (2009). "*The Cambridge Handbook of Earth Science Data*". Cambridge University Press, Cambridge, UK.

Jones, M. W., Smith, A., Betts, R., Canadell, J. G., Prentice, I. C. and Le Quéré, C. (2020). "Climate Change Increases the Risk of Wildfires". ScienceBrief. Accessed from https://tyndall.ac.uk/sites/default/files/wildfires_briefing_note.pdf

Lewis, S. L. & Maslin, M. A. (2018). "*The Human Planet*". Pelican Books, London.

Lovelock, J. E. (1979). "*Gaia. A New Look at Life on Earth*". Oxford University, Oxford.

Morris, P. W. G. (1994). "*The Management of Projects*". Thomas Telford, London.

Natural History Museum (2020). "Why is the Sea Salty?". Accessed from https://www.nhm.ac.uk/discover/quick-questions/why-is-the-sea-salty

NOAA (2020). "National Oceanic and Atmospheric Administration. US Department of Commerce". Accessed from https://www.noaa.gov/; https://www.climate.gov/news-features/understanding-climate/climate-change-atmospheric-carbon-dioxide

Pevsner, N. (2002). Neave, D. (ed.). *Yorkshire: York and the East Riding* (2 ed.). Yale University Press, London.

Rockström, J., Steffen, W., Noone, K., Persson, A., Chapin, F. S., Lambin, E., Lenton, T. M., Scheffer, M., Folke, C., Schellnhuber, H., Nykvist, B., De Wit, C. A., Hughes, T., van der Leeuw, S., Rodhe, H., Sörlin, S., Snyder, P. K., Costanza, R., Svedin, U., Falkenmark, M., Karlberg, L., Corell, R. W., Fabry, V. J., Hansen, J., Walker, B., Liverman, D., Richardson, K., Crutzen, P. and Foley, J. (2009). "Planetary Boundaries: Exploring the Safe Operating Space for Humanity". *Ecology and Society* 14/2, p. 32. [online] URL: http://www.ecologyandsociety.org/vol14/iss2/art32/

Scruton, R. (2012). "*Green Philosophy. How to Think Seriously About the Planet*". Atlantic Books, London.

Silvius, G., Schipper, R., Planko, J., van den Brink, J. & Kőhler, A. (2012). "*Sustainability in Project Management*". Gower, Farnham, Surrey.

Sturges, J. L. (1999a). "A Methodology for Environmentally-Aware Materials Selection in Construction". Proc. 15th Annual ARCOM Conference, Sept. 1999, Liverpool John Moores University, Vol. 1, pp. 355–362.

Sturges, J. L. (1999b). "Construction Materials Selection and Sustainability". Proc. 2nd International Conf. on Construction Industry Development, National University of Singapore, Vol. 1, pp. 297–304.

Taylor, T. (2011). "*Sustainability Interventions—For Managers of Projects and Programmes—With Some Serious Opportunities, Challenges and Dilemmas*". The University of Salford Centre for Education in the Built Environment. Salford, UK.

Urbina, I. (2019). "*The Outlaw Ocean*". The Bodley Head, London.

Washington, H. (2015). "*Demystifying Sustainability. Towards Real Solutions*". Earthscan Abingdon.

Wignall, P. B. (2019). "*Extinction. A Very Short Introduction*". Oxford University Press Oxford.

Wohlleben, P. (2017). "*The Secret Network of Nature*". The Bodley Head, London.

Wohlleben, P. (2018). "*The Hidden Life of Trees*". Greystone Books, Vancouver/Berkeley

World Commission on Environment and Development (1987). "*Our Common Future*" (Brundtland Report), Oxford University Press, Oxford.

WWF (2020). "The Impact of a Cotton T-Shirt: How Smart Choices Can Make a Difference in Our Water and Energy Footprint". Accessed from https://www.worldwildlife.org/stories/the-impact-of-a-cotton-t-shirt

11

Glossary and reference section

Melanie Smith

GLOSSARY

Adjacent Owner Owner of property affected by an adjacent Building Owner carrying out works affecting a party wall or nearby structure which comes under The Party Wall etc. Act (Gov.uk, 1996). Includes a person in receipt of, or entitled to receive, the whole or part of the rents or profits of land; a person in possession of land, otherwise than as a mortgage or as a tenant from year to year or for a lesser term or as a tenant at will; and a purchaser of an interest in land under a contract for purchase or under an agreement for a lease, otherwise than under an agreement for a tenancy from year to year or for a lesser term.

AFD Automatic fire detection

Air flow and pressure differential Air flow into and around a building can be analysed by measuring pressure differentials. Limited air flow could lead to stagnant air and moisture problems; excessive air flow causes more heat loss.

Air permeability The unintended leakage of air through fissures, gaps, cracks, etc., in the external envelope of a building. It is measured as the volume of air leakage per hour per square metre of external building envelope ($m^3/h{\cdot}m^2$) at a tested pressure of 50 pascals (Pa). See Design air permeability.

Air quality Many factors affect indoor air quality (IAQ) including ventilation, humidity, levels of chemicals, mould spores, and radon. As occupants have varying tolerances to different contaminants and irritants and there is not a definitive definition of IAQ, it is taken to be a subjective sliding scale, from there being no known contaminants at harmful concentrations, to building conditions that negatively impact human health and wellbeing.

Airtight layer Prevents the movement of air and may or may not act as a vapour control layer. Airtight vapour-permeable membranes are intended to reduce unwanted air infiltration (movement of unwanted external air through the construction) whilst minimising the risk of condensation by maximising vapour permeability.

Architectural disability When the physical design, layout, and construction of buildings and places confront people with hazards and barriers that make the built environment inconvenient, uncomfortable, or unsafe for everyone to use and even prevent some people from using it at all. The built environment can disable anyone by poor design.

Architrave Timber moulding around doors and windows.

ASHRAE American Society of Heating, Refrigerating and Air-Conditioning Engineers. Esteemed worldwide, it funds research, offers educational programmes, and develops technical standards to improve building services engineering, energy efficiency, indoor air quality, and sustainable development.

BER Building Energy Rating for residential or commercial. Scale of A to G with A-rated buildings labelled as the most energy efficient and G-rated the least. Discussed in Chapter 6.

Biophilia A love for nature, considered by some to be the missing part of sustainable design. The term "biophilia" refers to the adaption or design of a building to the environment, rather than the other way around. Biophilic design aims to create strong connections between nature and human-made environments which can have benefits for health and wellbeing.

Breathability The way water moves in relation to the building fabric. Breathability does not, as the term might imply, relate to an exchange of air. Instead it is the ability of a material to allow moisture vapour to pass through it. Breathability is based upon three essential mechanisms: vapour permeability, hygroscopicity, and capillarity.

Breather membrane Defined in BS 5250 (BSi, 2016), as a membrane with a vapour resistance less than 0.6 MNs/g. Breather membranes are positioned to the outside of the insulation acting as a weather barrier while still allowing moisture particles to escape from the inside.

Building Owner Owner of property carrying out works affecting a party wall or nearby structure which comes under The Party Wall etc. Act. Includes: a person in receipt of, or entitled to receive, the whole or part of the rents or profits of land; a person in possession of land, otherwise than as a mortgage or as a tenant from year to year or for a lesser term or as a tenant at will; and a purchaser of an interest in land under a contract for purchase or under an agreement for a lease, otherwise than under an agreement for a tenancy from year to year or for a lesser term.

Building pathology Short-hand term for a set of terms which are not definitively defined, including building pathology, building physics, deterioration of materials and elements over time, building performance, and defects.

Building works *Taken from UK legislation – The Building Regulations 2010 (Gov.uk, 2010):*

a. The erection or extension of a building
b. The provision or extension of a controlled service or fitting in or in connection with a building

c. The alteration of a building, service or fitting
d. Work required due to changing the building's use
e. The insertion of insulating material into the cavity wall of a building
f. Work involving the underpinning of a building
g. Work required due to a changing the building's or area's energy status (e.g. providing heating where there was none previously)
h. Work required due to renovating or replacing a thermal element
i. Work required due to a building's extension or the provision, increase or extension of building services (consequential improvements to energy performance)

Calorific value The amount of heat which can be released by a unit mass of fuel at a particular temperature and pressure.

Capillarity Upward movement of moisture in a material containing capillary pores (e.g. bricks). Height depends on potential pressure differences.

Cavity surface resistance See Surface resistance.

CDM UK's Construction (Design and Management) Regulations (Gov.uk, 2015).

Cess pit Sealed tank/container holding sewage. Requires periodic emptying. See also Septic tank.

Chartered Surveyor The description of Professional Members and Fellows of the Royal Institution of Chartered Surveyors (RICS) who are entitled to use the designation. The term is protected by law in many countries. The RICS grants the designation only following the successful achievement of a period of established learning and professional interview/assessment test of competence. Regulations require Chartered Surveyors to continue with acceptable levels of competence and carry insurance for the protection of the public.

Cold roof Roof construction where the insulation is placed between and over the ceiling joists. External air is designed to ventilate through the roof void. Blockage of this ventilation can result in condensation on the roof timbers. See also Warm roof.

Commissioning Going through the commissioning process for building services, to ensure that they are operating at the capacity specified and to the performance standard required.

Condensation Condensed water vapour from the air. The water vapour cools to its dew point especially when in contact with cold surfaces, turning into droplets of water. Interstitial condensation can occur within a building element, if the temperature profile within the material(s) reach the dew point of the transmitted water vapour. Condensation which remains in its liquid state can cause dampness and subsequently the development of mould and rot. 100% relative humidity at the surface is required for surface condensation to occur.

Conditioned space Space that is heated, cooled, and/or mechanically ventilated. Unconditioned space is naturally heated, cooled, and ventilated. For example, basements, garages, and roof voids can be unconditioned spaces.

Construction observation By observing the building at stages throughout construction any faults or mistakes that may be hidden when the building is complete can be identified, and if possible rectified.

Construction work *Taken from UK legislation – The Construction (Design and Management) Regulations 2015 (Gov.uk, 2015).*

The carrying out of any building, civil engineering or engineering construction work, including:

a. The construction, alteration, conversion, fitting out, commissioning, renovation, repair, upkeep, redecoration or other maintenance (including cleaning which involves the use of water or an abrasive at high pressure, or the use of corrosive or toxic substances), de-commissioning, demolition, or dismantling of a structure;
b. The preparation for an intended structure, including site clearance, exploration, investigation (but not site survey) and excavation (but not pre-construction archaeological investigations), and the clearance or preparation of the site or structure for use or occupation at its conclusion;
c. The assembly on site of prefabricated elements to form a structure or the disassembly on site of the prefabricated elements which, immediately before such disassembly, formed a structure;
d. The removal of a structure, or of any product or waste resulting from demolition or dismantling of a structure, or from disassembly of prefabricated elements which immediately before such disassembly formed such a structure;
e. The installation, commissioning, maintenance, repair, or removal of mechanical, electrical, gas, compressed air, hydraulic, telecommunications, computer or similar services which are normally fixed within or to a structure

Coping Stone/masonry/tile/metal capping to top of parapet, wall, or chimney. It protects the masonry below from premature deterioration due to excess moisture. It should therefore be weathered (i.e. have a slanted or curved top surface) to throw off the water in an appropriate direction.

Cornice A horizontal decorative moulding, especially that covering the junction between a wall and a ceiling.

Creep The tendency of a solid material to move slowly or deform permanently under the influence of persistent mechanical stresses.

Demise The extent of property described under the terms of a lease is called the demise. It is defined either descriptively, or by demarcation of a floor plan showing the boundaries or edges of the tenant's occupation and any limits of lease obligations. The demise is worth so much to the tenant and so much to the landlord.

Design air permeability The target value for air permeability of a building set at design stage and evaluated through a mandatory testing regime outlined in the UK's approved document ADL1A (Planning Portal, 2016). A default value is set in the UK's ADL1A which may be used for specific cases and in the absence of testing. Similar legislative guidance exists for other countries.

Design reviews Insights gained through testing of a building fed back to the design team, helping to avoid any problems that arose in a tested building and make improvements to the design and construction phases.

Dew point The temperature at which water (water vapour) condenses. Dependent on humidity and pressure.

Diffusivity See Thermal diffusivity.

Door set or doorset Door and its frame. Arguably its fixings as well.

Easement A right to make some limited use of land belonging to someone else, or to receive something from that person's land. Examples include rights of way or rights to light or support.

Element An element of a building refers to the assemblage or unit of a roof, over-site, floor, external wall, internal wall, or ceiling.

Elemental approach To surveys: considers the elements, or services or fittings, of the building in isolation from other elements, etc. Common in "tick box" survey proforma. Compare with Problem approach and Systemic approach.

Entropy A measure of the amount of disorder in a system. Gradual decline into disorder, for example the process of erosion, corrosion, or decay.

Environmental monitoring Recording environmental conditions within the building, including temperature, relative humidity, particulates, etc., to determine how effective building services are.

Exfiltration air Unplanned (not designed) air exchange or air movement, out of a building or space

External surface resistance See Surface resistance.

Feathers The linings separating different flues in a chimney. Often made of half brick, the mortar may have been compromised (through deterioration from soot and other products of combustion) so that the flues connecting different fireplaces can merge, enabling products of combustion to pass into different flues and enter rooms. Any compromise requires flues in use to be relined.

Fitting Of a building refers to a window, door, fixture, component, or other separate item. May be part of the building fabric, but not a building element.

Fixture An asset that is installed or otherwise fixed in, or to, a building or land so as to become part of that building or land in law (HMRC, 2016).

Flood resilience The ability of a building to resist damage as a result of flooding.

Flood resistance The ability of a building to resist the entry of flood water from the outside to the inside.

Fuel poverty Where a home occupier is unable to afford to heat their home to a level that is healthy and safe. Three elements in determining whether a household is fuel poor: household income, household energy requirements, and fuel prices. Energy-inefficient homes are a major factor. The UK Government

Gov.uk (2020) describes a household to be fuel poor if they have required fuel costs that are above average (the national median level), and were they to spend that amount, they would be left with a residual income below the official poverty line.

Going Horizontal distance between risers in a stair. Measured from the outer edge of the tread's nosing to the edge of nosing in the next higher tread. See Tread.

Hazard A dangerous phenomenon, substance, human activity, or condition that may cause loss of life, injury or other health impacts, property damage, loss of livelihoods and services, social and economic disruption, or environmental damage. See Chapter 8, which discusses disaster management.

Heat flux measurement Heat flux plates are placed on a building element, along with internal and external temperature sensors. Readings are logged over time and used to calculate a U-value for the building element. Care should be taken to avoid measuring elements in direct sunlight.

Height of a building *Taken from UK legislation – The Building Regulations 2010 (Gov.uk, 2010)*:

Measured from the mean level of the ground adjoining the outside of the external walls of the building to the level of half the vertical height of the roof of the building, or to the top of the walls or of the parapet, if any, whichever is the higher.

Holistic Not piecemeal. See Whole house approach.

Humidity The concentration of water vapour present in the air. **Absolute humidity** is the total mass of water vapour present in a given volume or mass of air. It does not depend on temperature. **Relative humidity** indicates a present state of absolute humidity relative to a maximum humidity given the same temperature. Relative humidity is expressed as a percentage, and depends on temperature and the air pressure. The same amount of water vapour results in higher relative humidity in cool air than in warm air. Relative humidity plays a role in human comfort, along with air temperature, radiant temperatures, air speed, metabolic rate, and clothing levels. **Specific humidity** is the ratio of water vapour mass to total moist air parcel mass. As temperature decreases, the amount of water vapour needed to reach saturation also decreases, so as the temperature lowers it can reach saturation point without adding any more moisture. At 100% relative humidity, the air is saturated and is at its dew point.

Hygroscopicity The capacity of a material to react to the moisture content of the air by absorbing or releasing water vapour. The material's ability to attract and hold water molecules.

Inertia See Thermal inertia.

Infiltration air Unplanned (not designed) air exchange or air movement, into a building or space.

Insolation The amount of solar radiation hitting a given area.

Integrity The quality or state of being whole and unimpaired, no gaps or fissures.

Internal surface resistance See Surface resistance.

Interstitial condensation Condensation occurring within the materials or layers of construction. Not necessarily associated with surface condensation and can therefore develop and continue unseen. See Condensation.

In-use monitoring Recording energy consumption and internal and external temperatures and conditions whilst the building is in use or simulating use conditions. Building performance under real conditions can be evaluated.

Leakage detection Using smoke from air current checkers to locate air leakage paths. During pressurisation testing leakage is more pronounced. During depressurisation air infiltration paths can be identified with infrared cameras when there is sufficient temperature difference.

Moisture closed A building or building material or system which does not allow moisture (as liquid water or water vapour) to move through it (Bristol, 2015).

Moisture open A building or building material or system which allows moisture (as liquid water or water vapour) to move through it (Bristol, 2015).

Moisture survey Carried out where moisture problems are suspected, using a protimeter to assess moisture content of elements. Remote sensors can be used to measure moisture content over time.

Mould Mould grows on surfaces where there is a source of moisture, food, oxygen, and acceptable temperature. Mould growth is detrimental to materials, aesthetics, and occupant health. The susceptibility of material to mould growth varies. BS5250 (BSi, 2016) states, "As a guide if the average humidity within a room stays above 70% for several days, the relative humidity at external wall surfaces will be above 80%, which is high enough to support the germination and growth of moulds".

Passiv Passiv, passivhaus and passive may be terms relating to a standard of buildings (designed AND constructed) "created to rigorous energy efficient design standards so that they maintain an almost constant temperature. Passivhaus buildings are so well constructed, insulated and ventilated that they retain heat from the sun and the activities of their occupants, requiring very little additional heating or cooling." (EST, 2019). Initially developed by Passivhaus Institute in Germany, certification is a rigid process, hence the use of other terms passiv and passive which may mean similar but not exact standards and performance.

Partial deconstruction Targeted deconstruction at areas identified as problematic. For example, removing bricks to inspect insulation in a cavity, lifting floorboards to inspect underfloor voids.

Post-occupancy evaluation (POE) Collecting information from building users and on in-use operation. Gives designers insights into building use and effectiveness when users interact with the building systems. Results can identify opportunities for reducing running costs, maintenance requirements, and repair work. See also Chapter 6.

Pre-design survey A survey (of a building, structure, site, etc.) to influence and inform the design decisions for building works.

Pre-design thermal retrofit survey An on-site anticipatory inspection and assessment of a building, which is undertaken to inform the design for a retrofitted energy efficiency measure, and must include (a) the nature and characteristics of the building, (b) the risks to that building and its occupants posed by the measure, (c) the risks to the efficiency of that measure posed by the building, and (d) the challenges required to mitigate these risks through and by the design.

Pre-installation survey A survey to assess the property based on the design decisions made. The assessment may influence design amendments for the building or structure.

Pressurisation testing Intentional ventilation is sealed off before a fan apparatus pressurises and then depressurises the building. Air flow rate and internal-external pressure difference is recorded and used to calculate the air permeability (and air leakage rate) of a building.

Problem approach To surveys: when conducting a survey, focussing on foreseeable problems common, or frequently found, in the age, type, condition, etc., of the specific property. Looking for damage or issues associated with or related to each problem suspected or found. Follow the problem pathways forwards and backwards to find the "source – pathway – receptor" of the problem. Works well in conjunction with systemic approaches. Compare with Elemental approach.

Psi-value This is the measure of linear thermal transmittance and is the measure of heat loss along a non-repeating thermal bridge calculated as shown in BR 497 (BRE, 2016). It is the rate of heat flow per degree temperature difference per unit length of a thermal bridge. It is measured in W/m·K and is a property of the thermal bridge. Psi-values or ψ-values are used to calculate the heat loss or gain through a non-repeating or geometrical thermal bridge.

Quality assurance A process of systematic checking, verification, calibration, commissioning, etc., all to ensure the required standard of quality of the product, service, or activity as determined at procurement. Such a standard may include all those features of the product/service/activity which are required by the customer/client. It includes the principles of Fit for Purpose and of Right First Time.

RdSAP The UK Government's reduced data version of SAP used as a lower-cost method of assessing the energy performance of existing dwellings (BRE, 2017).

Receptor As used in risk assessment. A receiver of the consequences from a hazard.

Retrofit The process of changing, upgrading, or improving an existing property's performance through practical or technical measures applied to that property.

Risk The probability of harmful consequences, or expected losses (deaths, injuries, property, livelihoods, economic activity disrupted, or environment damaged) resulting from interactions between natural or human-induced hazards and vulnerable conditions. Source: see Chapter 8, "Disasters and the built environment".

Rising damp Moisture rising from the ground in the walls of a building through capillary action until it reaches equilibrium.

Riser The vertical height of a step.

SAP "The Standard Assessment Procedure (SAP) is the methodology used by the Government to assess and compare the energy and environmental performance of dwellings. Its purpose is to provide accurate and reliable assessments of dwelling energy performances that are needed to underpin energy and environmental policy initiatives" (Gov.uk, 2014).

Septic tank Basic treatment system for sewage. Uses bacterial anaerobic action. Accumulating sludge may need occasional emptying. Ground water pollution may occur. See Cess pit.

Solid wall Masonry wall without a designed cavity.

Sorption The ability of a hygroscopic material to absorb or release water vapour from or into the air until a state of equilibrium is reached.

Spandrel Vertical plane below staircase, usually triangular.

Stakeholder A person or body who has an interest (financial) in the subject. This might be a client, an occupier, or another role.

Stoichiometric ratio The ratio of fuel to oxygen, in which the amount of oxygen present is just sufficient to completely oxidate the fuel. See Chapter 7, "Fire safety".

Stratification (air) The layering of differing (normally rising) air, for example from floor to ceiling. Air stratification is a layering effect that allows large air pockets with different core temperatures to remain intact.

Stratification cold The perception of feeling cold due to the effects of stratification. Within normal ambient conditions, temperature differences between a person's head and feet of around 3 degrees Celcius are noticeable to humans (ASHRAE, 2013). Stratification does not depend on the actual temperature of the floor, but dissatisfaction occurs with the temperature difference so that even a small amount of stratification can cause cool feet or warm head discomfort. It has been found that a stratification of up to 7 degrees Celcius may sometimes be acceptable, whilst at other points no stratification difference is acceptable.

Surface resistance The electrical resistance of a surface layer to a current; applies to surfaces in contact with air only and not in direct contact with another material. Surface resistance depends, not only on the material, but also on environmental conditions that may affect the values determined. The value of surface resistance can be derived from the procedure set out in ISO 6946:2007.

Each surface resistance incorporates external wind speed, emissivity of finishes, the mean radiative temperature coefficient, and pre-determined non-variables such as conductive/convective coefficient. Not every material will have the same resistance and not every site location experiences the same wind speed (source authors: Cormac Flood & Lloyd Scott). See also the section on "U-value calculation tools" below in "Reference section".

Survey (building) An assessment, observation, inspection, evaluation, etc., of a building, structure, or site. There are many types of survey (for example, measurement, cost, value, geographical, societal, use, legislative, etc.).

Surveyor (building) A person undertaking a survey, inspection, assessment, observation, etc., of a building. Additionally they may give advice on design, construction, cost, maintenance, and repair of properties, or assess damage or dilapidations. A person identifying as a surveyor may or may not be competent, experienced, or qualified. Fully qualified building surveyors in the UK are frequently, but not always, members of the RICS. Compare with Chartered Surveyor.

Sustainable development "Sustainable development is development that meets the needs of the present without compromising the ability of future generations to meet their own needs" (WCED, 1987).

System control checks Similar to commissioning building services, ensure that the controls, both automatic and those available to the user of the building, function as intended.

Systemic approach To surveys: considers and looks for interactions between the building's fabric, condition, elements, environs, ventilation systems, services, indoor air quality, etc. Systemic approaches work well in conjunction with problem-based approaches. Contrast with elemental approaches, which are one-dimensional.

Thermal bridge Part of an element or structure of lower thermal resistance that bridges adjacent parts of higher thermal resistance and which may result in localised cold surfaces on which condensation, mould growth, and/or pattern staining can occur (BRE, 2002). Thermal bridges are thermally weak points in the building envelope where the heat loss is greater than through the main building elements. There are two kinds of thermal bridges, repeating and non-repeating. Repeating bridges occur at regular intervals as part of a building element such as timber struts in timber-walled properties and wall ties. These are accounted for in U-value calculations, although their effect can be underestimated. Linear non-repeating thermal bridges occur at junctions of different elements. Psi-value calculations are required to take account of these.

Thermal bypass Heat transfer that bypasses the conductive or conductive-radiative heat transfer between two regions (Harrje, 1986). Even in an airtight building, there may be air movement through or around the insulation, in effect bypassing the insulation and its effect and making it perform less well. Thermal bypass can dramatically reduce the performance of insulation. See Mark Siddall's publication at https://www.glosfordsips.co.uk/wp-content/uploads/2016/03/Thermal_bypass-study.pdf

Thermal comfort See also Chapter 6. Subjective satisfaction with the thermal environment and assessed by subjective evaluation. However, some generalities apply. Low humidity can create discomfort, respiratory problems, and susceptibility to viruses, and can aggravate allergies. Humans can be comfortable within a wide range of humidities, from above 30% to below 70%, depending on the temperature, ideally 50–60%. Thermal equilibrium with the surroundings (thermal neutrality), when the heat generated by human metabolism is able to dissipate, maintains thermal comfort. The main factors that influence thermal comfort are those that determine heat gain and loss.

Thermal conductivity The theoretical rate at which a material conducts heat across a unit thickness. It is a measure of the rate at which temperature differences transmit through a material. The lower the thermal conductivity of a material, the slower the rate at which temperature differences transmit through it, and so the more effective the material is as an insulator. It is a value independent of the material's thickness. Expressed in W/(m·K). Sometimes referred to as k-value or lambda (λ) value.

Thermal diffusivity The speed with which heat soaks into a material. Thermal conductivity divided by the product of the density of the material and its specific heat capacity.

Thermal element *Taken from UK legislation – The Building Regulations 2010*
A wall, floor or roof (but not windows, doors, roof windows or roof-lights) which separates a thermally conditioned part of the building (the conditioned space) from:

a. The external environment (including the ground); or
b. In the case of floors and walls, another part of the building which is:
    i. unconditioned;
    ii. consists of a small conservatory, porch, covered yard, covered way, or a carport open on at least two sides; or
    iii. conditioned to a different temperature,

and includes all parts of the element between the surface bounding the conditioned space and the external environment or other part of the building as the case may be.

Thermal imaging survey Using infrared cameras to identify anomalies within building fabric or to locate thermal bridges and areas of thermal bypass. Can indicate missing or improperly fitted insulation. Must be used with expertise to avoid erroneous conclusions.

Thermal inertia Measure of time taken for a surface to heat up. Product of thermal conductivity, specific heat capacity, and density of the material.

Thermal resistance Thermal resistance determines the heat insulation property of a material. The higher the thermal resistance, the lower is the heat loss. Symbol: R. The ratio of the temperature difference between the two faces of a material to the rate of heat flow per unit area. Expressed in $(m^2 \cdot K)/W$. It is the inverse of thermal transmittance.

Thermal transmittance See U-value.

Three pillars of sustainability:

- Environmental sustainability
- Economical sustainability
- Social sustainability

Tracer gas measurement A traceable gas is released into the building. Detectors placed throughout the building record concentrations of the tracer gas over time, allowing air movement in the building to be analysed. By measuring decay rate, the rate of air loss at ambient pressures can be determined.

Tread Step in a staircase. Measured horizontally from the outer edge of the step to the vertical riser (i.e. distance incorporates any nosing). See Going.

Toxicology The science determining the limits of safety of chemical agents. There are three main routes for an agent to enter the human body: ingestion – oral route; skin absorption – dermal route; and inhalation. The risk relates to the dose. Toxicity is the ability of the agent to produce injury once it reaches a susceptible site in the body. Hazard of the agent is the probability that injury will be caused by the circumstances of use.

Traditional building A building built with construction methods and materials typical for that geographical area prior to industrialisation, off-site manufacturing, etc. In the UK, often considered to be typical construction prior to 1919. These usually present as constructed with moisture-permeable materials. For example, brick or stone solid-walled, or timber-framed with masonry or wattle and daub (or lath and plaster) infills.

Trickle vent Small purpose-provided opening in a window or building envelope intended to provide some ventilation to internal spaces when windows and doors are closed and ventilation fans are turned off.

U-value Thermal transmittance. The higher the value, the more heat energy is transmitted and the lower the insulating properties. Measured in $W/m^2 \cdot K$. See also Thermal conductivity. Maximum limits by building regulation in many countries.

Unconditioned space See Conditioned space. Part of a building that is not subject to heating, cooling, or similar control or intervention.

Unintentional consequences Consequences arising from changing one thing but causing problems elsewhere or for something else

Universal design The design of products and environments to be usable by all people, to the greatest extent possible, without the need for adaptation or specialised design. Intent of universal design is to simplify life for everyone by making products, communications, and the built environment more usable by as many people as possible at little or no extra cost. Universal design should benefit people of all ages and abilities.

Vapour control layer A material that can limit movement of vapour by diffusion, as well as air movement (sometimes called vapour checks or vapour barriers). Helps protect from the consequences of condensation, including interstitial. A vapour control layer (VCL) is positioned to the inside of the insulation in order to minimise the amount of warm moist air entering the construction. Compare Breather membrane.

Vapour permeability The measure of a material's ability to transmit vapour. The greater the permeability of a material, the more rapidly vapour passes through it. Compare Air permeability.

Vapour-permeable membrane

Taken from BS 5250 (BSi, 2016):

HR (high water vapour resistance) membrane: having a vapour resistance greater than 0.25 MNs/g.

LR (low water vapour resistance) membrane: having a vapour resistance less than 0.25 MNs/g.

Can also have type LR with air-permeable underlay which has water vapour resistance not more than 0.25 MNs/g combined with an air permeability of not less than 20 $m^3/h \cdot m^2$ at 50 Pa, which allows for the transfer of both water vapour and air.

Ventilation A designed system of air exchange for removing pollutants and excess moisture, for providing oxygen for fuel burning, and/or for maintaining good air quality within a building for occupants.

Warm roof Roof construction where the insulation is placed on the outside of the roof structure, removing the need for ventilation of the roof space at eaves level. Positioning of air barriers is important to avoid thermal bypass. See also Cold roof.

Whole house approach "Considers the house as an energy system with interdependent parts, each of which affects the performance of the entire system. It also takes the occupants, site, and local climate into consideration" (Bonfield, 2016). Also known as a holistic property approach.

REFERENCE SECTION

This section gives some useful reference data commonly required by building surveyors on the following matters:

Brick bond patterns
Brick sizes
Conversions factors
Dating the property
Geometric measurement equations
Greek alphabet
SI units and nomenclature
Thermal conductivity of common building materials
U-value calculation tools

Brick bond patterns

Common bond patterns for one brick thick solid-walled properties, particularly in the north-east of England, are English and Flemish garden wall bonds. The brickwork bonds were chosen, not to reduce the number of bricks used, but to produce a sound wall with minimal build time and cost, including the amount of mortar required. The more stretcher courses used, the faster and more cost-effective the build.

The "garden wall" versions commonly include three to nine courses of stretcher bond between each header course or each Flemish bond course. Stretcher bond courses

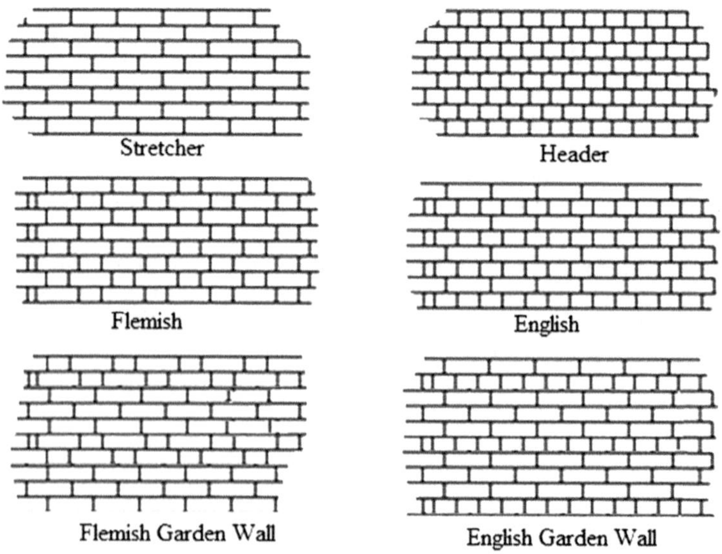

Figure 11.1 Brick bond patterns.

result in internal straight vertical joints, for the three, five, seven, or nine courses between header courses, parallel to the wall surface. Additionally, snapped headers were often used, with only a few full-sized headers to tie the two leaves together, to save money on facing bricks. As there are about one-third fewer headers in Flemish garden wall bond than in English garden wall bond the former will have the greater proportion of inner joints. This bond type can produce air movement paths.

There was a common practice of not filling all the unseen joints. This results in finger-sized (≈8–12 mm) air gaps in the solid wall construction. When not filled with mortar, the vertical joints create routes for air movement. This can result in the loss of warm air from inside the property or a thermal bypass. It has long been understood, by practice, that air passages in masonry interconnect.

Brick sizes

The bricks in existing UK properties vary depending on age, locality, and purpose.

Larger brick dimensions were manufactured when the Government introduced a brick tax in 1784. Brick producers started to make bigger bricks because the brick tax was paid per brick, so that bigger bricks meant that fewer bricks were required to cover the same area. Brick sizes tended to be around 3" in thickness. The brick tax was withdrawn in 1850 as it was considered to be detrimental to industrial development. Many brick manufacturers had already invested in machine-made bricks, in place of handmade, and so continued to produce these larger brick sizes.

Most bricks in the UK are now made to a standard size of 215 mm long, 102.5 mm wide and 65 mm high (215 mm × 102.5 mm × 65 mm) and laid with a nominal 10 mm mortar joint. Older standard bricks were in imperial measurements (UK), i.e. inches, denoted as ". One inch = 25.4 mm.The American inch is also 25.4 mm.

Table 11.1 Historic brick sizes

Style, period	**Common sizes of bricks**
Roman	(9" × 9" × 1") up to (18" × 12" × 1 ½")
Early medieval	(8 ¼" × 4" × 2") up to (10" × 5" × 2")
Late 15th century	9 ½" × 4 ½" × 2"
York standard, 1505	10" × 5" × 2 ½"
Brickmaker's Charter Statute brick, 1571	9" × 4 ½" × 2 ¼"
Clamp bricks	9" × 4 ½" × 2 7/10"
Edwardian	Narrow bricks 2" for Arts and Crafts styles
Inter-war, 1918–1939	Thinner bricks, e.g. 2 ½" generally
British Standard BS657, 1936	8 5/8" × 4 3/16" × 2" or … × 2 5/8" … or × 2 7/8"
British Standard BS3921, 1965	8 5/8" × 4 1/8" × 2 5/8"
British Standard BS3921, 1969	215 mm × 102.5 mm × 65 mm

Conversion Factors

Table 11.2 Conversion factors

1 J	2.7778×10^{-7} kW·h (kilowatt-hour)	1 kg m^2/s^2
1 W	1 J/s	1 kg m^2/s^3
1 kW·h	3.6×10^6 J (or 3.6 MJ)	
1 W·s	1 J	1 N·m
1 atmosphere	101.325 kPa (kilo Pascals)	
1 bar	100 kPa	0.986923 atmospheres
1 W	1 N·m/s	
1°C	1 K (Kelvin)	1.8°F
°F	(°C × 1.8) + 32	
°C	(°F – 32) ÷ 1.8	
1kg	9.81 N (Newtons)	
1 kg/m^2	9.81 N/m^2	
1 N·m/s	1 W	
1 N/m^2	1 Pa	0.102 kgf/m^2 (kg force per m^2)
1 Pa·m^2	1 N	
1 kN/m^2	0.14504 psi (pounds per sq inch)	
1 litre	0.001 m^3	@ 4°C, 1l water weighs ~ 1 kg
1 m^3	1000 l (litres)	
1 acre	0.40467 hectare	0.00404 km^2
1 hectare	2.47105 acres	0.01 km^2
1 m^2	10.7639 sq ft (square foot)	
1 sq ft	0.092903 m^2	
1 m	39.3701 in (inches)	3.281 ft (feet)
100 watts (incandescent light bulb)	25 watts (fluorescent/LED)	1500 lm (lumen)
1500 lm	(@70° apex angle) 1320 candela	(with surface area 1 m^2) 1 lux

Dating the property

Refer to Chapter 4, "Building pathology", especially Table 4.1. Generally, dating is conducted by comparison with other known and dated examples. Typical periods are given in Chapter 4.

If the property is historic or unusual, it may be listed or on a protection register:

For England, these properties can be found on the register at https://historicengland.org.uk/listing/the-list.
For Scotland, https://www.historicenvironment.scot.
For Wales, https://historicwales.gov.uk.
For Northern Ireland, https://www.communities-ni.gov.uk/topics/historic-environment or https://www.nidirect.gov.uk.
For Eire, a start can be made at https://www.buildingsofireland.ie/buildings-search.
For the Isle of Man, https://www.gov.im/categories/planning-and-building-control/registered-buildings-and-conservation-areas/is-my-building-registered.
For Jersey, https://www.gov.je/PlanningBuilding/ListedBuildingPlaces.

Other countries may have similar registers.

Desktop studies, particularly of Ordnance Survey (2020), and similar older, dated maps, can indicate when estates or individual buildings were constructed.

Books, such as those produced by Pevsner (2020), Pevsner Architectural Guides can also be a resource for older, interesting properties in the UK.

Geometric Measurement Equations

Table 11.3 Geometric measurement equations

Pi π	3.1416	
Diameter ϕ	Of a circle	2r = twice the radius
Radius r	Of a circle, arc, etc.	$\phi \times 0.5$ = half the diameter
Circumference	Of a circle, tree trunk, etc.	$\pi.d$ = pi × diameter
Area	Of a circle	$\pi.r^2$ = pi × radius squared
	Of a rectangle, room, etc.	length × width
	Of a triangle	base length × 0.5 perpendicular height
Volume	Of a rectangular block, building, etc.	length × height × width
	Of a cylinder	$\pi \cdot r^2 \cdot h$ = pi × radius squared × height
	Of a wedge, double-pitched roof, etc.	area of base × 1/2 perpendicular height
	Of a pyramid, hipped roof, etc.	area of base × 1/3 perpendicular height
Rafter length	Use Pythagoras $AB^2 = BC^2 + CA^2$ 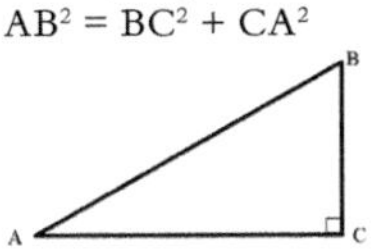	sloping length (hypotenuse) $= \sqrt[2]{(\text{height}^2 + \text{base span}^2)}$

Greek Alphabet

Table 11.4 Greek alphabet

Uppercase	Name	Lowercase	Uppercase	Name	Lowercase
Α	Alpha	α	Ν	Nu	ν
Β	Beta	β	Ξ	Xi	ξ
Γ	Gamma	γ	Ο	Omicron	ο
Δ	Delta	δ	Π	Pi	π
Ε	Epsilon	ε	Ρ	Rho	ρ
Ζ	Zeta	ζ	Σ	Sigma	σ ς
Η	Eta	η	Τ	Tau	τ
Θ	Theta	θ ϑ	Υ	Upsilon	υ
Ι	Iota	ι	Φ	Phi	φ
Κ	Kappa	κ	Χ	Chi	χ
Λ	Lambda	λ	Ψ	Psi	ψ
Μ	Mu	μ	Ω	Omega	ω

SI units and nomenclature

SI units are the current form of the metric system. It is the most widely used, international, system of measurement. There are seven base units (second, metre, kilogram ampere, kelvin, mole, candela), and other derived and supplementary units.

Table 11.5 SI units

Quantity name	SI name	units	comment
Time	Second	s	T (used in calculations)
Length	Metre	m	L (used in calculations)
Mass	Kilogram	kg	M (used in calculations)
Electric current	Ampere	A	I (used in calculations)
Thermodynamic temperature	Kelvin	K	θ (theta) (used in calculations)
Amount of substance	Mole	mol	

(continued)

Table 11.5 *(continued)*

Quantity name	SI name	units	comment
Luminous intensity	Candela	cd	
Area	Hectare	ha	10,000 m^2
Energy	Joule	J	$kg \cdot m^2/s^2$ or W·s or N·m or $Pa \cdot m^3$ 1 J = 2.7778 × 10^{-7} kW·h (kilowatt-hour) 1 kW·h = 3.6 × 10^6 J (or 3.6 MJ)
Force	Newton	N	$kg \cdot m/s^2$
Frequency	Hertz	Hz	1/second, i.e. 1 Hz = one per second
Heat flow	Kilowatt	kW	Q
Heat transfer		$W/m^2 \cdot K$	Kilowatt per m^2 per °C (or per kelvin) two-dimensional
Illuminance	Lux	lx	lm/m^2, i.e. one lux is one lumen per sq. m
Luminous flux	Lumen	lm	Total quantity of visible light emitted by a source per unit of time
Power	Watt	W	J/s
Pressure	Pascal	Pa	N/m^2 or $kg/m \cdot s^2$ or J/m^3 1 atmosphere = 101.325 kPa 1 bar = 100 kPa
Temperature	Celsius	°C	1°C= 1K °F = (°C × 1.8) + 32 °C = (°F – 32) ÷ 1.8
Temperature difference	Delta T	K	ΔT = the temperature difference across the sample
Thermal conductivity	k or λ (lambda)	W/m·K	Linear, a material constant
Thermal resistance	R-value	$m^2 \cdot K \cdot W^{-1}$	The thermal resistance of unit area of a material The reciprocal of thermal conductivity
Thermal resistivity	R	m·K/W	Linear, a material constant
Thermal transmittance	U-value	$W/m^2 \cdot K$	Rate of transfer of heat through matter Overall heat transfer coefficient

Table 11.6 SI prefixes

Term	Symbol	In relation to base SI unit	Common examples
Giga	G	× 1,000,000,000	Gt gigatonne (1 t = 1000 kg = 1 Mg)
Mega	M	× 1 000 000	Mt megatonne
Kilo	K	× 1000	km kilometre, kg kilogram
Hecto	H	× 100	hl hectolitre (1 l = 10^{-3} m^3)
Deca	Da	× 10	daP decapascal (audiology (rarely used))
Deci	D	/10	dB decibel (1 dB = 1/10 of 1 B (rarely used))
Centi	C	/100	cm centimetre
Milli	M	/1,000	mm millimetre
Micro	μ (mu)	/1,000,000	μm micrometre or micron
Nano	N	/1,000,000,000	nm nanometre

Thermal conductivity of common building materials

Be careful of the different terminologies (conductivity, resistance, transmittance, etc.) – some of these are the inverse of others.

Thermal conductivity, denoted as λ (lambda), measures how easily heat can travel through a **material** by conduction. The lower the figure, the better the thermal performance. It is measured in W/m·K.

Thermal resistance, denoted as R, measures a **construction's** resistance to heat travel by conduction due to the various thickness and conductivities of the materials that make up that construction.

$$\text{Thermal Resistance } (m^2.K/W) = \text{Thickness (m)}/\text{Conductivity (W/m.K)}$$

To build up the thermal resistance for more than one material:

$$\begin{aligned}\text{Thermal Resistance} = & \left(\text{Thickness A/Conductivity A}\right)\\ & +\left(\text{Thickness B/Conductivity B}\right)\\ & +\left(\text{Thickness C/Conductivity C}\right)\\ & +\cdots\text{etc.}\cdots\text{plus its internal and external surface resistances.}\end{aligned}$$

U-value of a plane (e.g. a wall, a floor, a roof):

U-value or **thermal transmittance** is the inverse of thermal resistance, i.e.:

$$1/\mathrm{R}\left(\mathrm{W}/\mathrm{m}^2.\mathrm{K}\right)$$

$$\text{U-value}\left(\mathrm{W}/\mathrm{m}^2.\mathrm{K}\right) = 1/\mathrm{R} = \text{Conductivity}\left(\mathrm{W}/\mathrm{m.K}\right)/\text{Thickness}\left(\mathrm{m}\right)$$

The U-value of the plane or element involves the building up, or adding up, of all the thermal transmittance values of each material that makes up that plane or element. This can be done by finding all the thermal resistances (including the internal and external surface resistances), adding them all up, then dividing 1 by the answer. Typical thermal conductivities of common building materials are given in Table 11.9.

Surface resistances: in simplified terms, there are particular resistances to the passage of heat at both sides, or surfaces, of a plane or element. These are referred to as **internal surface resistance (Rsi)** and **external surface resistance (Rse)**. Additionally, there is a separate **cavity surface resistance (Rcav)** for surfaces facing a cavity. These resistances are affected by various factors, including emissivity. BS EN ISO 6946 (BSi, 2017) gives figures for the surface resistances that can be used in calculations. See also Chapter 6.

Table 11.7 Surface resistances used in calculations

Surface resistances (examples of those used as standard)		
External surface resistance	0.04 m^2W/K	Rse
Internal surface resistance	0.13 m^2W/K	Rsi
Cavity surface resistance	0.18 m^2W/K	Rcav

Source: Cormac Flood and Lloyd Scott, see Chapters 6 and 11.

Table 11.8 Surface resistances of elements for simple calculations

Element type	**Heat flow direction**	**Rsi (internal) $m^2 \cdot K/W$**	**Rse (external) $m^2 \cdot K/W$**
Floor	Downwards	0.17	0.04
Wall	Horizontal	0.13	0.04
Roof	Upwards	0.10	0.04
Window	Horizontal	0.13	0.04

Source: BS EN ISO 6946.

Table 11.9 Typical thermal conductivities of building materials

Material	Typical thermal conductivity λ (lambda) in W/m·K (check with manufacturer)
Aerogel	0.014–0.03
Block light	0.380
Block medium	0.510
Bock dense	1.630
Brick	0.600–1.600
Cellular glass	0.041
Cellulose	0.035 (horizontal); 0.038–0.040 (vertical)
Concrete aerated lightweight	0.160–0.38
Concrete medium	0.510–1.28
Concrete dense	1.130–1.800
Cork	0.038–0.040
Cross laminated timber	Depends on insulation and thickness
Earthwool	0.032–0.038
Expanded clay aggregate	0.100
Expanded polystyrene	0.030–0.038
Extruded polystyrene foam	0.033–0.035
Flax	0.038–0.060
Foamed glass	0.075–0.080
Glass	0.96
Glass mineral wool	0.035–0.04
Granite	1.700–4.000
Gypsum board	0.17
Hardboard (high density)	0.15
Hemp	0.038–0.060
Hempcrete	0.060–0.110 not used below ground level
Limecrete	0.060–0.110 not used below ground level
Limestone	1.26–1.33
Marble	2.070–2.940
Phenolic foam	0.020–0.023

(continued)

Table 11.9 *(continued)*

Plaster gypsum	0.160–0.500 typically 0.460
Plasterboard	0.160–0.210
Plywood	0.160
Polyisocyanurate foam (PIR)	0.020–0.026
Polyurethane foam (PUR)	0.020–0.026
Render (sand/cement)	0.500
Rock mineral wool	0.032–0.044
Sandstone	1.500–1.700
Screed (sand/cement)	0.410
Sheep's wool	0.038–0.042
Slate	2.100
Spray foam	0.025–0.027
Steel	Carbon 45; stainless 15
Straw	0.055–0.080 perp to fibres
Timber softwood (commonly used)	0.013–0.140
Timber hardwood (commonly used)	0.120–0.190
Woodfibre	0.038–0.050

U-value calculation tools

Calculating the U-value of plane elements of a building (e.g. walls, roofs, and floors) is a relatively simple process and can often be done without the need for thermal modelling software. Simple elements consisting of relatively homogenous layers can be calculated using the methods laid out in BS EN ISO 6946 (BSi, 2017). Simple calculation methods require minimal data inputs, e.g. construction thicknesses, material thermal properties (thermal resistance), and surface resistance. Where specific material properties are unknown, reference values can be used, for example from BS EN ISO 10456 (BSi, 2007). Further guidance for more complicated construction can be found in BR-443, "Conventions for U-value calculations" (BRE, 2006).

Simple U-value calculation tools are available, such as the BRE U-value calculator (BRE, 2020), BuildDesk (2020), etc. These require information about the thicknesses and properties of each part or layer of a building element, about the perimeter of floors, and about the fraction of timber in a stud wall. This kind of tool calculates U-values automatically, saving time for the user and reducing the risk of potential calculation errors.

REFERENCES

ASHRAE (2013). *Standard 55-2013—Thermal Environmental Conditions for Human Occupancy*. American Society of Heating, Refrigerating and Air-Conditioning Engineers, Inc. Atlanta, Georgia.

Bonfield, P. (2016). Each Home Counts: An Independent Review of Consumer Advice Protection, Standards and Enforcement for Energy Efficiency and Renewable Energy. Commissioned by Secretaries of State for the Department of Energy and Climate Change (DECC), now part of the Department for Business, Energy and Industrial Strategy (BEIS), and the Department for Communities and Local Government (DCLG).

BRE (2002). *BR 262 Thermal Insulation: Avoiding Risks: A Good Practice Guide Supporting Building Regulations Requirements*. Building Research Establishment, UK.

BRE (2006). *BR 443:2006 Conventions for U-Value Calculations*. 2006 edition. Author Brian Anderson. BRE, Scotland; Watford.

BRE (2016). *BR 497 Conventions for Calculating Linear Thermal Transmittance and Temperature Factors*. 2nd edition. Authors T. Ward, G. Hannah and C. Sanders. BRE, Watford.

BRE (2017). RdSAP 2012 version 9.93 (19 November 2017) Appendix S: Reduced Data SAP for existing dwellings. Available from: https://www.bre.co.uk/filelibrary/SAP/2012/RdSAP-9.93/RdSAP_2012_9.93.pdf

BRE (2020). BRE U-Value Calculator. Available from: https://projects.bre.co.uk/uvalues

Bristol (2015). Bristolian's Guide to Solid Wall Insulation. Bristol City Council. Available from: https://warmupbristol.co.uk/content/solid-wall-insulation [Accessed 7 July 2016].

BSi (2007). *BS EN ISO 10456 Building Materials and Products. Hygrothermal Properties. Tabulated Design Values and Procedures for Determining Declared and Design Thermal Values*. British Standards Institution, London, UK.

BSi (2016). *BS 5250:2011+A1:2016 Code of Practice for Control of Condensation in Buildings*. British Standards Institution, London, UK.

BSi (2017). *BS EN ISO 6946 Building Components and Building Elements. Thermal Resistance and Thermal Transmittance. Calculation Methods*. British Standards Institution, London, UK.

BuildDesk (2020). *Software for U-Value Calculation*. Available from: http://www.builddesk.co.uk

EST (2019) Passivhaus: what you need to know. Energy Savings Trust. Available from: https://energysavingtrust.org.uk/passivhaus-what-you-need-know/

Gov.uk (1996). *Party Wall etc. Act 1996*. Available from: https://www.legislation.gov.uk/ukpga/1996/40/contents

Gov.uk (2010). *The Building Regulations 2010*. UK Statutory Instruments 2010. No. 2214. Available from: https://www.legislation.gov.uk/uksi/2010/2214/contents/made

Gov.uk (2014). *Standard Assessment Procedure.* Department for Business, Energy & Industrial Strategy. Available from: https://www.gov.uk/guidance/standard-assessment-procedure

Gov.uk (2015). Available from: https://www.hse.gov.uk/construction/cdm/2015/index.htm

Gov.uk (2020). Fuel Poverty Statistics. Available from: https://www.gov.uk/government/collections/fuel-poverty-statistics

Harrje, D. T., Dutt, G. S., and Gadsby, K. J. (1985). Convective Loop Heat Losses in Buildings. *ASHRAE Transactions* 91(2): pp. 751–760.

HMRC (2016). HMRC Capital Allowances Manual. Her Majesty's Revenue and Customs, UK Government. Available from: https://www.gov.uk/hmrc-internalmanuals/capital-allowances-manual/ca26025

Ordnance Survey (2020). Ordnance Survey Limited, Southampton. Available from: https://www.ordnancesurvey.co.uk

Pevsner (2020). *Pevsner Architectural Guides.* Pevsner Guides, Yale University Press, London.

Planning Portal (2016). Part L—Conservation of Fuel and Power. Approved Document L1A: Conservation of Fuel and Power in New Dwellings. Available from: https://www.planningportal.co.uk/info/200135/approved_documents/74/part_l_-_conservation_of_fuel_and_power

WCED (1987). *Our Common Future.* World Commission on Environment and Development, Brundtland Commission. Oxford University Press, Oxford.

Index

Page numbers in bold denote tables, in italic denote figures

AAR (alkali aggregate reaction) 115, 116, 132
absorption 64, 120, 150
access 4, 23, 47, 49, 51, 51, 58, 84–85, 141, 165, 191, 194, 222, 232, 240, 248, 271, 299, 303, 318
access for fire services 248
accessible and adaptable 84–85, 103, 161–162, 248
accidents 76, 77, 316
acid 16, 89, 109–110, 112, 290
acoustic performance 152, 154–155, 167
acoustics 21, 45, 57, 62–64, 133, 152, 154–155, 167
Actinomycetes 106
adaptive thermal comfort 179–180
Adjacent Owner 59
aerosol particle releases 312, 315, 323, 324
affordability 265–266, 268
aggregate reaction 115
air exchange 150, 174
air infiltration 19
air leakage 150, 162, 172–174
air movement 138, 176
air permeability 172–173
air quality 94, 173; *see also* indoor air quality (IAQ)
air temperatures 26, 175–176, 175, 179, 202, 204
air velocity 175–176, 175, 179
airborne sound 62, 64, 153–155
air-flow 137
airtightness 4, 94–95, 108, 139, 155, 169, 172, 174, 192
algae 112, 315
algal growth 98, 104
alkalinity 100, 116–117
Alternaria 106
alternative construction materials 329
aluminium 99–101, 100, 163, 241, 244
Anglo-Saxon 90
anodic index 99, 100
anthropometric measurements 85
Antrodia 107
Approved Documents 5, 46, 48, 49, 85, 87, 109, 227, 239–240, 248–250
architectural disability 82–83, 83
architectural periods 89, 90
architrave 156–157
Art Nouveau 92
Arts and Crafts 92
asbestos 69, 93, 132, 276, 290
ASHRAE 175, 177, 217
Aspergillus 106
ASR (alkali silica reaction) 115, 116, 119
ASSI (area of special scientific interest) 300
Aureobasidium 106
automatic fire detection (AFD) 163, 241, 249
autonomic responses 178

Bacillus 106
bacteria 106–107, 133, 172, 255; actino- 106; aerobic 106
Baroque 91
barriers 18–19, 27, 27, 29, 64, 83, 94, 103–104, 123, 128, 129, 136, 138, 147, 166, 200, 210, 212, 218, 222, 233, 240, 245, 267, 270, 271, 281, 294–295, 295
basements 19, 20, 43, 44, 91, 104–105, 106, 127, 144–146, 151, 213, 222, 231, 248, 277, 282
bearing capacities 34, 37, 41, 43, 151, 228, 240, 242, 282, 283, 284
BER 148, 182–183, 185, 188, 190–191, 193–195, 198, 196, 197, 304
bi-metallic corrosion 99, 99, 101
biodiversity 312–313, 318–319, 323, 334–335, 338
biophilia 165
boundary walls 57, 61; brick 29, 92, 95, 206, 208, 209
BPE (building performance evaluation) 4, 169, 199, 200–202, 201
breathability 150
BREEAM 182–185, 186, 187, 197–198
BREEAM categories and weightings 186
brick 26, 29, 31, 33, 35, 90, 90–93, 95, 96, 98–99, 101, 110–111, 110, 119–120, 119, 132, 144–145, 155, 163–164, 206, 207, 208, 209–214, 242–243, 305–306, 336–339
Brutalist 93, 113
BSI 87, 177, 179, 181, 298, 306
Building Acts 46, 91
building envelope 12, 16, 18, 21, 170, 173, 188, 204, 216–217, 304–306
building form 11, 14, 21
Building Notice 48
building owner 5, 59, 60, 61, 80, 183, 190, 197, 201
building pathology 9, 89, 90, 96–97, 170, 203, 206, 276, 279
building performance 1, 3, 4, 95, 182, 199, 200, 203, 216
building performance simulation 216
building physics 139
building regulation 3, 4–5, 29, 45–46, 47, 48, 62–63, 84, 87, 93, 102, 109, 148, 152, 163, 167, 191, 227–228, 233, 235, 239, 272, 293, 304–305
building retrofit 134, 144–145, 216
building services 73, 81, 160, 201, 203
building standards 29, 48, 190, 217, 239
building survey 6, 96
building surveying 1–2, 2, 4, 6, 9, 9, 65, 82, 88, 169, 170, 291
building surveyors 1, 3, 5, 9–10, 45, 81–82, 89, 160–161, 247–248, 276
building works 4–5, 45–46, 47, 48, 66, 143, 235

calcium chloride additive 132
calcium silicate brickwork 132
calorific value 251
capillarity 150
capillary action 105, 305–306
carbon dioxide (CO_2) 96, 110, 116–117, 143, 191, 193, 195, 198, 222, 242, 284, 290, 293–294, 293, 301–302, 310, 312–314, 319, 323–324, 324, 325–326, 328–329, 333, 338
carbonation 93–94, 100, 115, 116–117
cavity barrier 233, 240
cavity trays 29, 104, 146, 295, 295
cavity wall ties 130
cavity walls 27, 47, 92–95, 100, 110, 120, 130, 148–149, 206–208, 207, 209–212, 305
CDM 45, 76, 77–78, 78
cellar walls 19
cellars 19
Chaetomium globosum 106
change of use 47, 50, 52, 85, 152, 157, 291
Chartered Surveyors 71, 88, 88, 276–277
chemical attack 108, 110, 123, 132, 329
chloride attack 115, 116

chlorides 100, 116–118, 290
chlorofluorocarbons (CFCs) 263, 314, 317, 324
Cladosporium 106
Class 0 228
classification of fires 222
clay bricks 119, 119, 132, 242, 336, 338–339
climate and construction 301, 304
climate change 9, 111, 122, 164, 182, 200, 217–218, 252, 276, 287–288, 296–297, 296, 301–305, 312–313, 318, 323, 324, 329
Climate Change Committee (CCC) 296, 303
clothing 77, 175, 176–177, 179, 230, 319
Code of Practice 88, 148, 299
cold bridge 169
combined foundations 36
combustibility 94, 226–227, 245
combustible 94, 221, 224–225, 234, 236, 241, 245, 247, 290
combustible cores 132
combustion of flammable liquids 221
combustion of solid materials 220
commercial condition survey 74
Commissioning 201, 202
common furniture beetle 111–112
community-based organisations (CBOs) 268
compaction 40, 41, 94, 113–114, 123
compliance audits 74
concrete 18–19, 20, 26, 29, 31, 34–35, 37, 39–40, 92–95, 97, 100, 110, 113–119, 115, 119, 123, 132–133, 141, 151–153, 151, 207, 242–244, 281, 284, 306, 321, 326, 329–331, 330, 333, 333, 338
concrete brickwork 132
condensation 18–19, 23–24, 26–28, 93–94, 98, 102–103, 108, 111, 113, 136, 138, 139, 144–146, 149–150, 171, 173, 203–204, 218
condition survey 6, 45, 73–75, 75, 81, 96
conditioned space 169
Coniophora puteana 107
conservation areas 45, 56
conservation of energy 29
construction materials and sustainability 326
construction types 207, 209–212
construction work 5, 78–79
containment 238–239, 291
contaminants 5, 42, 277, 290, 292, 294
contaminated land 289–291, 292, 293
contractor 5, 65–66, 79, 160, 297, 327
contracts 4, 26, 45, 65–67, 81, 109, 327
Control of Substances Hazardous to Health (COSHH) 299
coping 256
correction factors 217, 304
corrosion 26, 93–94, 97, 99–100, 99, 101, 109–110, 113, 115, 116–117, 119, 132, 136, 210, 329
cost reimbursement contracts 66
cracking 21, 26, 31, 99–100, 109–110, 115, 120, 122, 121–122, 124, 124–126, 132–133, 136, 242
cracks 18, 24–26, 32, 96, 110, 117, 120–121, 124, 126–127, 130–131, 131–132, 153, 173, 228, 242, 282, 296
creep 26, 43, 121, 250
crypto-florescence 98, 110–111, 119
crystallisation 97–98, 101, 110–111, 119
CWI (cavity wall insulation) 94, 133, 142–143, 146, 148–149

damp 18–19, 20, 69, 91–93, 94–95, 95, 98, 100–102, 101, 102, 104–105, 109–112, 133, 134, 136, 138, 144–146, 203, 217, 330, 337; *see also* rising damp
data capture approach 8
data collection 7, 8, 75
deathwatch beetle 111–112
decay 4, 8, 98–99, 103, 106–107, 111–112, 133, 135, 137, 141, 206, 224–225, 286, 336
decay fungi 106–107, 133

Decent Homes 45, 68
defect management 96, 96
defective premises 58
Defective Premises Act 1972 58, 72
defects 8, 26, 58, 74–75, 89, 95–98, 111–112, 115, 115, 118, 131, 139, 206, 217, 295
defects in retrofit 143, 144–146
deleterious materials 131, 132, 276
demise 73–74, 289
Department for International Development (DFID) 257–258
desalination 117
design and build contracts 66–67
Design Pyramid 83, 84
designer 7, 29, 78–79, 140, 152, 157, 160, 200–201, 230, 271, 279, 297, 338; lead 161–162
desktop surveys 297
dew point 26–27, 103
diagnosis 8, 112
diagnostic approach 8
differential settlement 30–31, 123, 124, 124, 280, 280
dilapidations 4, 6, 45, 68, 70–73, 74, 75, 88, 289
Dilapidations Protocol 73
diminution 71–72
disabilities 54, 82–83, 83, 85, 186
Disability Discrimination Act 82
disaster management 259–260; cycle 259, 259, 261
disaster risk 256–257, 257, 261
disasters 9, 252–254, 256–262, 257, 265–267, 331; post- 10, 253, 256, 258–259, 262, 264–270, 269, 270, 271
Donoghue v. Stevenson 65
door set 155–156, 241
DPC 5, 20, 29, 29, 92–95, 104–105, 106, 120, 136, 140, 305, 337
drainage 4, 49, 69, 81, 92, 105, 112, 124, 126, 280, 288, 297, 305
draughts 19, 20, 94–95, 102, 109, 135, 136, 141–142, 144–146, 148, 176, 178
dry lining 106
dry rot 107, 111, 113
due diligence 4, 6, 45, 76, 81
durability 24, 89, 96, 105, 113, 135, 139, 305, 330–331, 335
dust explosions 221
duty holders 78, 78–79
duty of care 58, 65, 81
dwarf walls 20

early post war 93–94
easement 59
eaves 14, 16, 17, 18, 140–141, 144, 247
economical sustainability 263, 352
Edwardian 92
effects of a fire: physiological and behavioural 219, 229
efflorescence 26, 98, 110–111
egress 23, 85
electrochemical methods 117
electronic communication 49, 166
element 7, 8, 11, 14, 16, 17, 18–19, 21–23, 30, 47, 53–54, 62, 64–65, 70, 73, 75, 96–97, 109, 138, 141–142, 150, 153, 157, 162, 164, 166, 172, 192, 194, 194, 200, 202–203, 204, 205, 208, 213–214, 217, 228, 238, 240, 243–244, 252, 256, 258, 262, 277, 290, 295, 304–305, 310, 314
Elizabethan 91, 164
energy efficiency 4, 23, 30–31, 47, 138, 143–145, 147–148, 182–183, 185, 188, 190, 193, 195, 197, 199, 216, 218, 302–303, 314, 337
energy losses 169, 171
energy-inefficient homes 303
entropy 8–9, 89, 135, 137, 206
envelope walls 16
environmental issues 276–277
environmental protection 48, 58
Environmental Protection Act 1990 62, 289, 298–299
environmental sustainability 263, 271
EPC (energy performance certificate) 3, 5–6, 188, 189
Equality Act 45, 82

erosion 14, 21, 24, 26, 104, 109, 127, 280, 287–288, 296, 329, 335
escape behaviour 230–232
EU Directive on the Energy Performance of Buildings (EPBD) 144
European standards 87
evacuation lift 85
evacuation time 236, 237, 250
EWI (external wall insulation) 94, 140, 140, 142–143, 149, 213
exfiltration 169, 173
expansion 24–25, 31, 34, 98, 100, 109–110, 119, 119, 132, 241, 243–244, 281, 306
exposure 11, 12, 15–16, 17, 18–19, 24, 62–64, 89, 109, 117, 149, 191, 193, 229, 230, 242–243, 256, 258, 281, 293, 294, 304
external fire spread 219, 245
external walls 12, 19, 29, 47, 61, 93–94, 102, 104, 109, 145, 148–150, 155, 213, 216–217, 233, 236, 245, 247, 295
extinguishment 222–223, 238–239

fashion 135, 164–165, 167, 232, 319
fashion in buildings 164–166
feathers 142
final 71, 97, 162, 187–188, 247, 321
finishes 17, 18, 21, 31, 95, 101, 106, 115, 120, 135, 154, 164, 166, 208, 209, 211, 217–218, 298, 305, 337
fire alarm 233, 236–237, 249
Fire Authority 160–161, 235, 250
fire doors 157, 158, 161, 163, 238, 240–241
fire engineering 219, 226–227, 248–250
fire legislation 238
fire load 226
fire precautions 81, 219, 233–234, 238
fire prevention 219
fire progression 232
fire propagation 213, 227–228
fire protection 132–133, 136, 154–156, 219, 238, 238, 240
fire protection measures 238, 238
fire resistance 155–157, 158, 162–163, 228, 228, 235, 238, 240–241, 243–245, 246, 247
fire safety 2, 24, 45, 49, 80, 80, 85, 155, 158, 160–162, 160, 219, 226, 230, 231, 233, 235, 241, 245, 246
fire safety in occupied building 45, 76, 80
fire smoke plume 224
fire triangle 219, 220, 222
fitting 47; ill- 143, 155, 202
fixture 23, 135, 296
flammable 221, 225, 293; non- 293
flammable gases 220–222, 227
flashover 224–225, 230
flashpoint 221
flood damage 296, 296
flood resilience 298
flooding 255, 296–298, 296
flora and fauna 298
flues 98, 141–142
forensic 170
formaldehyde 69, 132–133
foundation movement 118, 122
foundation types 32, 33, 127
foundations 5, 30–39, 31–32, 34, 41, 42–43, 57, 61, 93, 109, 118, 119, 122, 123, 124, 127, 151, 277–282, 283, 285–286, 298, 332–333; heat transfer 31, 42
frass 112
freeze-thaw 17, 18, 149
fresh water 312–313, 319, 323, 324, 325
frost damage 98, 104, 111, 149, 213, 337
frost heave 123, 279, 281
fuel poverty 147, 303–304
full planning 52
fungal attack 91, 111
fungi 8, 20, 69, 91–92, 102–103, 106–108, 111–112, 133, 255, 300, 335–336
Fusarium 106

galvanic corrosion 99, 99
gas protection system 295
gases 98, 109, 111, 163, 183, 220–223, 227, 229–230, 230, 234, 241, 263, 290, 293, 294–295, 301, 314, 331
geometric thermal bridges 19
George IV 92
Georgian 56, 90, 92, 164, 244
glass 43, 108, 115, 133, 142, 148, 158, 160–161, 163, 165, 233, 244–245, 330; fibre- 64, 210
Gloeophyllum 107
going 59, 101, 135, 232, 236
Gothic 91
Grenfell Tower 68, 236
ground floors 18, 20, 28, 39, 43, 104–105, 144–146, 148, 162, 213, 282
ground stabilisation 42
ground treatment 41–42, 280
ground water 19, 113, 115, 116, 280
Guidance Note 66, 71, 76, 81, 87, 88, 277, 325

HAC (high alumina cement) 132
hazard 22–24, 53, 62, 68, 69–70, 81, 83, 94, 108, 112, 118, 221, 223, 235, 241, 253–254, 255, 256–258, 260, 262, 264–266, 289, 293–294, 299, 320, 331
hazard assessment 292
hazard identification 292
health and safety 46, 68, 74, 76–77, 78–79, 81, 162, 272, 296
heat loss 11–12, 17, 19, 20, 30, 42, 109, 134, 136, 143, 146, 147, 150, 171–173, 176–178, 190, 192, 209
heating, ventilation and air-conditioning (HVAC) 193, 194
heave 123, 124, 126, 129, 279, 281, 285
height of a building 248
High-Tech 94
Historic England 3, 56, 56, 138, 142–143, 149–150, 160
holistic 8, 89, 160, 200, 259, 263–264, 271
holistic 8, 89, 160, 200, 259, 263–264, 271
house longhorn beetle 111
household income 266, 303
Housing Act 1988 73
Housing Act 2004 73
Housing Health and Safety Rating 68, 69
Housing Health and Safety Rating System (HHSRS) 68, 69
HSE 45, 63, 77–78, 78, 80, 87, 108
human comfort 94, 135
human conditions 82, 83
human conflicts 253
human respiration 173
humidity 20, 22, 24–25, 104, 106–107, 173, 175–176, 175, 179, 332; absolute 176; high 69, 106, 132; relative (RH) 106–107, 150, 217, 304
hydrodynamic action 296
hydrostatic action 296
hygroscopic 105–106, 141–142, 149–150
hygroscopicity 105–106, 141–142, 149–150
hygrothermal behaviour 203, 205
hygrothermal changes 216
hygrothermal modelling 138, 203
hygrothermal simulation 205

ICC contracts 67
ignitability 226
impact sound 62, 153–154
inadequate consolidation 123
inclusivity 45, 82–83, 83
incompatible materials and technologies 135
indoor air quality (IAQ) 108, 171
Industrial Revolution 2, 4, 95, 313
infestation 25, 111–112, 255
infiltration 169, 173, 288; *see also* air infiltration
information paper 88, 299
insects 25, 92, 101, 108, 111–112, 255, 300, 323, 334–335

in-situ measurements 216
instability 210, 287–289
insulation *see* CWI; EWI; IWI; MMMF; SWI; thermal insulation; TIWI; UF
insulation positioning 26
insurance 6, 66, 76, 81, 122, 123, 248, 257, 262, 276, 291, 296, 297–298
integrity 1, 20, 25, 105, 133, 142, 155–157, 158, 160, 228, 240, 244, 247, 278–280
inter-dependence of life 323
interest 1, 4, 9, 48, 52, 54, 61–62, 71, 165, 238, 269, 326, 332, 334; historic 52, 55–56; special 52, 54–56
interim 71
internal surface resistance 217, 304
internal walls 105, 150, 155, 298
internally displaced people (IDPs) 253
International Journal of Building Pathology and Adaptation 138
international non-governmental organisations (INGOs) 267–268
international standards 87, 179, 181
interstitial condensation 22, 26, 27, 92, 94, 103, 171, 205, 218
inter-war 93
invasive weeds 298
investigation time 236
IWI (internal wall insulation) 2, 94, 140, 140, 142–143, 144, 149–150, 213–214

Jacobean 91
JCT contracts 66

King's Cross Underground 234
Kingspan Ltd 150

Lakanal House 235
land water resources 318
landfill 53, 95, 286–287, 291, 293–294
landlord 3, 4, 45, 58, 69, 71–74, 297
Landlord and Tenant (Covenants) Act 1995 73
Landlord and Tenant 4, 45, 68, 71
Landlord and Tenant Act 1927 72
Landlord and Tenant Act 1954 72
Landlord and Tenant Act 1985 72
landslip 122, 123, 287
lateral restraint 130
law 1–2, 6, 54, 70–72, 77, 80, 82, 190, 276, 299–300, 311; common 58, 62, 65, 71
Law of Property Act 1925 72
lawful development certificates 52
leaching 116
lead 69, 100, 101, 133, 244, 263, 290
Leasehold Property (Repairs) Act 1938 72
least developed countries (LDCs) 253–254, 256–257, 262, 264–270, 269, 271
LEED 182–185, 187, 187, 197–198
Legionella 108
legislation 4–5, 45–46, 48, 50, 56–57, 58, 59, 61, 63, 68, 72, 74, 74, 76–77, 87, 191, 238, 319, 327
lessee 4, 71
lessor 4, 71
letter of intent 66
Leucogyrophana 107
lichen 112
lime leaching 98
limited combustibility 227
listed buildings 51, 52, 54–56, 56, 154, 157, 160–162
long-term defects 115
low carbon 135, 172, 174
lump sum contracts 66

made-up ground 277–278, 283, 286–287
masonry 16, 26, 29–30, 29–30, 33, 35, 90, 90, 98–100, 101, 104–105, 110–111, 120, 130, 153, 155, 206–207, 207, 211
masonry walls 29–30, 30, 33, 206
means of escape 85, 94, 161, 219, 230, 238, 245, 246
measured survey 74

measurement contracts 66
measurements 2, 4, 67, 85, 86, 169, 172, 177, 180, 183–184, 193, 199, 207, 229
Medieval 90, 164
membranes 18, 20, 95, 101, 105, 295, 330
metabolic rate 175, 176–177, 179
methane 290, 293–295, 293
Misrepresentation Act 65
MMC (modern methods of construction) 94, 104, 166
MMMF (machine made mineral fibre) 133
models for surveys 6
modern construction 89–90, 95
moisture accumulation 28, 205
moisture content 34, 97, 107, 111, 138, 150, 306
moisture movement 120, 137, 137, 141–142, 332, 335
monitoring 96, 96, 117–118, 129, 131, 199, 201, 294
moss 104, 112
mould 22–25, 27, 79, 94, 101, 103, 106, 133, 149, 171, 205
mouldings 158, 337
movement 19, 24–26, 31–32, 34, 86, 90, 94, 109, 113, 118–122, 119, 123, 124, 125, 127, 129–131, 132, 137–138, 137, 138, 141–142, 157, 158, 163, 176, 227, 231–232, 236–237, 240, 253, 255, 278–279, 281, 284–285, 332, 335
mundic 133

natural capital 310–311, 319
natural disasters 9, 252–254, 265–267
natural hazards 254, 255, 256, 258, 264, 266
nature and development of fire 219
near-zero energy building (NZEB) 43, 193
NEC contracts 67
nickel sulphide impurities 133
nitrogen pollution 312–314, 324, 324
noise control 45, 57, 62, 64
non-combustible materials 226–227, 241, 247
Non-Domestic Energy Assessment Procedure (NEAP) 191, 193, 194, 195
Norman 90

occupier's liability 58
ocean acidification 312–313, 323, 324
orientation 82, 109, 193, 216–217, 284
outline planning 51–52
overloading 120, 122, 280, 288
ozone layer 312–314, 317, 323–324, 324
ozone layer damage 312, 314, 317, 323–324, 324

pad foundations 32, 35–36, 36
Palladian 91
participatory design 268
party wall 4, 11–12, 45, 57, 59–61, 60, 61, 74, 88, 91, 130, 150, 213–214, 247–248
Passiv 30, 30
Passivhaus 43, 43, 211
pathology 89; *see also* building pathology
PCBs (polychlorinated biphenyls) 133, 290
penetrating damp 104, 109, 144–146
Penicillium 106
performance assessment 4, 169
performance gap 200, 202, 202
personal factors 174–176
photochemical effects 108
pile foundations 32, 38, 43
piles 38–40, 38–39, 151–152, 333
planned maintenance surveys 74
planning 29, 45, 48, 50–52, 51, 53, 55–56, 59, 63, 73, 75, 93, 140, 163, 185, 213, 217, 254, 260, 262, 271–272, 290, 298–299, 305, 320
Planning Policy Guidance Notes (PPGs) 50
Planning Policy Statements (PPSs) 50

plaster 5, 31, 91, 97, 101, 106, 120, 141, 155, 157, 159, 208, 209, 211–212, 244, 296
PMV (predicted mean vote) 179–180
POE (post-occupancy evaluation) 201–202
policy drivers 143
polymers in construction 329–330
Poria 107
post disaster reconstruction 262, 267
Postmodern 94
post-occupancy 197, 201
Practice Statement 88
pre-acquisition survey 75
Pre-Action Protocol 73
precipitation 217, 296, 304–305, 310
pre-design thermal retrofit survey 143
primary environmental variables 174–175
principal contractor 79
principal designer 79
principles of sustainable operation 324
professional indemnity insurance (PII) 6, 66, 81, 276
professional statement 88
project management 320
property investment 4–5
protected species 289, 299

qualitative design review (QDR) 250
Queen Anne 91
quinquennial inspections 74, 96

RAAC (reinforced autoclaved aerated concrete) 133
radiant temperature 175–176, 175, 179
radiation 16, 30, 69, 108–109, 169, 176, 216, 222–223, 227, 234, 247, 294, 304, 317, 335
radon 20, 69, 293–295, 293
raft foundations 32, 37, 38, 39, 43
rain 11, 14, 16, 17, 21, 23–25, 95, 97, 99, 104, 109–110, 113, 140, 149, 206, 209, 216–218, 277, 280, 287–288, 305–306, 318, 337
rainwater 28, 92, 96, 98, 104, 111, 217, 297, 330
rainwater goods 92, 98, 104, 330
RdSAP 6, 188
re-alkalisation 117
receptor 7, 291, 292, 298
reduced energy use 135, 160
reflection 8–9, 96, 96
refuges 85
refurbishment 23, 45, 63, 133, 134–136, 152, 155, 157, 164–167, 183–185, 200, 235
Regency 92
reinforced autoclaved aerated concrete (RAAC) 133
reinstatement 71, 297
remediation 291–292, 294, 319
remedies 101, 101, 106, 106
reporting 79, 96, 96, 292
reserved matters 51–52
residential condition survey 73
residential survey 74
resilience 16, 17, 253, 257–258, 258, 260–261, 298, 312, 323
respiratory problems 69, 107, 133, 229
response times 237, 301
retrofit 2, 3, 5–6, 7, 9, 17, 23, 94, 136, 136, 140, 143–145, 144–146, 147–150, 160, 162, 164, 166, 203, 206, 208, 212, 216–218, 236, 247; *see also* building retrofit; thermal retrofit
reversion 71–73
RICS 1–2, 3, 5, 45, 65, 71, 76, 81, 87–88, 88, 96, 277, 299
rights of light 59
rights over land 45, 57, 58
rising damp 20, 98–99, 105–106, 106, 144–145, 218
risk 4, 7, 8–10, 14, 19, 22, 26–27, 28, 65–66, 68, 69, 73, 78–79, 80–82, 83, 87, 92, 99, 99, 103, 108–109, 114, 118, 128, 131, 132–133, 136, 138, 139–140, 143, 149–150, 157, 161, 166, 171, 203–204, 206, 213, 217, 253–254, 256, 258, 260–262, 276–277, 288–289, 291, 292, 293,

296–298, 300, 305–306; *see also* disaster risk
risk assessment 6, 7, 9, 63, 68, 76, 80, 234, 292, 297
risk estimation 292
risk reduction 80, 256–258, 288
robust details 63
rocks 133, 242, 255, 277–278, 282, 282–283, 284–285, 287–288, 294, 300, 363
Rococo 91
rodents 112, 300
Roman 2, 90, 113
Romanesque 90
Ronan Point 233
roof 8, 12, 14, 16, 17, 18, 22–23, 30, 47, 62, 89, 91, 93, 98–101, 104, 109–110, 112–113, 119, 130, 132, 136, 138, 140, 141, 144–146, 160–162, 165, 190, 192, 194, 204, 214, 223, 247–248, 288
roof structure 16, 17
roof timbers 8, 14
roof void 94, 102–103, 247–248, 300
root barriers 128
routes of escape 237
rules of conduct 88
running costs 188, 202

safe methods of work 76
SAP 5–6, 182–183, 185, 188, 190, 197–198, 207
schedule of condition 73–75
schedule of repair 75
sea water 117, 314–315, 318, 324
seasonal soil variation 32, 34
selection of materials 338
Serpula lacrymans 107
settlement 30–31, 93, 122, 123, 124, 124, 151, 264, 268, 278–280, 280, 284–286
settlement and movement 279, 285
shrinkage 31, 94, 111, 115, 119, 120, 132, 137, 284, 296, 332
shrinkage cracking 110, 120, 121
significant fires 232, 233
Simplified Building Energy Model for Ireland (SBEMie) 193
site inspections 4–5, 96
site investigation 292, 294
smoke obscuration 229
social sustainability 263, 266, 311, 326
soil characteristics 284
soil erosion 127, 287–288
soils 31–32, 34, 41, 42, 122, 123, 124–125, 127, 128, 277–279, 281–282, 283, 284–289, 294, 300, 335
solar 11–12, 16, 17, 21, 23–25, 30, 109, 113, 171, 173, 192, 194, 203
solar energy 11, 23
solar radiation 16, 108–109, 216, 335
solid floors 18–19, 20, 141
solid wall 27–28, 28, 95, 104, 110, 148–150, 171, 208, 209, 213
sorption 218
sound transmission 153–154
SPAB 138
space standard 165
spandrel 20, 213
spontaneous fracturing 133
spontaneous ignition 221
SSSI (sites of special scientific interest) 299–300
stability 23–26, 31, 42, 91, 118, 122, 130, 142, 151, 155, 228, 244, 278–279, 281, 284, 331
Stachybotrys 106
stack effect 173
stakeholder 187, 201, 253, 259, 303, 328
Stardust Disco 223, 234
STBA 138
steel 14, 26, 34–35, 37, 39–40, 79, 92–93, 95, 99–100, 100, 104–105, 113–114, 115, 116–118, 124, 136, 152, 163, 165–166, 206, 210, 241, 243–244, 281, 318, 326, 329–331, 330, 333, 333, 338
steel and concrete production 326, 331
stock condition survey 73, 75, 96
stoichiometric ratio 251, 328
stonework 90, 212, 242

strata 29, 30–31, 33–35, 36, 37–39, 41–42, 42, 151–152, 277–278, 280, 280, 282, 283, 286–288
stratification cold 171
Streptomycetes 106
strip foundations 29, 32–35
Stuart 91
subsidence 25, 120, 122, 123, 124, 125–126, 128, 255, 288–289, 296
sulphate attack 26, 97–98, 101, 110–111, 114, 115, 116, 119
sulphate resistance 116
Summerland Leisure Complex 223
sun 21, 24, 30, 108, 188, 193, 218
Sun Valley Poultry 235
surface condensation 19, 24, 94, 149, 203, 218
surface flooding 287–288
surface resistance 202, 217, 304
surface spread of flame 156, 227–228, 227, 234
surface water 19, 289–290, 297
survey (building) *see* building survey
surveyor (building) *see* building surveyors
sustainability 9–10, 49, 81, 134, 182–185, 197–199, 216, 261–266, 268, 271, 276–277, 310–312, 319–320, 322, 326, 327–328, 334, 338
sustainability assessments 182, 184, 187, 198, 263
sustainability checklist 325
sustainable construction 9, 263–266, 271
sustainable design 165, 265, 271
sustainable development 46, 257, 262–265, 271
Sustainable Energy Authority of Ireland 182, 188
SWI (solid wall insulation) 148–149

tall buildings 15, 16, 331–332, 334
TDD (technical due diligence) 81
Technical Handbooks 46, 47, 48, 49, 240, 248
technology 1, 2, 4, 89, 97, 113, 166, 189, 181, 199
temperature 10, 18, 21–22, 24–26, 29, 69, 94, 96, 99, 102–103, 106–107, 115, 119–120, 119, 131, 132, 147, 150, 171–173, 175–176, 175, 178–180, 191, 192, 202, 204, 216–217, 221, 224–226, 229, 230, 241–244, 251, 255, 282, 296, 301, 304–306, 314, 318, 326, 337
temperature control 169
temperature factor 204, 205
tenant 3, 4, 45, 68, 70–73, 289
terminal 71, 73
testing 4, 8–9, 63, 81, 96, 96, 105, 108, 117, 156, 169, 201, 229, 292, 294, 337
thermal analysis 216
thermal bridge 19, 20, 28, 28, 30, 42, 144–146, 150, 169–172, 204, 204
thermal bridging 29, 94, 140, 142, 169–171, 190, 202–204, 211, 295, 295
thermal bypass 12, 94, 172
thermal comfort 23, 68, 147, 160, 174–181, 175
thermal conductivity 150, 170, 241–243, 251, 306, 333, 333
thermal diffusivity 241, 243, 251, 333, 333
thermal element 47
thermal envelope 170, 172
thermal equilibrium 174, 177
thermal inertia 251
thermal insulation 18, 20, 22, 26, 27, 29, 93–94, 133, 149, 154, 170, 172, 228–229, 235, 247
thermal judgement scale 179, 179
thermal modelling 202, 204
thermal movement 90, 109, 118, 120, 132, 158
thermal performance 139, 140, 145, 147–150, 155, 169, 171, 199, 206–207, 209–212, 216–218, 304–306
thermal resistance 103, 172, 179
thermal retrofit 2, 6, 94, 206
thermal stratification 94
thermal transmission 169

thermal upgrades of external fabric 140
thermoregulation 174, 176–178
thin internal wall insulation (TIWI) 143
three pillars of sustainability 263–264, 303
timber 8, 14, 18–19, 20, 25, 27, 39–40, 69, 90, 90–95, 97–98, 102–105, 108, 111–112, 115, 119–121, 119, 130, 138, 141, 143, 154, 156–157, 158–160, 164–165, 211–212, 241–242, 263, 296, 298, 300, 321, 325, 329, 332–335, 333, 338–339
timber as a construction material 166, 207, 207, 211, 331–332
timber decay 111, 135, 141
timber infestation 111
timeline of surveying 2, 3
toughened glass 133, 244
toxic moulds 108, 149
toxic substance releases 312, 316, 323, 324
toxicology 220
traditional 6, 22, 30, 34–35, 83, 90, 130, 135, 143, 163, 166–167, 172, 207, 208, 211, 216, 268, 271, 289, 337
traditional building 55, 208
traditionally constructed buildings 18–19, 89–90, 137, 137
traits 1
tree removal 123, 124, 127, 279, 286
trees 13, 14–15, 58, 123, 124–125, 127–128, 127–129, 279–281, 285–286, 288, 319, 333–336, 338
trees and vegetation 281
trees as part of the life support system 310, 334
Trichoderma viride 106
trickle vent 64, 103
Tudor 91
typical defects in retrofit 143, 144–146

UF (urea formaldehyde) 133
ultra violet light 108, 304
underpinning 47, 89, 151–152, 151
uneven ground movement 124, 125
Unfair Contract Terms Act 65
unhydrated cement 116
unintentional consequences 22
universal design 83
University of Leicester Engineering Building 160
unknowns 135, 138, 140, 226
upgrade risks 22
upgrading 21–23, 29–30, 134–135, 136, 137, 139–141, 143, 152, 154–157, 158–160, 161–163, 241
upgrading acoustic performance 152
upgrading fire protection 155
upgrading thermal performance 139
upgrading timber floors 157, 159–160
uses classes 50, 52, 53
U-value 42–43, 139, 146, 149–150, 172, 194–195, 202–203, 206–208, 207, 209–212, 216–218, 304, 306
U-value calculation 216

Valley Parade Football Stadium 234
vapour control layer (VCL) 149
vapour permeability 150
vapour pressure 102–103
vasoconstriction 178
vasodilation 178
ventilation 4–5, 21–22, 30, 49, 64, 81, 91–94, 95, 102–103, 105, 106, 107–108, 132–133, 136, 137, 137, 138, 139, 141–142, 144–146, 148, 150, 164–165, 172–174, 190, 191, 192, 193–194, 195, 202, 203, 213–214, 224–226, 248, 294–295
Victorian 89–90, 90, 92, 95, 167, 324–325
visitable 84
VOC (volatile organic compounds) 69, 107, 293–294, 293, 316
volatiles 119, 290
vulnerability 132, 252–254, 256–258, 260, 262

walls *see* boundary walls; cavity walls; cellar walls; dwarf walls; envelope

walls; external walls; internal walls; masonry walls; party wall; solid wall
water *see* fresh water; ground water; rainwater; surface water
water vapour 20, 26, 28, 29, 102, 111, 149–150, 310
wet rot 111, 113
wheelchair user dwelling 85
whole house approach 147–148
Wildlife and Countryside Act 1981 298–299
William and Mary 91
wind 11–12, 12–13, 14–15, 15–16, 21, 23–25, 98, 104, 108–109, 110, 113, 140, 149, 173, 216–218, 255, 277, 287, 304–306, 329
wind and building form 14
wind effects 173
windows 16, 22, 30, 47, 64, 70, 91, 93, 95, 102–104, 109, 140, 141–143, 144–146, 148–150, 155, 164–167, 172, 178, 188, 192, 194, 205, 213–214, 231, 247, 297, 330–331
woodworm 91, 111, 300
Woolworths, Manchester 223, 233, 236
worker 63, 76, 78–79, 95, 107, 186, 311

For Product Safety Concerns and Information please contact our EU representative GPSR@taylorandfrancis.com Taylor & Francis Verlag GmbH, Kaufingerstraße 24, 80331 München, Germany

Batch number: 10725831

Printed by Printforce, the Netherlands